An Introduction to Tropical Rain Forests

Relaxing in the jungle, French Guiana. Reproduced by courtesy of Mary Evans Picture Library.

An Introduction to

Tropical Rain Forests

Second Edition

T.C. WHITMORE

UNIVERSITY PRESS

Great Clarendon Street, Oxford 0X2 6DP

Oxford University Press is a department of the University of Oxford.
It furthers the University's objective of excellence in research, scholarship,
and education by publishing worldwide
in Oxford New York
Athens Auckland Bangkok Bogotá Buenos Aires Calcutta
CapeTown Chennai Dares Salaam Delhi Florence Hong Kong Istanbul
Karachi Kuala Lumpur Madrid Melbourne Mexico City Mumbai
Nairobi Paris São Paulo Singapore Taipei Tokyo Toronto Warsaw

with associated companies in Berlin Ibadan

Oxford is a registered trade mark of Oxford University Press
in the UK and in certain other countries

Published in the United States
by Oxford University Press Inc., New York

First published 1998
Reprinted 1999

A catalogue record for this book is available from the British Library

Library of Congress Cataloging in Publication Data

ISBN 0 19 850148 X (Hbk)
ISBN 0 19 850147 1 (Pbk)

Printed in Great Britain
on acid-free paper by The Bath Press, Bath

Preface to second edition

There has been a change since 1989 in the perception of tropical rain forests. Faltering steps are being taken to attempt to contain impact on them, and indeed all the world's forests. The catalyst was the 1992 United Nations Conference on Environment and Development at Rio de Janeiro. Chapter 10 has been completely rewritten and expanded into two chapters to reflect the numerous national and international moves aimed to make exploitation sustainable. This includes forest certification and so-called 'eco-labelling' of timber. The most recent global assessment of human impact on tropical forests by FAO is summarized. Timber exploitation continues to spread, even though the other values of forests as reservoirs of biological diversity and for local people are now more widely accepted. The virgin Eastern rain forests are nearing exhaustion and moves into tropical America are beginning. Rain forests are increasingly being reduced to fragments in anthropogenic landscapes. The nature of biodiversity, its fate, and steps to be taken to describe and conserve it, have recently come under the spotlight. These new developments are all described. Topics of particular concern, and an agenda for action, are highlighted. These aspects are perhaps those that the layman hears most about, and perhaps which interest him most. Their interpretation rests on biology and ecology which, as before, are described in the first nine chapters. Here, change has been mainly an incremental increase in knowledge and understanding, with a few important breakthroughs, though there are no major paradigm-shifts to report.

Information is brought up to date throughout, attempting to incorporate the main points from what seems to be an increasing torrent of publications. Some of the highlights are as follows. Chapter 2: the northern limit of rain forest is now increased to 28°, and the eastern flank of the Andes is confirmed as the richest region for both plant and animal species; the record is now 307 tree species on one hectare. Chapter 3: there is better understanding of flowering and fruiting patterns in seasonal climates. Chapter 6: the past history of rain forests and their biogeography is clearer, especially in America and Asia. Chapter 7 on forest dynamics, is substantially rewritten and expanded. After over 20 years of intensive research the framework is firmly in place, continuing studies are now beginning to fill it in. The complexity of the microclimate of canopy gaps and the adjacent understorey has been unravelled. This forms the basis for a better understanding of shade-tolerance, the ecophysiological basis of which can now also be described. Interest in low-impact logging, exploitation within the ecological limits of the forest concerned, is burgeoning, and the necessary procedures are described. The near ubiquity of community-scale disturbances to rain forest, many of them human, has been increasingly realised. Their signs are analysed. Chapter 8: advances in understanding the site conditions that determine the occurrence of both heath and upper montane forest are described.

The book, as before, aims to introduce tropical rain forests to a wide audience, to attempt to convey something of their fascination and importance, the scientific challenges they still pose, and what is really happening to them. Even at this level there is much new information that clarifies or extends what was known 8 years ago, and there are important political changes too. It has been a challenge to cover developments without substantially increasing the length.

To enhance the usefulness of the book as a starting point for deeper enquiry, more numerous references to other books and reviews are included. A fuller description is given of silviculture.

Cambridge T.C.W.
March 1997

Preface to first edition

The aim of this book is to provide an introduction to the world's tropical rain forests for a broad audience, to describe their structure and functioning, their value to man, and what he is doing to them. Examples are drawn from all parts of the humid tropics. Today there is more research being conducted in tropical rain forests than ever before. Some is driven by curiosity, some by the desire to harness these forests to mankind's needs. What generalizations can be made? To what extent are there real differences between rain forests in different places, and what are the current frontiers of knowledge? Is present-day concern about man's impacts on tropical rain forests justified? The book seeks to provide an answer to these questions, at a simple level accessible to all who want to know something about these grand forests.*

In lands where tropical rain forests occur man has lived in closest dependence on them since time immemorial. Europeans became aware of them over two millennia ago (Chapter 1). Increasing knowledge since the Renaissance with the voyages of discovery and then the colonial era revealed that there are in fact many different kinds of tropical rain forest (Chapter 2). Plants exist in a luxuriance and a diversity of bizarre forms undreamed of in temperate latitudes (Chapter 3). Animal life is also rich and diverse (Chapter 4). Modern science continues to unravel the many kinds of complicated interdependence of plants and animals, for example, in flower pollination and seed dispersal (Chapter 5). Tropical rain forests have waxed and waned in extent through geological time, and the present patterns of species' distributions are a result of these historical events. The former idea that these great forests have survived immutable 'since the dawn of time' is a romantic fallacy, as investigations of the last three decades have shown (Chapter 6). The forests are also continuously changing at the other end of the time scale, the life span of an individual tree. The elucidation of forest dynamics (Chapter 7) has been the other major breakthrough of recent years. We now know a great deal about the ecology of individual tree species and the particular requirements for growth of their seedlings in canopy gaps formed by the death of big trees. Silviculture, the manipulation of forests by man to favour tree species of his choice, is dependent on understanding these innate characteristics. Tropical rain forests can be a sustained source of timber, renewed by regrowth after felling, so long as (and it is a vital proviso) humans work within the limits of their natural dynamics. Tropical rain forest nutrient cycles (Chapter 8) are also now reasonably well understood, with enough detailed studies made for tentative generalizations to be possible. The old ideas of a closed cycle and with nearly all the nutrients in the plants have not survived.

* To aid comprehension some of the more technical or specialized terms are defined in a glossary at the end of the book, and some concepts are discussed more fully in the text notes

Shifting agriculture is now well-known to be a sustainable form of farming, suitable for infertile soils. Sustainable human utilization of forest lands for crops or trees depends, as with silviculture, on working within the natural limits of the nutrient cycle.

One of the driving forces of science is the puzzles found in nature. Of these one of the greatest is to understand the phenomenal species richness of tropical rain forests. The extremes so far discovered are a 100 m^2 plot in Costa Rica on which grew 233 species of vascular plants (including 73 tree species, mostly as seedlings) (Table 2.3), and a 10 000 m^2 plot in Peru with 580 trees of over 10 cm in diameter, of which every second stem was a different species (Fig. 2.27). In both samples the study plot did not contain all the species in its locality. Many different factors already mentioned contribute to this hyper-richness; they are brought together in Chapter 9.

The book ends with a consideration of human impact on tropical rain forests, Chapter 10. This final chapter starts with a historical sketch of increasing intervention through historical time. Present-day impacts are described, and it is shown that pan-tropical generalizations are simplistic and misleading. The very real causes for concern are analysed. These have led to the increasing public unease, and this in turn has triggered a response in governments, international institutions, and businesses which, as this book goes to press, is suddenly accelerating. Because of this response, the book is able to end on an optimistic note. All is not yet lost of the world's tropical rain forests. There are still big gaps in our scientific understanding which need urgently to be addressed. But mankind has the ability to discover how these self-perpetuating dynamic ecosystems work and then turn that knowledge to advantage. The present book aims to review the scientific basis for good decisions. If human societies can then exercise the self-restraint to utilize rain forests by working within their natural limits the cries of doom and gloom of the late eighties will have served a useful purpose. This is a big 'if'.

Why have tropical rain forests always been so thrilling to biologists? What is their excitement that drives scientists throughout the world forward when for most the pay is so poor and the support of their employing institutes and universities minimal? Is present public concern about rain forests justified? Are they really disappearing, and if they are, does it matter? Are they so fragile that their sustainable utilization is impossible? These are the questions to which the following pages are addressed.

Witney, Oxon
June 1989 T.C.W.

Acknowledgements

There seemed to be a need for a book that looked at all aspects of tropical rain forests and mankind's impact upon them, yet in a concise text, written so as to be readily understood by the concerned layman as well as the non-specialist college student. Here, I have attempted this daunting task, but the wider the subject matter the further the scientist strays outside his own special field. The opinions and interpretations remain my responsibility but I have been dependent on assistance from a large group of people, unfortunately too numerous to mention individually by name. I do mention especially the help of S.C. Chin, R.T. Corlett, P. Lucas, I. Polunin, Hugh Tan, and I.M. Turner, who made valuable comments on drafts of the whole text and those who likewise commented on one or a few chapters, namely C.P. Burnham, N.M. Collins, W.K. Gong, T.J. Lowery, N. Sizer, and N. Tamin. For all this assistance I express my deep thanks.

The following acknowledgements are made for the figures: Ambio (Fig. 10.14); American Museum of Natural History Novitates 2387 12 March 1987 (Fig. 2.10); P.S. Ashton (Fig. 1.5); *Biological Conservation* (Fig. 10.15); *Biotropica* (Fig. 10.20); British Ecological Society (Figs. 2.32, 3.26, 8.5); N.D. Brown (Fig. 7.12); E. Brunig (Fig. 2.11); Butterworth (Figs. 6.10, 6.11, 6.12); Cambridge University Press (Fig. 6.1); Lord Cranbrook (Figs. 2.8, 2.9); D.J. Chivers (Fig. 5.10); Ding Hou (Fig. 3.13); J. Dransfield (Figs. 3.14, 5.2); Elsevier (Figs. 4.3, 4.5); *Evolution* (Fig. 6.17); Gustav Fischer Verlag (Figs. 2.1, 2.19); Forest Research Institute Kepong (Figs. 3.34, 3.35, 3.37, 3.40); A. Gentry (Fig. 1.6); C. Huxley-Lambrick (Fig. 5.11); World Conservation Monitoring Centre (Fig. 4.1); International Palm Society (Figs. 6.2, 6.9); Junk (Figs. 7.3, 7.13, 10.9); J.A. MacKinnon (Fig. 4.4); Malayan Nature Society (Figs. 5.14, 5.15, 7.1); *Malaysian Forester* (Fig. 3.25); S. Mayo (aroids on Figs. 1.8, 2.2, 3.17, 9.1); *Nature* (Fig. 7.38); *Photochemistry and Photobiology* (Fig. 7.2); I. Polunin (Figs. 3.22, 3.36); H. Rijksen (Fig. 8.2); Rijksherbarium Leiden (Fig. 6.15); *Science* (Fig. 5.7); Springer Verlag (Figs. 2.21, 8.8); M.D. Swaine (Figs. 3.31, 3.33, 7.34, 10.6, 11.6); J. Tan (Figs. 4.6, 10.19); *Taxon* (Fig. 10.17); Rosemary Wise (Figs. 5.14, 5.15).

Readers of my earlier publications will recognize some of the illustrations. In particular I have borrowed generously from my study of the Eastern rain forests (Whitmore 1975, 1984*a*) which was written to a greater depth and for a more specialist audience than the present work.

I thank Professor Gloria Lim and Associate Professor S.C. Goh who enabled me to spend one semester at the National University of Singapore in an ideal environment, and to the undergraduates who took courses A203 and A303 in 1988 and were unwitting guinea-pigs for most of the text. Ivan Polunin provided the intimate contact with living rain forest, which was an indispensable source of inspiration. Christine Brotherton gave invaluable assistance, especially in chasing references.

Finally, my wife typed the whole book in all its drafts and has made the whole enterprise possible.

Many friends and colleagues have continued to contribute ideas and criticisms which have helped to clarify my understanding. In particular, parts of this new edition were read and commented on by S.R. Aiken, R. Borchert, L.A. Bruijnzeel, P.D. Hardcastle, R. Morley, J. Proctor, E.V.J. Tanner and I.M. Turner.

I am most grateful to Professor G. Glatzel who provided me with the chance to try out the complete revision of human impact, which forms the two final chapters, in a course of lectures at the Institute of Forest Ecology, Universität für Bodenkultur, Vienna, and to the students who came to listen.

The following acknowledgements are made for new figures: CAB International (Fig. 10.26); *Conservation Biology* (Fig. 11.5); *Forest Ecology & Management* (Elsevier Science) (Fig. 7.16); Forestry Department HQ, Peninsular Malaysia (Fig. 10.16); the late T. Inoue (cover photograph); Institute of Hydrology (Fig. 10.30); *Interciencia* (Fig. 10.3); *Journal of Ecology* (Fig. 7.21); *Journal of Tropical Ecology* (Cambridge University Press) (Fig. 7.21); K.T. Lee (Fig. 11.7); *Oryx* (Fauna & Flora International) (Fig. 10.27); World Wide Fund for Nature (Fig. 10.25). Although every effort has been made to trace and contact copyright holders, in a few instances this has not been possible. If notified the publishers will be pleased to rectify any omission in future editions.

Contents

Explanatory notes

REFERENCES AND FURTHER READING

This book is an introduction to tropical rain forests and the text aims to discuss the main topics. The reader who wants to discover more can do so via the notes at the back of the book which are referred to by superscript numbers in the text. In the text notes are named original research papers on which particular points are based. In addition, for every chapter these text notes include a selection of books and review papers which give fuller information than there is space for here. The sources of data included in Figures and Tables are given in their captions. All the papers referred to in text notes, Figures, and Tables are cited together at the end of the book.

Knowledge of tropical rain forest fills a great warehouse. The references cited have been carefully chosen to help the reader peer through its windows and discover the huge store of knowledge arranged within its walls.

UNITS, SYMBOLS, AND ABBREVIATIONS

Throughout this book SI units are used. The convention is followed that values such as metres per year or stems per hectare are shown as m $year^{-1}$, stems ha^{-1} not m/year, stems/ha.

$>$ means 'greater than'
$\geqslant$ means 'equal to or greater than'
$<$ means 'less than'
$\leqslant$ means 'equal to or less than'

B.P. means years Before Present
c. = circa, about
M = mega, 10^6
n = nano, 10^{-9}
s.l. means 'in the broad sense'
$ means $US

1

An introduction to tropical rain forests

Rain forests have crossed a threshold of perception. Reports on television, radio, or in the press of another piece of destruction, or a new message of gloom for the planet, have become commonplace.[1] The public firmly believes that something nasty is happening down on the Equator, even that the once vast Amazon rain forests have all but disappeared. Man's present-day impact on tropical rain forests is, however, just the last stanza of a saga stretching back into the past beyond the beginning of written history.

European knowledge of tropical forests began when Alexander the Great crossed the Khyber Pass in 327 BC, into the Punjab, to establish the eastern limits of his short-lived empire on the banks of the Indus.[2] 'His army saw mangrove swamps (which upset conventional views on trees), jackfruit, mangoes, bananas, cotton, and banyans—which upset everybody's views on what roots are supposed to do.'[3] These bizarre findings were incorporated in the *Enquiry into plants* of Theophrastus, philosopher and pupil of Plato and Aristotle, to become part of the general knowledge of plants, copied, corrupted, and not improved upon for nearly two thousand years until the great voyages of discovery of the sixteenth and seventeenth centuries and the subsequent European colonial expansion.

The word jungle, still often in use, comes from the Hindi jangal, a reference to the dense impenetrable forest and scrub around settlements.

Tall stories percolated back to Europe from the early visitors (Fig. 1.1). For example, in the East, poisoned arrows were discovered to be tipped with the sap of the upas tree,[4] accounts of whose identity and preparation mingled fact and fable. Of this the great Dutch naturalist G.E. Rumpf wrote in 1750:

> Under the tree itself no plant, shrub or grass grows—not only within its periphery but, even, not within a stone's throw of it; the soil is sterile, dark and as if burned. Such poisonousness does the tree exhibit that from the infected air birds perching on the branches are stupefied and fall dead, and their feathers strew the soil. So caustic were the branches sent to me in a stout bamboo vessel that when the hand was placed on the vessel, a tingling was produced such as one feels on coming out of the cold into the warmth. Everything perishes which is affected by its exhalation, so that all animals avoid it, and birds seek not to fly over it. No man dare approach it unless his arms, legs and head be protected by clothes.[5]

Osbeck, on a voyage from Sweden to China, stopped in east Java and, on 20 January 1752, saw a tree with flowers on its trunk. Cauliflory is unknown in northern Europe. He believed he had found a leafless parasite, and called it *Melia parasitica* (Fig. 1.2) naively commenting:

> A small herb of barely a finger's length growing on tree trunks. It is so rare that so far as is known no one ever saw it before.[6]

Fig. 1.1. Early European travellers brought back exaggerated tales about tropical rain forests. This engraving from *Flora Brasiliensis* of von Martius, 1840 (plate IX) is a scene in the Atlantic coast forest of Brazil.

Fig. 1.2. Flowers borne on the trunk of *Dysoxylum parasiticum*, something unknown in northern Europe. When the Swedish botanist Osbeck saw this species in Java he thought the flowers were a leafless parasite. Solomon Islands.

With Colonial penetration scientific specimens began to flood back to the museums of Europe. At first the plants were the weeds of open places, many of which have wide occurrence. Indeed, when Linnaeus made his great synthesis of the world's plants, the *Species plantarum* of 1753, he believed from this evidence that the tropics had a rather species-poor and uniform flora.

The tropics had a powerful influence on the development of biology in the nineteenth century. Biogeography and ecology are both founded on the journeys in South America of the German Alexander von Humboldt, in the Andes (where he recorded how vegetation changes with climate) and in the lowland rain forests of Venezuela. He travelled with the Frenchman Aimé Bonpland. They arrived at Cumana, Venezuela, on 16 July 1799 and the effect of the tropical environment led von Humboldt to write home:

> What trees! Coconut trees 50–60 feet high; *Poinciana pulcherrima*[7] with a foot high bouquet of magnificent bright red flowers; pisang and a host of trees with enormous leaves and scented flowers, as big as the palm of a hand, of which we knew nothing ... . We rush around like the demented; in the first three days we were unable to classify anything; we pick up one subject to throw it away for the next. Bonpland keeps telling me he will go mad if the wonders do not cease.

Perhaps even more important for the development of biology was the stimulus the tropical rain forest gave to the minds of Charles Darwin and of Alfred Russel Wallace in their independent expositions of the theory of evolution by natural selection. Darwin, as a young man 22 years old, went as naturalist on the voyage of the *Beagle*, whose first tropical landfall was Salvador on the Atlantic coast of Brazil (Fig. 10.11). He went ashore on 29 February, Leap day, 1832 and has recorded:

> Delight ... is a weak term to express the feelings of a naturalist who, for the first time, has wandered by himself in a Brazilian forest. The elegance of the grasses, the novelty of the parasitical plants, the beauty of the flowers, the glossy green of the foliage, but above all the general luxuriance of the vegetation, filled me with admiration ... . The noise from the insects is so loud, that it may be heard even in a vessel anchored several hundred yards from the shore... . To a person fond of natural history, such a day as this brings with it a deeper pleasure than he can ever hope to experience again.

Wallace spent five years travelling in South America (1848–52) and then eight more (1854–62) in the Malay archipelago where he discovered the two distinct faunas of the region, epitomized by the boundary Wallace's Line named after him (Chapter 6). He too was impressed by the richness of the forests:

> If the traveller notices a particular species and wishes to find more like it, he may often turn his eyes in vain in every direction. Trees of varied forms, dimensions and colours are around him, but he rarely sees any one of them repeated. Time after time he goes towards a tree which looks like the one he seeks, but a closer examination proves it to be distinct. He may at length, perhaps, meet with a second specimen half a mile off, or may fail altogether, till on another occasion he stumbles on one by accident.[8]

Richness in species was one of the vivid discoveries of these nineteenth century explorer naturalists. It is now believed that about half the world's species occur in tropical rain forests although they only occupy about seven per cent of the land area. Herbs familiar in Europe have woody relatives which gives a whole new dimension to taxonomy:

> Nearly every natural order of plants has here *trees* among its representatives. Here are grasses (bamboos) of 40, 60, or or more feet in height, sometimes growing erect, sometimes tangled in thorny thickets, through which an elephant could not penetrate. Vervains[9] form spreading trees with digitate leaves like the horse-chestnut.[10] Milkworts,[11] stout woody twiners ascending to the tops of the higher trees, and ornamenting them with festoons of fragrant flowers not their own. Instead of your periwinkles[12] we have here handsome trees exuding a milk which is sometimes salutiferous, at others a most deadly poison, and bearing fruits of corresponding qualities. Violets[13] of the size of apple trees. Daisies (or what might seem daisies) borne on trees like alders.[14]

Tropical rain forest is certainly very different from the vegetation of northern Europe familiar to these naturalists, and few were able to resist recording their impressions in lyrical prose, or to exaggerate (Fig. 1.3). In perhumid climates, on normal tropical soils and at its grandest, as in the western Malay archipelago, it is, to use von Humboldt's phrase, forest piled upon forest, the top-most trees 45 m or occasionally even taller (Figs. 1.4, 1.5, 1.6), often as solitary

Fig. 1.3. The orang-utan of Sumatra and Borneo is in fact docile and shy and even if provoked is more likely to flee than to attack. This, the frontispiece to A.R. Wallace's *Malaya Archipelago*, gave the European reader the thrill he was expecting.

emergents which stand head and shoulders above a billowing continuous canopy, many shades of green. Within the canopy[15] there are trees of all different heights, which sometimes locally occur in layers or strata (pp. 29–30), with crowns of many shapes. Trunks are mostly slender with only a minority exceeding a metre in girth. The trunks may be buttressed (p. 54), and the bark variously sculptured and coloured (p. 57). But the forest is more than just a collection of trees, as has been vividly described by E.J.H. Corner:

On its canopy birds and butterflies sip nectar. On its branches orchids, aroids and parasitic mistletoes offer flowers to other birds and insects. Among them ferns creep, lichens encrust, and centipedes and scorpions lurk. In the rubble that falls among the epiphytic roots and stems, ants build nests and even earthworms and snails find homes. There is a minute munching of caterpillars and the silent sucking of plant bugs. On any of these things, plant or animal, a fungus may be growing. Through the branches spread spiders' webs. Frogs wait for insects, and a snake glides. There are nests of birds, bees and wasps. Along a limb pass wary monkeys, a halting squirrel, or a bear in search of honey; the shadow of an eagle startles them. Through dead snags fungus and beetle have attacked the wood. There are fungus brackets nibbled round the edge and bored by other beetles. A woodpecker taps. In a hole a hornbill broods. Where the main branches diverge, a strangling fig finds grip, a bushy epiphyte has temporary root, and hidden sleeps a leopard. In deeper shade black termites have built earthy turrets and smothered the tips of a young creeper. Hanging from the limbs are cables of lianes which have hoisted themselves through the undergrowth and are suspended by their grapnels. On their swinging stems grows an epiphytic ginger whose red seeds a bird is pecking. Where rain trickles down the trunk filmy ferns, mosses and slender green algae maintain their delicate lives. Round the base are fragments of bark and coils of old lianes, on which other ferns are growing. Between the buttress-roots a tortoise is eating toadstools. An elephant has rubbed the bark and, in its deepened footmarks tadpoles, mosquito larvae and threadworms swim. Pigs squeal and drum in search of fallen fruit, seeds and truffles. In the humus and undersoil, insects, fungi and bacteria and all sorts of animalculae participate with the tree roots in decomposing everything that dies.[16]

During the twentieth century knowledge of tropical rain forests developed in two streams.[17] On the other hand academic scientists continued to collect and identify the plants and animals, and to describe forest structure. Many of these studies were made on short visits of a few months' duration. They tried to comprehend the nature of forest variation from place to place. This was in accord with the preoccupation of ecologists in temperate countries at that time with the nature of climax communities and with succession. This phase of study culminated with the publication of a masterly synthesis by P.W. Richards of the whole field up to about 1940.[18]

Fig. 1.4. Lowland evergreen dipterocarp rain forest in profile (Brunei), with *Dipterocarpus crinitus* right centre and *Shorea curtisii* to its left. Without the scale object the reader would not realize the huge size of these trees.

Independently, colonial foresters began to delimit blocks of forest to be preserved from felling for agriculture, to control the utilization of forests for timber production, and to develop silviculture. This last involved the application of the centuries old European knowledge about what is today sometimes called 'gap-phase dynamics', namely the ability of different species to regenerate after different degrees of canopy disturbance.

In recent decades the two separate streams have merged to lead to the new synthesis which is a major part of the present book.

Apart from the growing knowledge of tropical rain forests in the Western scientific tradition it must not be forgotten that within the tropics man has lived close to Nature and in intimate contact with tropical forests for millennia. The forests yielded all the products needed for his life, and he learned how to grow crops on inherently infertile rain forest soils, by shifting agriculture, moving the fields every 2 or 3 years and allowing forest regrowth to restore site fertility. This discovery was made independently in all parts of the tropical zone. His numbers were never large.

Fig. 1.5. Profile diagram of lowland evergreen dipterocarp rain forest, Brunei, Ridge crest plot 60 × 7.5 m, all trees over 4.5 m tall shown. Mature phase forest except for extreme right hand end. (From Ashton 1965, in Whitmore 1984*a*, Fig. 1.6; see latter for species' identification.)

The era of European exploration of the world followed by the Industrial Revolution led to increasing human impact, which has increased continually till the present day (Figs. 1.7, 1.9). The evolution of medical knowledge in the West, plus the development of powerful drugs, has this century removed most of the health hazards of the humid tropics so that death rates have diminished and life expectancy increased, both dramatically. This has added the new pressure on tropical forests of much higher and rapidly increasing human populations (Fig. 10.10). The technological development of reliable machines for road building and log hauling, of chain saws for tree felling, and of bulldozers for land-clearance since World War II have made it possible to remove tropical rain forests on a scale and at a rate that was previously

Fig. 1.6. Emergent kapok, *Ceiba pentandra* var. *caribaea*; riverine forest near Iquitos, Peru. Subsequently all felled to supply a short-lived plywood industry.

This species is one of the biggest and commonest emergents of the Amazonian rain forests (Gentry and Vasquez 1988) and occurs also in Africa (Fig. 3.31).

Fig. 1.7. Lowland semi-evergreen rain forest penetrated by logging road, with S. Coutinho (1985). Lower Amazon, Jari, Brazil.

Fig. 1.8. Giant herbs, here the aroid *Colocasia gigantea* at Langkawi, Malaya, are a distinctive feature of the lowland humid tropics.

impossible. The forest frontiers have been rolled back. What seemed limitless forests a few decades ago are now seen as finite and vulnerable. Even only forty years ago, when my own Odyssey began, tropical rain forests seemed boundless. I descended the Amazonian flank of the Andes in Ecuador on muleback to investigate montane forest zonation, a journey which now takes an hour by bus. Then I travelled widely through the Solomons in the western Pacific by schooner, and collected plants never seen by science on sparsely inhabited islands where the rain forest came down to the wild coconuts leaning over the sandy beach. Later I explored the eastern part of the Malay peninsula, poling up rivers by prahu to the head of navigation before walking for several days to the peaks cresting the watershed.[19] Now one flies to those lands from London in only a day or so and can next day be in the patches of jungle which remain, arriving by dump truck along a muddy logging road. There is more general and scientific interest in tropical rain forests than ever before and this new ease of access is part of the reason. So let us now turn in Chapter 2 to a close examination of the tropical forest zone and then continue with analysis of its plants, animals, dynamics, and present status.

Fig. 1.9. Lowland evergreen dipterocarp rain forest in profile, beside a newly built telecommunications tower access-track, Malaya. Emergent *Canarium* and *Shorea* but no other distinct strata. Note man.

2

What are tropical rain forests?

2.1. TROPICAL MOIST FORESTS AND THEIR CLIMATES

The German botanist A.F.W. Schimper was one of the great nineteenth century naturalists. In. 1898 he wrote a monumental book, translated in 1903 into English as *Plant geography upon an ecological basis*, which built upon the hundred years and more of European scientific discoveries in the tropics. To Schimper we owe the term tropical rain forest (*tropische Regenwald*) for the forests of the permanently wet tropics. He recognized in total four major sorts of woody vegetation in the tropics. Rain forests, then in progressively drier and more seasonal climate, monsoon forests, savanna forests, and thorn forests. In still drier climates non-woody vegetation, tropical grasslands, and deserts, were recognized. This classification is still useful as an outline framework in all parts of the tropics. For most purposes Schimper's great groups provide too coarse a classification of vegetation and it is useful to have more finely defined classes. These are called formations, and in all six of Schimper's groups several can be recognized. A vegetation formation is defined on its structure and on the physiognomy of its component plants. For tropical forests the structural properties include the height of the trees, whether they tend to have their crowns in layers, and the presence of different kinds of climbers and epiphytes. The physiognomic properties include whether the trees are buttressed, their crown shape, the nature of the leaves, (size, shape, thickness, margin), whether the forest is evergreen or, if not, how strongly deciduous, and where on the trees flowers and fruits are borne. Using these criteria different forest formations can be defined and are found to occur in many different places.

Amongst the various tropical rain forest formations are the most structurally complex and diverse land ecosystems that have ever existed on earth, with the greatest numbers of co-existing plant and animal species. In species richness they are only rivalled by coral reefs. These lofty forests are the apex of creation.

Climate

Constant high temperature is characteristic of tropical climates, and climates in which the mean temperature of the coldest month is 18 °C or over, is often used as a definition. This excludes some tropical montane areas and a difference of less than 5 °C between the mean temperatures of the warmest and coldest months is sometimes used as an alternative definition. Rainfall is the second most important factor, and its amount, and even more important its distribution through the year, define different tropical climates. Night is the winter of the

tropics, because the diurnal range of mean daily temperature exceeds the annual range and is greater in the drier months.

Rain forests develop where every month is wet (with 100 mm rainfall or more), or there are only short dry periods which occur mainly as unpredictable spells lasting only a few days or weeks. Where there are several dry months (60 mm rainfall or less) of regular occurrence, monsoon forests exist. Outside Asia these are usually called tropical seasonal forests. They are slightly less lofty but with more even canopy top, and the bigger trees are mostly deciduous, although perhaps only briefly. They have fewer dependent climbers and epiphytes, and are less species rich.

The blanket term tropical moist forest was coined in 1976[20] to cover both rain and monsoon/seasonal forests, Schimper's first two great groups, which together comprise the tall, closed-canopy forests of the wet tropics. Tropical moist forest has proved a useful grouping for analyses of the rate of forest disappearance (which will be discussed in Chapter 10), because the exact boundary between rain and seasonal forests is hard to pinpoint and because individual nations seldom distinguish between them. To the biologist, however, there are major differences, and this book is about tropical rain forests, those which occur in the everwet (perhumid) climates, with only passing mention of monsoon forests.

The climates in various parts of the rain forest belt are shown in Fig. 2.1 using the useful pictorial device of Climate Diagrams devised by Walter and Lieth.[21] Note the distinction, even within the rain forest belt, between places with no dry season (e.g. much of the Indo-Malayan forests and the western Amazon), and those where there is a slight dry season (e.g. West Africa and the eastern Amazon). One savanna forest site is also shown (Tamale in Ghana) to demonstrate the much lower and less evenly distributed rainfall of the seasonal tropics.

The Climate Diagrams depict average climate. For plants and animals extremes are often more important than means, even when only of rare occurrence. For rain forests unusually prolonged dry periods are of particular significance, for example as triggers for flowering (section 3.3), or because drought can actually kill plants or make the forest flammable (section 7.8).

Forests, because of their stature, have internal microclimates that differ from the general climate outside the canopy. These are discussed more fully in section 7.1. In general terms, it is cool, humid, and dark near the floor of a mature patch of forest, progressively altering upwards to the canopy top. Different plant and animal species have specialized to the various forest interior microclimates, as will be seen below.

Occurrence of tropical rain forests

Tropical rain forests occur in all three tropical land areas[22] (Fig. 2.1). Most extensive are the American or neotropical rain forests, about half the global total, 4×10^6 km^2 in area, and one-sixth of the total broad-leaf forest of the world.[23] These occur in three parts, of which the largest lies in the Amazon and Orinoco basins. Second is a block which lies across the Andes on the Pacific coasts of Ecuador and Colombia, extending northwards through middle America as far as Veracruz in southernmost Mexico (19°N). The Atlantic coast of Brazil has a third block, a strip less than 50 km wide on the coastal mountains, extending beyond the tropics to the vicinity of Rio de Janeiro (*c.* 23°S). Today only about 12 per cent remains[24] (Fig. 10.11). It merges southwards into subtropical rain forest of simpler structure and different flora, but in a way we shall never now be able to define. Brazil is the nation with more rain forest than any other.

The second largest block of tropical rain forest occurs in the Eastern tropics, and is estimated to cover 2.5×10^6 km^2. It is centred on the Malay archipelago, the region known to botanists as Malesia. Indonesia[25] occupies most of the archipelago and is second to Brazil in the amount of rain forest it possesses. The Eastern rain forests extend beyond Malesia into the Pacific and southwards as a narrow broken

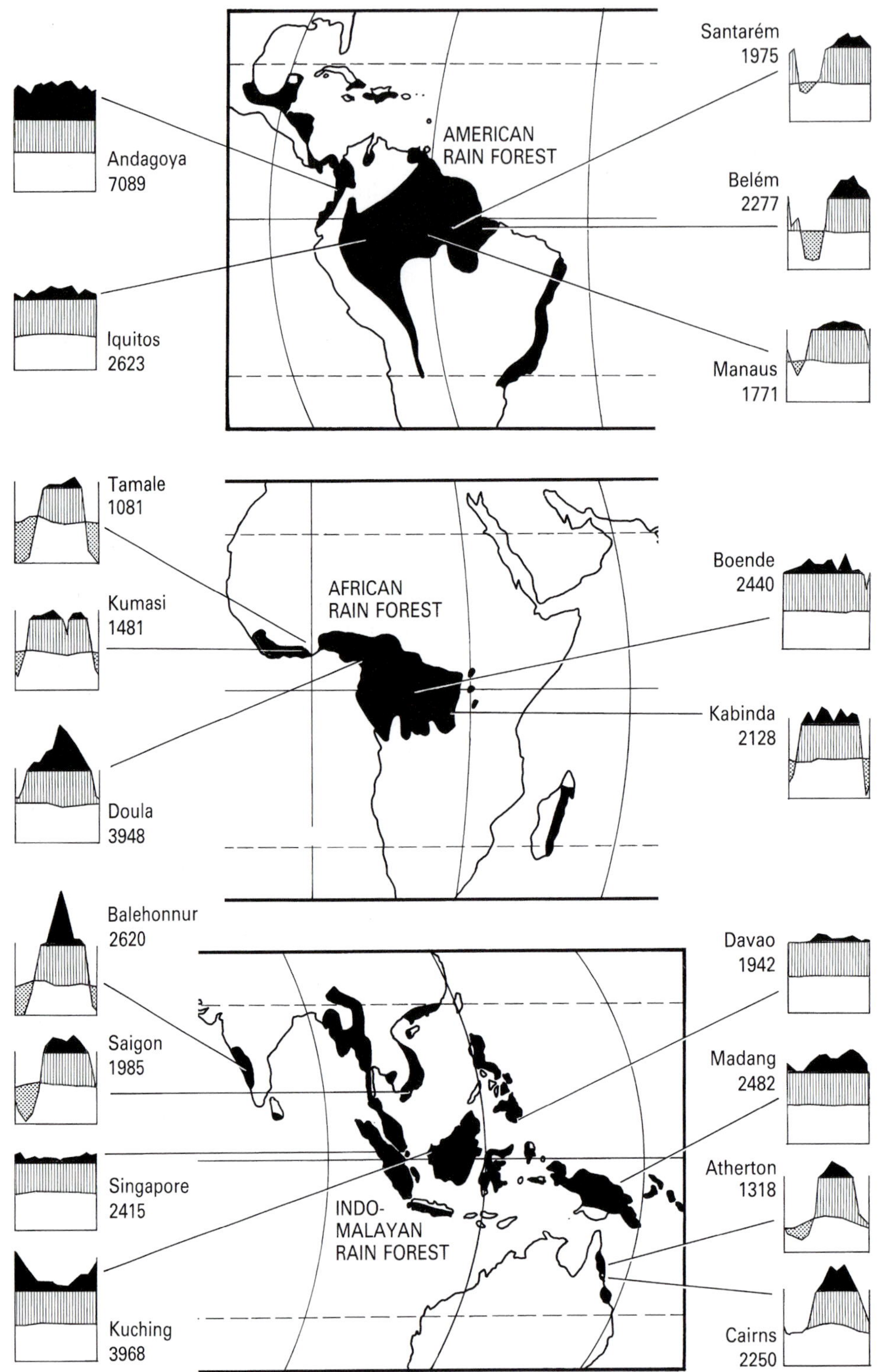

AMERICAN
RAIN FOREST
Andagoya
7089
Iquitos
2623
Santarém
1975
Belém
2277
Manaus
1771
AFRICAN
RAIN FOREST
Tamale
1081
Kumasi
1481
Doula
3948
Boende
2440
Kabinda
2128
INDO-
MALAYAN
RAIN FOREST
Balehonnur
2620
Saigon
1985
Singapore
2415
Kuching
3968
Davao
1942
Madang
2482
Atherton
1318
Cairns
2250

coastal strip in Queensland, northeast Australia, in a similar position to the Atlantic coast forests of Brazil. As in Brazil, this forest extends beyond the tropics, southwards into New South Wales. Close study of the merging zone has shown that towards their temperate limit tropical rain forest formations become increasingly restricted to the wettest sites and the deepest, most friable, and most fertile soils.[26] The Malesian forests extend northwards up the Malay peninsula into continental southeast Asia, Burma, Thailand, and Indo-China. Again, at the boundary they probably occur as a mosaic and occupy the 'best' sites, though we lack detailed studies. There is also an ill-defined change to subtropical rain forest. Between 92° and 98°E below the great south wall of the Himalaya, in southern China, Assam and upper Burma tropical rain forest reaches further away from the Equator than anywhere else, attaining 28°N.[27] There are further outliers, in southwest Sri Lanka centred on Sinharaja, and also on the Western Ghats of India where seasonality is strong.

Africa has the smallest block of tropical rain forest, 1.8×10^6 km^2. This is centred on the Congo basin, reaching from the high mountains at its eastern limit westwards to the Atlantic Ocean, with outliers in East Africa. It extends as a coastal strip into West Africa, but woodlands reach the coast at the so-called Dahomey Gap.[28] There are tiny patches of rain forest on the east coast of Madagascar and in the Mascarenes. Outside the Congo core the African rain forests have been extensively destroyed.

2.2. THE FOREST FORMATIONS

From a satellite or high flying aeroplane the earth is seen to have a sombre, dark, blue-green girdle about the Equator, just glimpsed through breaks in the cloud. There is considerable variation from place to place in this rain forest mantle due to the distinct forest formations. A forest formation, as described above, is recognized by a particular combination of vegetation structure and physiognomy regardless of flora. Convergent evolution has occurred because different species in the three land areas have evolved similar responses to particular environments. In many cases we still do not understand what facet of the environment evokes a particular response, but sometimes we do, as is discussed further in relation to features of montane and heath forests in sections 8.4, 8.5.

The forest formations occupy different physical habitats and these are mostly sharply bounded, but where this is not so there is a merging zone.

The formations can be grouped for convenience according to the main physical characteristics of their habitats (Table 2.1). This is an arbitrary arrangement but gives a useful hierarchy. The naming of vegetation types is always problematical. In the case of tropical rain forest formations the names (listed in Table 2.1) reflect site and the structure and physiognomy of the formation.

The first division is between climates with a dry season and those that are perhumid. The

Fig. 2.1. (Left) Tropical rain forest, distribution and climate. (After Whitmore 1983, 1984*a*; Walter and Lieth 1967.)

The Climate Diagrams show dry periods as dotted and rainy periods as hatched (or where monthly rainfall exceeds 100 mm shown black at 1/10 scale). Long-term mean annual rainfall (mm) is shown in figures. Evergreen rain forest occurs in perhumid climates, i.e. virtually all of the Indo-Malayan region, the upper Amazon and Pacific coast South America. It is replaced by semi-evergreen rain forest around the fringes of Indo-Malaya, and in the lower Amazon and in Africa, including the Congo basin, where dry periods occur irregularly scattered through the year and do not show up on these diagrams because they are based on long-term means.

Table 2.1 The formations of tropical moist forests

Climate	Soil water		Soils	Elevation		Forest formation
Seasonally dry	Strong annual shortage					Monsoon forests (various formations)
	Slight annual shortage					Rain forests: **Semi-evergreen rain forest**
Everwet (perhumid)	Dryland		Zonal (mainly oxisols, (ultisols)	Lowlands		**Lowland evergreen rain forest**
				Mountains	(750) 1200–1500 m	**Lower montane rain forest**
					(600) 1500–3000 m (3350) m	**Upper montane rain forest**
					3000 (3350) m to tree line	Subalpine forest
			Podzolized sands	Mostly lowlands		**Heath forest**
			Limestone	Mostly lowlands		Forest over limestone
			Ultrabasic rocks	Mostly lowlands		Forest over ultrabasics
	Water table high (at least periodically)	Coastal salt-water				Beach vegetation Mangrove forest Brackish water forest
		Inland fresh-water	Oligotrophic peats			**Peat swamp forest**
			Eutrophic (muck and mineral) soils	±Permanently wet		**Freshwater swamp forest**
				Periodically wet		**Freshwater periodic swamp forest**

Those shown bold are discussed in the text

second division is a crude measure of soil water availability, and distinguishes swamp from dryland forests. The third division is on soils and, within dryland forests, distinguishes those on parent materials with atypical properties—peat, quartz sand, limestone, and ultrabasic rocks—from the widespread 'zonal' soils, mainly ultisols and oxisols. Finally there is a division of the forests on zonal soils by altitude.

In this scheme a few formations are defined by physical habitat as well as vegetation structure and physiognomy, for example, forest over limestone. The overall result is a set of easily recognized forest formations with memorable names that say something about the most distinctive characteristics of the formation. It is a pragmatic scheme of rain forest classification and the formations can be recognized throughout the humid tropics although many different regional names have been applied. For example, in Brazil fresh-water swamp forests are called varzea or igapo.

This book is only a primer or introduction to rain forests and descriptions are given just to the more extensive forest formations.

Tropical lowland evergreen rain forest

Description

(Figs. 1.4–1.7, 2.2–2.4) This is the most luxuriant of all plant communities. It is lofty, dense, evergreen forest 45 m or more tall, characterized by the large number of tree species which occur together. Gregarious dominants (consociations) are uncommon and usually two-thirds or more of the upper-canopy trees are of species individually not contributing more than 1 per cent of the total number. This formation is conventionally regarded as having three tree layers (see

Fig. 2.2. Lowland rain forest with luxuriant aroids; *Monstera deliciosa* (cheeseplant, left) and *Philodendron* sp. (right), showing their cord-like pendent roots. (Kerner and Oliver 1895, fig. 423.)

pp. 29–30): the top layer of individual or grouped giant emergent trees, over a main stratum at about 24–36 m, and with smaller, shade-dwelling trees below that. Ground vegetation is often sparse and mainly of small trees; herbs are patchy. Some of the biggest trees have clear boles of 30 m and reach 4.5 m girth, and may be deciduous (Fig. 3.20) or semi-deciduous without affecting the evergreen nature of the canopy as a whole. Boles are usually almost cylindrical (Fig. 3.29). Buttresses (Figs. 3.31–3.33) are common. Cauliflory (Fig. 1.2) and ramiflory (Fig. 5.2) are common features. Pinnate leaves (Fig. 2.5) are frequent; leafblades of mesophyll size (Fig. 3.26) predominate. Big woody climbers (Fig. 3.15), mostly free-hanging, are frequent to abundant and sometimes also bole climbers (Fig. 3.17). Shade and sun epiphytes (Figs. 3.21, 9.1) are occasional to frequent. Bryophytes are rare.

All other rain forest formations differ from this in having simpler structure, sometimes with fewer life forms, and fewer species.

Habitat and occurrence

Tropical lowland evergreen rain forest occurs in perhumid lowland climates where water stress is intermittent or absent, from sea-level to *c.* 1200 m on dryland sites. It is the main lowland formation of the Eastern tropics and also occupies western Amazonia plus the Pacific coast of South America. In Africa it is restricted to three small blocks near the coast of West Africa, between Guinea and Liberia and Cameroon and Gabon.[29] On the Climate Diagrams (Fig. 2.1), all sites with this formation have no regular annual dry season.

Tropical semi-evergreen rain forest

Description

Semi-evergreen rain forest is a closed, high forest in which the biggest emergent trees sometimes attain 45 m in height. It includes both evergreen and, in the top of the canopy, deciduous trees, in an intimate mixture but with a definite tendency to gregarious occurrence. Deciduous trees may comprise up to one-third of the taller trees, though not all are necessarily leafless at the same time. The number of species is high, but less so than in evergreen rain forest. Buttresses continue to be frequent and occur in both evergreen and deciduous species. Bark tends to be thicker, and rougher and cauliflory and ramiflory rarer. The stature tends to be slightly less than evergreen rain forest with emergents as scattered individuals which are sometimes rare. The canopy is sometimes locally stratified. Big woody climbers tend to be very

Fig. 2.3. Lowland evergreen dipterocarp rain forest on hill slopes, Selangor, Malaya, at about 800 m elevation.
The giant emergent trees 50 m or more tall are *Shorea curtisii*, clumped on spur ridges, and showing the cauliflower-like crowns typical of many Dipterocarpaceae.

Fig. 2.4. Lowland rain forest, Caroni basin, Venezuela.
This forest has no emergent trees, the canopy top is fairly flat, and the crowns of the numerous species are nested closely together but mostly with a small surrounding space.

Fig. 2.5. Pinnate leaves are a common feature of lowland rain forest; *Canarium sp.* Guandong, China.

abundant. Bamboos are present. Epiphytes are occasional to frequent and include many ferns and orchids.

Habitat and occurrence[30]

Where there is a regular annual period of moisture stress, either due to rainfall seasonality or particular soil conditions, this replaces the previous formation. It occupies most of the African rain forest block, including the whole Congo basin. It occurs today as a narrow fringe around the main Eastern rain forest block, and also forms the outlier in India, and most of the Australian tropical rain forest (Fig. 2.6). It was probably more extensive in parts of continental southeast Asia in the past but has been reduced to its present extent by human activities. This is probably the formation that occurs in middle America. The eastern and southern Amazon (1.8×10^6 km^2, about half of the forested area) do not have a rain forest climate. The dry season is extremely strong and may last five months. Semi-evergreen rain forest, rather than a deciduous monsoon forest, occurs because the very deep soils act as a reservoir and are tapped for water by roots that reach down to 8 m depth.[31] This is probably the formation that occurs in middle America. The northernmost rain forests below the Himalaya, as well as the south China Xishuangbana rain forests, are mainly the evergreen formation.[32] The occurrence of a dry season is clearly shown in the Climate Diagrams of Fig. 2.1, except in the Congo basin where dry periods are so irregular in occurrence that they do not show up on these long-term mean records.

Fig. 2.6. The rain forest margin, northeast Queensland, Australia. Lowland semi-evergreen rain forest in the foreground sharply merging into wet sclerophyll forest with emergent *Eucalyptus grandis* in the background. The view is looking westwards, away from the coast, to increasingly dry climates.

The montane rain forests

Description and habitat[33] (Figs. 2.7–2.10)

As one climbs a tropical mountain one successively encounters forests of different structure and physiognomy (Table 2.2). The most dramatic change, which usually occurs over a short distance, is from mesophyll-dominated forest with an uneven billowing canopy surface to a lower, more even, often pale-coloured, microphyll-dominated canopy, of more slender trees, usually with gnarled limbs and very dense subcrowns. This is upper montane rain forest. It is clearly distinctive both from the air and to the traveller on foot. This formation is often only 10 m tall or less, and its shorter facies are sometimes called elfin woodland. The trees may be heavily swathed in bryophytes and filmy ferns and the formation is then known as mossy forest (though-liverworts dominate). Peat often forms, sometimes with the bog moss *Sphagnum*. On small mountains upper montane rain forest abuts on lowland rain forest, but on bigger ones an intermediate formation, lower montane rain forest, occurs and this has a broad ecotone against the lowland formation. On the highest peaks, upper montane rain forest itself is replaced upwards by a shorter more gnarled formation with even tinier leaves (nanophylls) known as subalpine forest, of which good descriptions exist for New Guinea, but which is much more extensive in the Andes. The tree line on the biggest mountains is at *c.* 4000 m. It is often depressed by fire. Above it, beyond the climatic limits of trees, occurs alpine vegetation, an edaphically determined mosaic of shrub heath, moss tundra, fern meadow, and grassland, called paramo in the Andes. These treeless landscapes extend up to the snowline at *c.* 4500 m.

On large mountains all the forest formations occur to higher elevation than on small ones and outlying spurs. This is the so-called *Massenerhebung* or mass elevation effect, first described for the Alps. In the tropics the compression and

Table 2.2 Characters of structure and physiognomy used to define the principal montane forest formations

Formation	Tropical lowland evergreen rain forest†	Tropical lower montane rain forest	Tropical upper montane rain forest
Canopy height	25–45 m	15–33 m	1.5–18 m
Emergent trees	Characteristic, to 60(80) m tall	Often absent, to 37 m tall	Usually absent, to 26 m tall
Pinnate leaves	Frequent	Rare	Very rare
Principal leaf size class of woody plants‡	Mesophyll	Mesophyll	Microphyll
Buttresses	Usually frequent and large	Uncommon, small	Usually absent
Cauliflory	Frequent	Rare	Absent
Big woody climbers	Abundant	Usually none	None
Bole climbers	Often abundant	Frequent to abundant	Very few
Vascular epiphytes	Frequent	Abundant	Frequent
Non-vascular epiphytes	Occasional	Occasional to abundant	Often abundant

From Whitmore (1984a) table 18.1
†Included for comparison
‡Following Raunkiaer (1934)

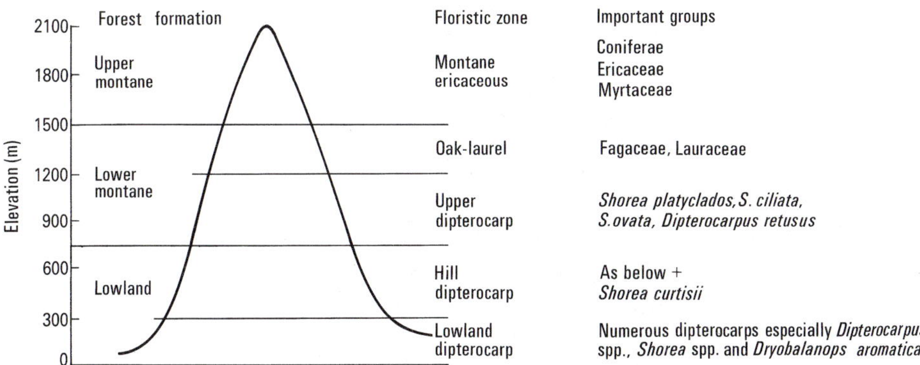

Fig. 2.7. Forest zones on the main mountains of Malaya. (Whitmore 1984*a*, fig. 18.1.)

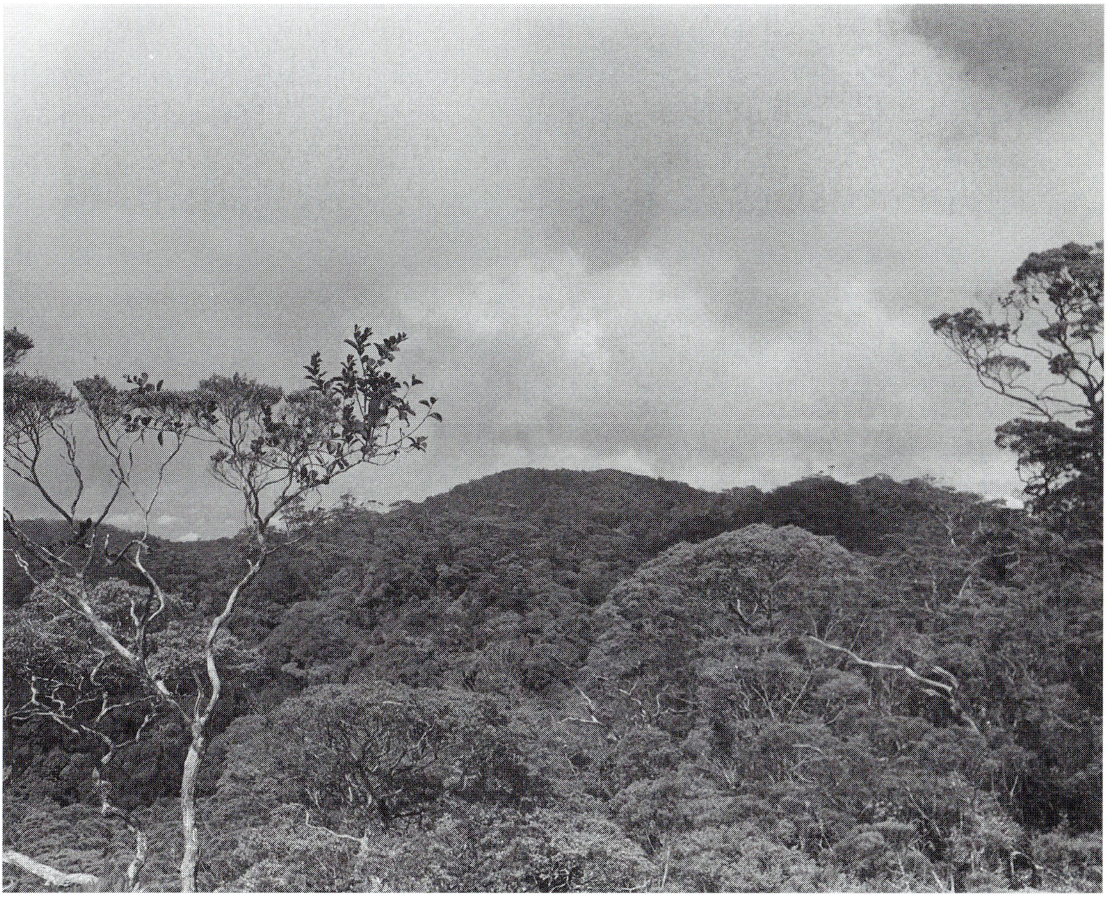

Fig. 2.8. Upper montane rain forest, Mount Benom, Malaya.
Note flattish canopy top, and trees with dense subcrowns of crowded foliage and small leaves. The big trees are *Leptospermum flavescens*; their gnarled limbs are another distinctive feature of this forest formation.

Fig. 2.9. *Leptospermum flavescens* with an undergrowth of low bushes, ferns, and sedges. A dry, open facies of upper montane rain forest; Mount Benom, Malaya. Lord Cranbrook for scale (1967).

Fig. 2.10 Upper montane rain forest, Mount Nokilalaki, Sulawesi, 2275 m elevation. The trees are swathed in filmy ferns and bryophytes. This facies is often loosely called 'mossy forest', though the bryophytes are mostly liverworts.

depression of the forest belts on small mountains seems mainly associated with lowering in the level at which cloud habitually forms.

The changes in forests and their environment with elevation are analysed further in section 8.5.

Occurrence

The Eastern tropics are very mountainous and many descriptions exist of the extensive montane forests and their zonation.[34] There are fewer descriptions for the Andes, although the complete series of formations occurs on both flanks and over large areas. The fullest studies have been made on Jamaica in the Caribbean where zonation is compressed. Montane rain forests are least extensive in Africa, where they occur in Cameroon and at the eastern fringe of the Congo river basin. Africa has the most extensive seasonally dry tropical montane woody vegetation (forests, bushlands, thickets, and shrublands), whose description is beyond the scope of this book.[35]

Heath forest

Description (Figs. 2.11–2.13)

Even the botanically inexperienced casual wanderer will notice the change when he enters the kerangas forest ... the storey formed by large saplings and small poles predominates and forms a tidy and orderly but forbidding phalanx which is dense and

Fig. 2.11. Heath forest with very even canopy top and small crowns, interdigitated with lowland evergreen dipterocarp rain forest in interior Sarawak.

Fig. 2.12. (Below left) Open stunted heath forest at Bako, Sarawak. The tree with feathery foliage massed as dense subcrowns is *Casuarina nobilis*. See Fig. 2.13.

Fig. 2.13. (Below right) *Calophyllum incrassatum* in the Bako heath forest, showing its characteristic shiny, vertically-held leaves. Many species in heath forest have thick, leathery leaves (sclerophylls) like these.

Figs. 2.12, 2.13 both show species with adaptations to minimize foliage water loss and heat load (see section 8.4).

Fig. 2.14. Mangrove forests are a forest formation that occurs on accreting coasts in all parts of the tropics. They have several peculiar features of structure and physiognomy, for example stilt roots (seen here on *Rhizophora apiculata* in Malaya), and viviparous reproduction. They are floristically much poorer than most other forest formations, with one family (Rhizophoraceae) predominant amongst the trees. Mangrove plants have specialized physiological mechanisms to enable them to live in salt water. Mangroves are disappearing fast as land is reclaimed for building (e.g. Singapore), or for prawn ponds or salt pans. They are the breeding ground for the marine life on which coastal fisheries depend and it is being increasingly realized that their destruction has serious repercussions. Fuller accounts can be found in Clough (1982) on salt balance, Lugo and Snedaker (1974) on ecology, and Tomlinson (1986) on taxonomy.

often difficult to penetrate. The canopy is low, uniform and usually densely closed with no trace of layering ... brownish and reddish colours prevail in the foliage of the upper part of the canopy and (the forest) ... appears considerably brighter.

Thus wrote E.F. Brunig, foremost student of the Asian heath forests, about the Bako National Park, Sarawak.[36] He also recorded the predominance of microphylls over mesophylls, and that many leaves are sclerophyllous, often held obliquely or vertically, often in dense clusters, and either waxy grey or highly reflective. Big woody climbers are rare but not slender wiry ones. Epiphytes are common. Myrmecophytes (Fig. 5.11) and insectivorous plants may be abundant, including in Asia the pitcher plants *Nepenthes* (Fig. 4.7).

Habitat and occurrence

Heath forest occurs on soils developed from siliceous sand, either coastal alluvium or weathered sandstones. These become podzolized. They are low in sesquioxides and so are poorly buffered, and are highly acidic (pH less than

Fig. 2.15. Steep limestone hills, here in Malaya in the form of tower karst, are extensive in the Eastern tropics and Caribbean. Limestone carries a distinctive forest formation with several very different facies; for example, sheltered enclosed valleys are rich in herbs (especially Gesneriaceae) and epiphytes; and exposed summits, which are prone to desiccation, have a stunted xeromorphic forest with resemblances to the heath forest formation. For fuller details see Whitmore (1984*a*).

4.0). They are frequently of coarse texture and freely draining. The streams draining heath forest are black, or tea-coloured when viewed by transmitted light (due to the presence of organic colloids), and are usually acid with a low cation content.

The most extensive heath forests are in the upper reaches of the Rio Negro (appropriately named) and Rio Orinoco in South America. In Brazil they are known as campina, campinarana, caatinga Amazonica, or campina rupestre. There has been intensive study of them

at the famous San Carlos research centre in southern Venezuela, but they only represent 6 per cent of the Brazilian Amazon rain forests. The description of the Guyana heath forest dominated by wallaba, *Eperua falcata*, by T.A. W. Davis and P.W. Richards, is one of the classic works of rain forest ecology.[37] There are extensive heath forests in Kalimantan, Sarawak, and Brunei in Borneo, where they are called kerangas. The other principal occurrences are small areas in Malaya, and on coastal sands in Africa in Gabon, Cameroon, and Ivory Coast.

Heath forest is perhaps the most strikingly distinctive lowland rain forest formation. There has been much debate on the extent to which its special features are correlated with either nutrient status or with water relations. The extremely acid soil is probably toxic to the roots of many plants including crops. These topics are discussed further in section 8.4.

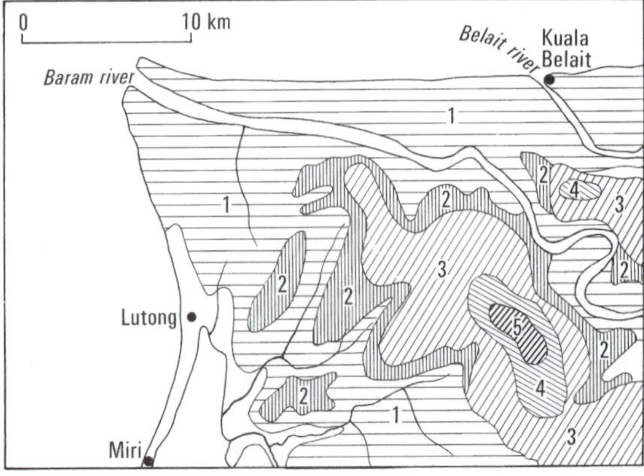

Fig. 2.16. The highly developed peat swamp forests of Sarawak are domed and bear a concentric series of different forest types (1–5) from edge to centre. (Whitmore 1984*a*, fig. 13.5).

A core taken through the peat at the dome centre contains pollen of the successive communities, with mangrove at the base. This demonstrates that as the dome gets higher by peat accumulation each community is replaced by the next in sequence.

Peat swamp forest

Description

This is a forest formation defined more on its special habitat than on structure and physiognomy. The Eastern peat swamp forests have a limited and distinctive flora.

Habitat and occurrence

In parts of Sumatra, southern Thailand, Malaya, Borneo, Mindanao, and west New Guinea a physiographic setting exists which has favoured the formation of peat. Since sea-level rose at the end of the last Glacial maximum (section 6.2), the rivers have deposited silt as levées and on flood plains. Swamps developed behind the levées and became less saline as their soil level was raised by continuing silt deposition at times of high river level. Mangrove became replaced by inland species whose litter failed to decay in the still salty, high sulphide, waterlogged conditions. This litter turned into peat which has continued to form up until the present day. The result is that now there are domed peat swamps up to 20 km across which are oldest in the centre. These carry up to six different forest types; developed concentrically on the dome (Fig. 2.16), each occurs at a particular height above the water table. The innermost type is an open, stunted forest. The outer forests reach 50 m tall and are a valuable timber resource.[38] Here some silt has been washed in. Further from the margin all the nutrient input has been from rainfall and the peat is extremely oligotrophic and acid. The peat, which is semi-liquid, reaches 13 m deep below the most developed domes. Agriculture is impossible on peat over 1–3 m deep and forests should be retained. Unfortunately, Indonesia already has failed examples of the ill-conceived conversion of deep peat to agriculture.[39] Lowland peat swamp forests in rain forest climates also occur in the Americas, including the pegass swamps of Guyana and parts of the Amazon basin (e.g. Tupinambara island east of Manaus), as well as in Africa (Kenya, Uganda and Congo) but these have not received detailed study. The total areas of tropical peat are 22×10^4 km^2 in Asia,

7×10^4 km^2 in America, and 3×10^4 km^2 in Africa[40] and most of it is under rain forest.

Freshwater swamp forests

Description (Figs. 2.17, 2.18)

Like peat swamp forests these are defined mainly on habitat. They are a very diverse assemblage of forest types flooded by river water, hence with soils richer in plant nutrients than peat swamp forest, and of fairly high pH (6.0 or more). Where flooding is periodic, either daily, monthly, or seasonally, further dimensions of variability are introduced, a group called periodic swamp forests. There may be a few centimetres of peat, or an organo-mineral (muck) soil may occur.

Habitat and occurrence

The Amazon, which has annual floods and is also influenced by tides to some 900 km from the mouth, has very extensive and diverse freshwater permanent and periodic swamp forests, known as varzea and igapo respectively, for white water (silt-laden) and black water rivers. Where unstable banks of alluvium occur, which are moved by every annual flood, extensive ephemeral grasslands develop during the low-water season. It is surprising to the visitor to see this mighty river and its main tributaries flanked by meadows and not by forest. The species occurring in swamp forests along blackwater rivers differ from those along more eutrophic white or clear waters. The alluvial plains of Asia once carried extensive swamp forests but little now remains as these have mostly been cleared for wetland rice cultivation. The Congo basin is about one-third occupied by swamp forests, many disturbed by man, and little studied.

Fig. 2.18. Freshwater swamp forest. Papua New Guinea.

Fig. 2.17. Seasonal freshwater swamp forest. Papua New Guinea.

2.3. FOREST MAINTENANCE—THE GROWTH CYCLE

The discussion so far has been about mature forests. The different formations all have distinctive structure related to which are particular physiognomic features and dispositions of epiphytes and climbers. But the canopy is not all

Fig. 2.19. Forest canopy gap formed by the fall of a single large tree. (Hallé *et al.* 1978, fig. 107.)
The medieval French word chablis is sometimes used to describe the canopy hole, plus the associated damage, and the fallen mass of branches, leaves, and twigs.

like this, because trees are mortal and eventually die. In reality, the canopy is in a continuous state of flux, with gaps (Fig. 2.19) developing from one of many causes, varying in size from tiny to huge. Gaps are colonized by seedlings which grow up to become saplings and then poles before attaining maturity. In order to analyse this dynamic nature of the forest canopy it is convenient to recognize a forest growth cycle consisting of gap, building, and mature phases, which are arbitrary subdivisions of the continual process of growth.[41] It follows that a forest consists of a mosaic of patches at different phases of the growth cycle (Fig. 2.20). This pattern in space reflects the processes of forest maintenance which continue through time. Where a tree dies of old age, its crown slowly dies back and then the limbs and finally the bole disintegrate. In this case a fourth, degenerate phase to the growth cycle exists. Commonly, however, a tree dies suddenly, struck by lightning or blown over (Fig. 2.21), or snapped off by wind. Wind may blow over several trees. In an extreme case, windthrow has created long narrow corridors to over 8 km long in Sarawak peat swamp forest. On steep slopes landslides occur, perhaps as a result of an earthquake. Volcanoes create mud flows (lahars) or ash or lava flows. Between 10–20° north and south of the Equator cyclones occur and these periodically clear huge swathes of forest, in the Caribbean, Bay of Bengal region, northeast Philippines, Queensland, and Melanesia. These various causes create canopy gaps of greatly different size and some places are more prone to extensive destruction than others. The forest structural mosaic varies from very fine to extremely coarse.

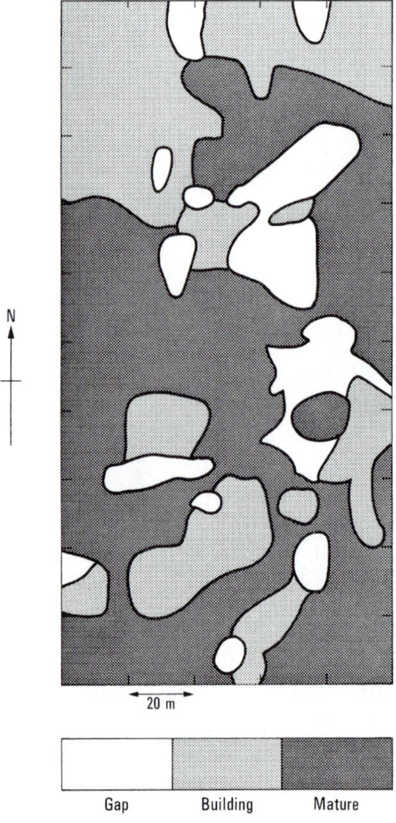

Fig. 2.20. Canopy phases on 2 ha of tropical lowland evergreen dipterocarp rain forest at Sungai Menyala, Malaya, 1971.
Long narrow gaps result from windfall of single moribund giant trees. Big gaps result from multiple windfall increasing an originally small gap. The extensive area at the north end of building phase forest results from regrowth after partial clearance in 1917. A plan like this is a valuable aid to interpreting forest structure and dynamics.

The Dutch forester F. Kramer created in the lower montane rain forest of west Java artificial

Fig. 2.21. View along a gap created by a single treefall. Note man to right of the upturned roots and soil (the so-called root plate). Sulawesi.

gaps of 0.1, 0.2, and 0.3 ha, equivalent to circles of 36, 51, and 62 m in diameter.[42] In the smallest gaps seedlings grew up which had established below the canopy, but in the larger two sizes they were replaced by a new set of different species which were not present before. This is what always happens as gap size increases, and the two kinds of species are called climax and pioneer, respectively.[43] They are discussed fully in section 7.2. Their essential characteristics are that climax species can germinate and their seedlings establish below a forest canopy, so these species can persist in the same place, the seedlings growing up after a gap develops. But if the gap is too big the climax species are replaced by pioneer species which germinate and grow fast after gap formation. The two essential features of pioneer species are that they need full light for both germination and seedling establishment. Their seedlings are therefore not found below canopy shade. They cannot therefore perpetuate themselves in the same place. Below pioneer trees, climax species establish and as the pioneers die off, one by one or in small groups, canopy gaps develop and the next growth cycle is based on these climax species. There is a floristic shift from one suite of species to the other, and this is what is termed succession (Figs. 2.22, 2.23).[44]

Any forested landscape is likely to have a patch somewhere which is recovering from a landslide or from multiple windthrow as a forest of pioneers, so this patch is in a state of succession. Elsewhere climax species will occur. These, as a class, are self-perpetuating, so their regrowth in gaps is cyclic replacement rather than succession. The landscape is in a state of dynamic equilibrium, not changing as a whole although small parts are in continual flux. It has aptly been termed a shifting mosaic steady state.[45]

Forests differ in the gap-forming processes to which they are prone. For example, Papua New Guinea[46] is a land of cyclones, earthquakes, volcanic eruptions, and periodic fires, all of which cause catastrophic destruction of big swathes of forest. The mosaic of structural phases is coarse and forests of pioneers are widespread. It is estimated that 8–16 per cent per century of the land surface of Papua New Guinea is disturbed by landslides (section 7.8), and 2 per cent of Panama, another earthquake-prone country.[47] Malaya by contrast has very little catastrophic destruction from these natural causes; fine structural mosaics and stands of climax species are widespread.

Gap-phase dynamics and the two ecological groups of species, pioneer and climax, will be described more fully in Chapter 7. It seems that most of the world's forests work according to this model.[48] Strongly seasonal tropical forests, in which fire is a major factor, are probably an

Fig. 2.22. Forest on the north coast of Kolombangara, Solomon Islands, dominated by overmature trees of the light-demanding climax species *Campnosperma brevipetiolatum* (CAMB) which is not regenerating itself. (See Whitmore 1974, fig. 2.3 for full species names.)

This forest resulted from massive disturbance, and unless that is repeated it will change in composition to resemble that of Fig. 2.23, whose species are already present in the lower part of the canopy.

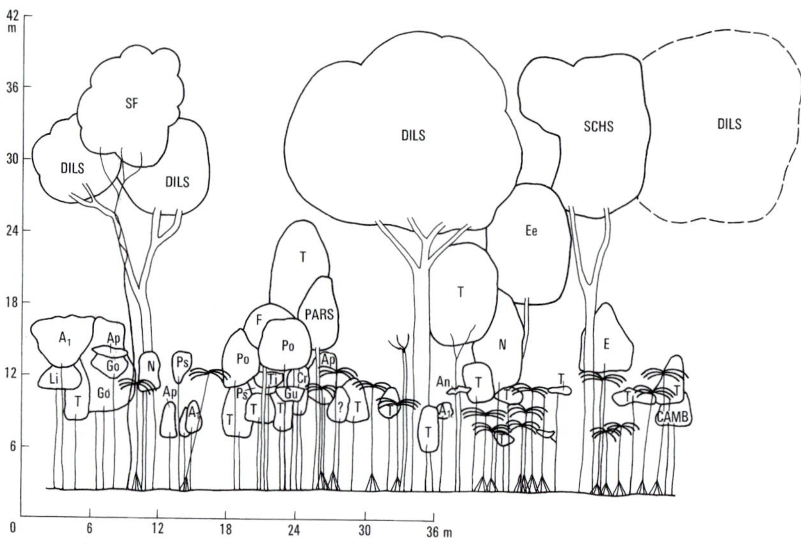

Fig. 2.23. Forest on the west coast of Kolombangara, Solomon Islands, dominated by the slow-growing, shade-tolerant climax species *Dillenia salomonensis* (DILS) and *Schizomeria serrata* (SCHS) which are regenerating themselves. (Whitmore 1974, fig. 2.4 and 1984*a*, fig. 17.20; see either source for full species list.)

exception. It is more likely that after disturbance resprouting rather than regeneration from seed is important; but we have no detailed ecological knowledge on how they work, which is still a serious deficiency awaiting an ecologist to apply modern approaches and insights.

Canopy layers

In this book the term forest canopy is used for the whole plant community above the ground. It is commonly described as being layered or stratified and this is a useful aid to description or analysis although, because the forest is dynamic with patches at all stages of the growth cycle, stratification is a simplification and abstraction.

A mature lowland rain forest has trees of many sizes. The tallest stand head and shoulders above the general level of the canopy and are known as emergents. They occur either alone or as groups. Single kapok trees (*Ceiba pentandra*), which reach immense size in Amazonia, are a common and conspicuous sight viewed from the

Fig. 2.24. Profile showing mature (ends) and building phases of the lowland evergreen dipterocarp rain forest at Belalong, Brunei. Plot area 60 × 7.5 m, all trees over 4.5 m tall shown. (Ashton 1964 in Whitmore 1984*a*, fig. 2.1; see latter for species names.)

Dipterocarps shown hatched; note how these still have tall, narrow, youthful monopodial crowns in the building phase, which change to sympodial, broader than deep, and with several large limbs in the mature phase.

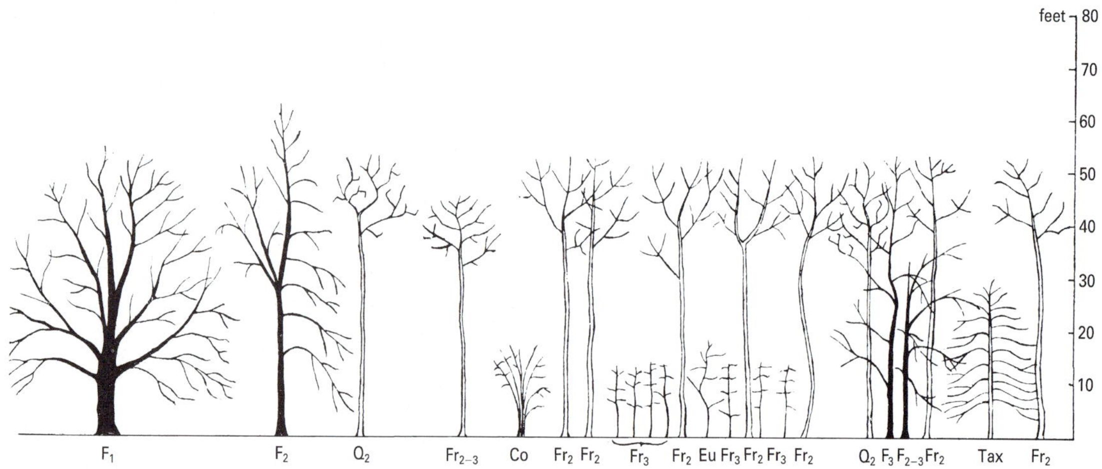

Fig. 2.25. Profile diagram of temperate deciduous forest, Sussex, England. F. *Fagus sylvatica* (beech); Co, *Corylus avellana* (hazel); Fr, *Fraxinus excelsior* (ash); Q, *Quercus robur* (oak); Tax, *Taxus baccata* (yew). (After Watt 1924, fig. 3.)

river (Fig. 1.6). By contrast, the rough, billowing canopy of a west Malesian forest results from groups of emergent species of Dipterocarpaceae (Figs. 2.3, 2.24). Strata have usually been depicted by a profile diagram, a side-view of a strip usually *c.* 60 × 6m. What such a strip depicts depends on where it lies with respect to the mosaic of structural phases as can be seen from Figs. 2.20 and 2.24. Before their use for tropical rain forests, profile diagrams were used to describe the structure of the forests of south England, which have a very simple structure, just one layer of trees over a shrub and a herb or ground layer (Fig. 2.25). It is probably by analogy to temperate forests that European scientists sought to see strata in tropical rain forests.

In addition to structural layering, which becomes confused by the growth cycle, different tree species habitually mature at different heights, and whole genera and even some families reach up to different parts of the canopy. For example, most Ebenaceae and Euphorbiaceae are small trees, most Burseraceae, Lecythidaceae and Sapotaceae reach the top of the canopy, and amongst emergents are many Leguminosae and nearly all rain forest Dipterocarpaceae. Far Eastern Myristicaceae are nearly all small trees, but in the neotropics there are larger canopy-top species, e.g. the important timber producing genera *Dialyanthera* and *Virola*. Young individuals, or trees dwarfed because of unfavourable growth conditions, obscure such layering and there is no evidence that habitual mature height falls into a small number of discrete classes. Stratification is most prominent in species-poor forest where groves occur of one or a few species, which is why semi-evergreen lowland rain forest is more conspicuously stratified than evergreen (p. 15). Crown form (p. 51) changes from monopodial, deeper than broad, to sympodial, broader than deep, as a big tree matures (Fig. 2.24). Some small tree species remain monopodial to maturity, e.g. Annonaceae, Ebenaceae, and Myristicaceae. It follows that layering of crown shape also occurs, and this is also easily seen on a profile diagram. The forest microclimate alters upwards through the canopy (Fig. 7.1). This probably triggers the metamorphosis of crown (section 3.2); it also leads to yet another kind of stratification, that of epiphytes and climbers (Fig. 3.18).

2.4. FLORISTICS

The humid tropics are extremely rich in plant species. Of the total of approximately 250 000 species of flowering plants in the world, about two-thirds (170 000) occur in the tropics. Half of these are in the New World south of the Mexico/US frontier, 21 000 in tropical Africa (plus 10 000 in Madagascar) and 50 000 in tropical and subtropical Asia, with 36 000 in Malesia. A few plant families are confined to humid tropical climates, e.g. Myristicaceae, the nutmegs; others are strongly concentrated there with a few temperate outliers, e.g. Annonaceae, the soursop family, Musaceae, the bananas, and Ebenaceae, the ebonies (whose temperate species provide the fruits called persimmon).[49]

There are similarities, especially at family level, between all three blocks of tropical rain forest, but there are fewer genera in common and not many species.[50] All three regions have abundant Leguminosae of subfamily Caesalpinoideae. Other big families include Annonaceae, Euphorbiaceae, Lauraceae, Moraceae, Myristicaceae, Rubiaceae, and Sapotaceae. On small areas Annonaceae, Euphorbiaceae, and Rubiaceae are nearly always among the ten most species-rich families in all three regions. America is characterized by numerous Lecythidaceae, the Brazil nut family, with 11 genera, and about 120 species. The most distinctive and unique feature of western Malesia is the abundance and species richness of Dipterocarpaceae.[51] Borneo for example has 287 species and 9 genera, and in many places most of the big forest trees belong to this single family. Conifers have many species in the East (Fig. 2.26) and are found at all altitudes. So far only one conifer has been found in the lowland rain forests of Africa and another in tropical America.

Fig. 2.26. The conifer *Podocarpus neriifolius* is a common and valuable timber tree found in lowland and lower montane rain forests throughout Malesia. Note the big, permanent, radial limbs on this mature specimen.

In flora Africa has been called 'the odd man out';[52] there are fewer families, fewer genera, and fewer species in her rain forests than in either America or Asia. For example, there are 18 genera and 51 species of native palms on Singapore island,[53] as many as on the whole of mainland Africa (15 genera, 50 species); Africa has only 4 species of bamboo, and Mt. Kinabalu (4101 m) in northern Borneo has almost the same number of ferns as the African continent.

There are also differences within each rain forest region. Not all species occupy the whole available area despite the absence of physical barriers; for example *Theobroma*, the cocoa genus, is confined to northwest Amazonia; and many species found in Sumatra do not reach New Guinea. These patterns are thought to have historical causes which will be discussed in Chapter 6. Other patterns are believed to relate to climate. For example, the rain forest flora of Ghana, herbs

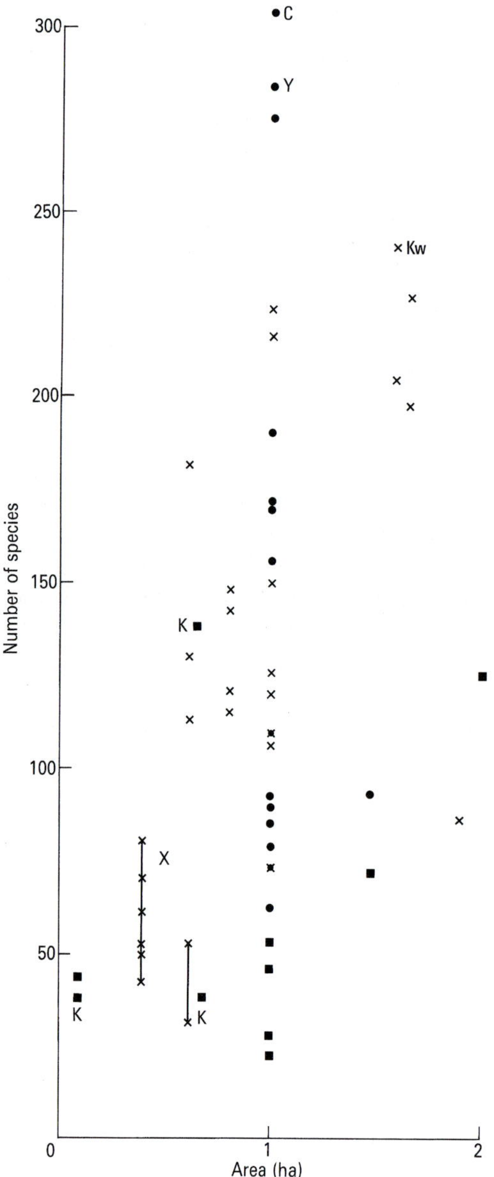

Fig. 2.27. Species richness among trees of 0.1 m in diameter and over on small plots in tropical lowland rain forest: ● America; × Eastern tropics; ■ Africa. Lines connect sample plots that lie close together. (Data of Whitmore 1984*a*, fig. 1.5; Gentry 1988*b*; Whitmore and Sidiyasa 1986; Whitmore *et al.* 1987; Valencia *et al.* 1994.)

Species numbers rise with increasing plot area as is shown on Fig. 2.28. In addition to this basic pattern, the African forests are species-poor except for one at Korup (K). Most of the Eastern tropics are species-rich but the richest forests of all are in America where, however, some are also very species-poor. The Yanamono (Y) and Cuyabeno (C) plots lie in the upper Amazon near the Andes (see Fig. 6.16b) and this region from Colombia to Peru is now believed to be richest in the tropics for birds, frogs and primates, as well as flowering plants, exceeding even Borneo. (Kw, Kalimantan, Wanariset-Samboja.)

and trees, alters clinally from west to east across the country, a distance of *c.* 300 km,[54] in parallel with increasing climatic seasonality.

To obtain a more precise picture of differences in species richness, Fig. 2.27 shows the numbers of tree species of 0.1 m in diameter or larger on small plots. The numbers vary from 23 ha^{-1} in Nigeria to 307 ha^{-1} in the Ecuadorian Amazon at Cuyabeno.

On the richest rain forest plots every second tree on a hectare belongs to a different species. It is difficult to conceive a forest much richer than the Ecuadorian plot, where 76 per cent of the species present had only one or two trees. This plot had 10 per cent of the entire tree flora of Amazonian Ecuador (and 16 per cent amongst trees ≥ 5 cm diameter).

Species-area curves have been constructed for several forests, and species numbers continue to rise over several hectares (Fig. 2.28).

On a slightly more extended scale a single 50-ha plot at Pasoh, Malaya, contained *c.* 830 species, 20–30 per cent of the total tree flora of the country down to the size measured (10 mm diameter).[55] An enumeration of 6.6 ha spread across 5 × 2.5 km of forest at Bukit Raya, Sarawak, had 711 species over 0.1 m in diameter, which is about half the trees of this size found in all of Sarawak.

To set these figures in context, the whole of Europe north of the Alps and west of Russia has only 50 indigenous tree species and eastern north America has 171.[56]

One component of the extreme species richness of many rain forests is the local endemics,

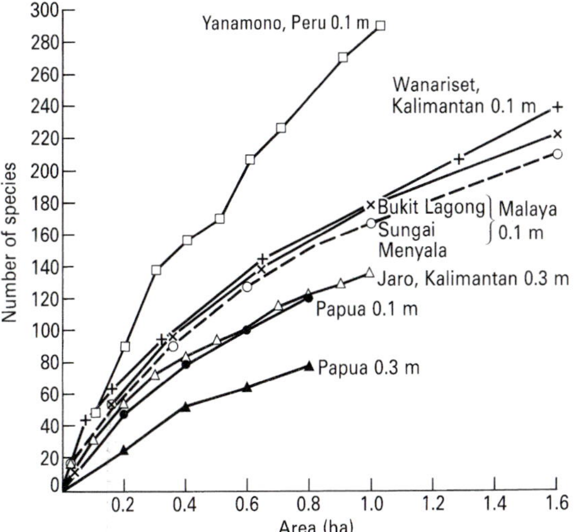

Fig. 2.28. Species area curves for tropical lowland evergreen rain forests. (After Whitmore 1984*a* and Gentry 1988*b*).

The Yanamono forest is the second richest yet found, every second tree on the hectare plot was a different species. These curves were mostly made by adding together the number of species found on contiguous subplots. See also Figs. 2.29, 6.16*b*.

namely species of very limited range.[57] For example, many trees collected in Perak, northwest Malaya, in the late nineteenth century have never been found elsewhere and, because the lowlands were soon afterwards largely converted to rubber plantations, are almost certainly extinct. The monospecific genus *Burkilliodendron album*, known from one collection from a limestone hill in Perak, has never been seen again and its habitat has been used for road metal. By contrast, although the broadleaf forest cover of England is reduced to *c.* 4 per cent, no species of woody plant has been lost from the flora.

Despite the great interest everyone has in the amazing plant species richness of tropical rain forests, there are very few total species counts. On a single 100 m^2 plot in wet lowland rain forest in Costa Rica, 233 vascular plant species were recorded (Table 2.3). This is by far the richest plant community ever enumerated on Earth. In three forests in western Ecuador, sample areas ten times bigger (0.1 ha), and each from ten non-contiguous subplots, had totals of 365, 173, and 169 species in perhumid, dry, and slightly seasonal forest respectively.[58] Even if the trees are excluded, all these four samples are amongst the richest plant communities ever encountered with 153, 251, 125, and 105 non-tree species, respectively. The Rio Palenque forest of western Ecuador has a total of 1030 species in 80 ha, one-quarter of them endemic; the whole of the British Isles have 1380 seed plant species, which helps put this figure in perspective.

The most nearly comparable species-rich plant communities outside the tropics are the Mediterranean heathlands of the Cape of South Africa (known as fynbos) and of southwest Australia. Here total species numbers are similar to trees alone in rain forest. Chalk grassland in England is extremely rich on a small scale and may have 32–33 species in a 0.5 m^2 quadrat, but the total flora of one site of a few hectares is only 50–55 species.[59]

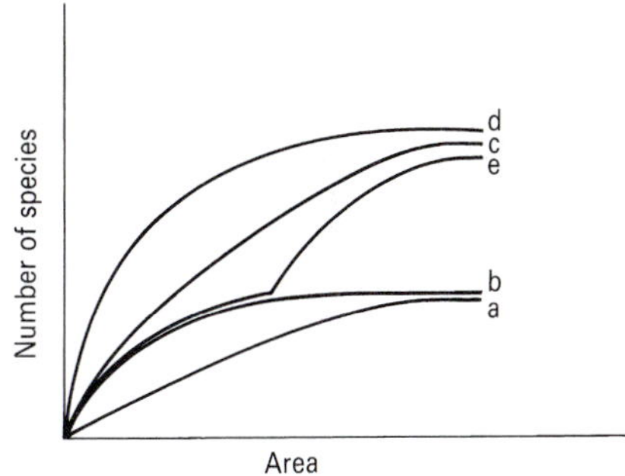

Fig. 2.29. The species-area curves of the different kinds of species diversity.

The two species-poor communities (a) and (b) have low alpha diversity and low and high beta diversity, respectively. Communities (c) and (d) by contrast are species-rich and have high alpha diversity and low and high beta, respectively. Community (e) differs from all the others in having two distinct constituent parts, namely it has higher gamma diversity; the parts each resemble (b). See also Fig. 2.28.

Table 2.3 Numbers of species and individuals in the different synusiae on a 100 m^2 plot in evergreen rain forest near Las Horquetas, Costa Rica

	Individuals	Species	
		Total	Cumulative total
Vascular plants:			
(a) Independent (free-standing) plants			
Plants ≤ 1 m tall	1349	132	132
(of which tree seedlings)	(566)	(64)	
Plants 1–3 m tall	144	36	140
(of which trees)	(134)	(5)	
Plants ≥ 3 m tall	38	18	144
(of which trees)	(35)	(4)	
(b) Dependent plants			
Climbers			
bole	233	21	165
woody (free-hanging)	68	20	174
Epiphytes	339	61	233
(of which aroids)	(90)	(17)	
(of which bromeliads)	(49)	(8)	
(of which ferns)	(87)	(9)	
Total all vascular	2171		233
Bryophyta:			
Liverworts	?	25	
Mosses	?	7	
Total all Bryophyta			32

From Whitmore *et al.* (1986) table 1
The count was made by a team of 11 people, removing one synusia at a time, and took a total of 192 man-hours. At this rate it would take 10 man-years to extend the plot to a full hectare.

These various data show that meaningful discussions of species richness must specify scale.[60] For example, we may usefully compare richness within rain forests by counting tree species on plots of *c.* 1 ha. This within-community diversity has been called alpha diversity. At the other extreme we can record species richness of a whole landscape made up of several communities, and this has been called gamma diversity. The fynbos is very rich with 8500 species on 89 000 km^2. It is made up of a mosaic of different floristic communities, each of which has rather few species. That is to say fynbos has low alpha and high gamma diversity. Within a single floristic community species replace each other from place to place. This gives a third component to richness, known as beta diversity. For example, within lowland rain forest there are differences in species within a single community between ridges, hillsides, and valleys. Figure 7.38 demonstrates how in Amazonian Peru species composition changes as the forest gets older and this is another example of beta diversity. The species richness of chalk grassland is due to high alpha diversity, beta diversity is low.

These different kinds of diversity can be shown on species-area curves (Fig. 2.29). Where numerous species occur on a small area the curve rises steeply. Where the minimum area is large the curve continues to rise a long way before flattening. These represent high alpha and beta diversity, respectively. A change in slope of the curve reflects transition to a different community, so if the curve has a series of steps before it flattens out this shows its full richness contains a component of gamma diversity.

2.5. NATURE OF THE TROPICAL RAIN FOREST COMMUNITY[61]

Now that the main kinds of variation in tropical rain forests have been described we can consider a question that has been very much discussed by tropical ecologists: namely, whether tropical rain forest is one huge floristic association varying haphazardly from place to place or whether distinct communities exist.[62] It has been shown in this chapter that there are in fact numerous kinds of variation. It is possible to arrange them roughly into a hierarchy of diminishing importance.

Biogeography

Variation can only operate on the species that are present, and over-riding all other reasons for it is the availability of flora. For example, dipterocarps dominate the rain forests of western Malesia and give those forests their unique characteristics which are referred to repeatedly throughout this book.

Disturbance

The influence of massive disturbance is the second most important factor. Forests regrowing after a cyclone or human destruction, for example, are dominated by pioneer or near-pioneer species (Fig. 2.22)[63] and because trees live a century or more (Fig. 7.35) rare catastrophes can have long-lasting effects.

Habitat

Major physical habitats bear different forest formations (Table 2.1) and this third cause of variation is of similar importance to major disturbance.

Variation within the formation

Topography

Further down the hierarchy, once these three factors have been allowed for, comes variation within a formation. This tends to be continuous, without sharp boundaries. It results from various causes. One kind is linked to geology, which manifests itself in various ways relating both to topography and to the chemical and physical properties of the soil. Within a rain forest formation it is common for some species to be associated with different topographic situations, especially with valleys or ridge crests.

Fig. 2.30. *Homonoia riparia*, a wiry shrub of swiftly flowing rocky rivers, is a typical rheophyte, with linear-lanceolate (willowlike) leaves called stenophylls. Yunnan, southwest China.

Rheophytes are a highly distinctive synusia or life-form community. They are commonest in the tropics, especially in Malesia where they reach greatest abundance in northern Borneo. The life-form is an adaptation to a very difficult habitat, an otherwise empty ecological niche, which has been colonized by only a few unrelated plant families. Van Steenis (1981) gave a full account of these peculiar plants.[64]

Fig. 2.31. The dipterocarp *Shorea curtisii* and the big, prickly, stemless bertam palm *Eugeissona tristis* commonly grow together in Malaya.

Fig. 2.32. (Right) Species patterns on a 50 ha plot at Barro Colorado Island, Panama. Trees over 0.2 m in diameter are shown. (a) *Trichilia tuberculata*, the commonest big tree, is ubiquitous. (b) *Poulsenia armata* is largely confined to steep slopes. (c) The pioneer species *Cecropia insignis* occurs in clumps not related to topography but which have developed in canopy gaps. (Hubbell and Foster in Sutton *et al.* 1983.)
The distribution of trees within a rain forest results from many factors which may interact and are not always easy to discover.

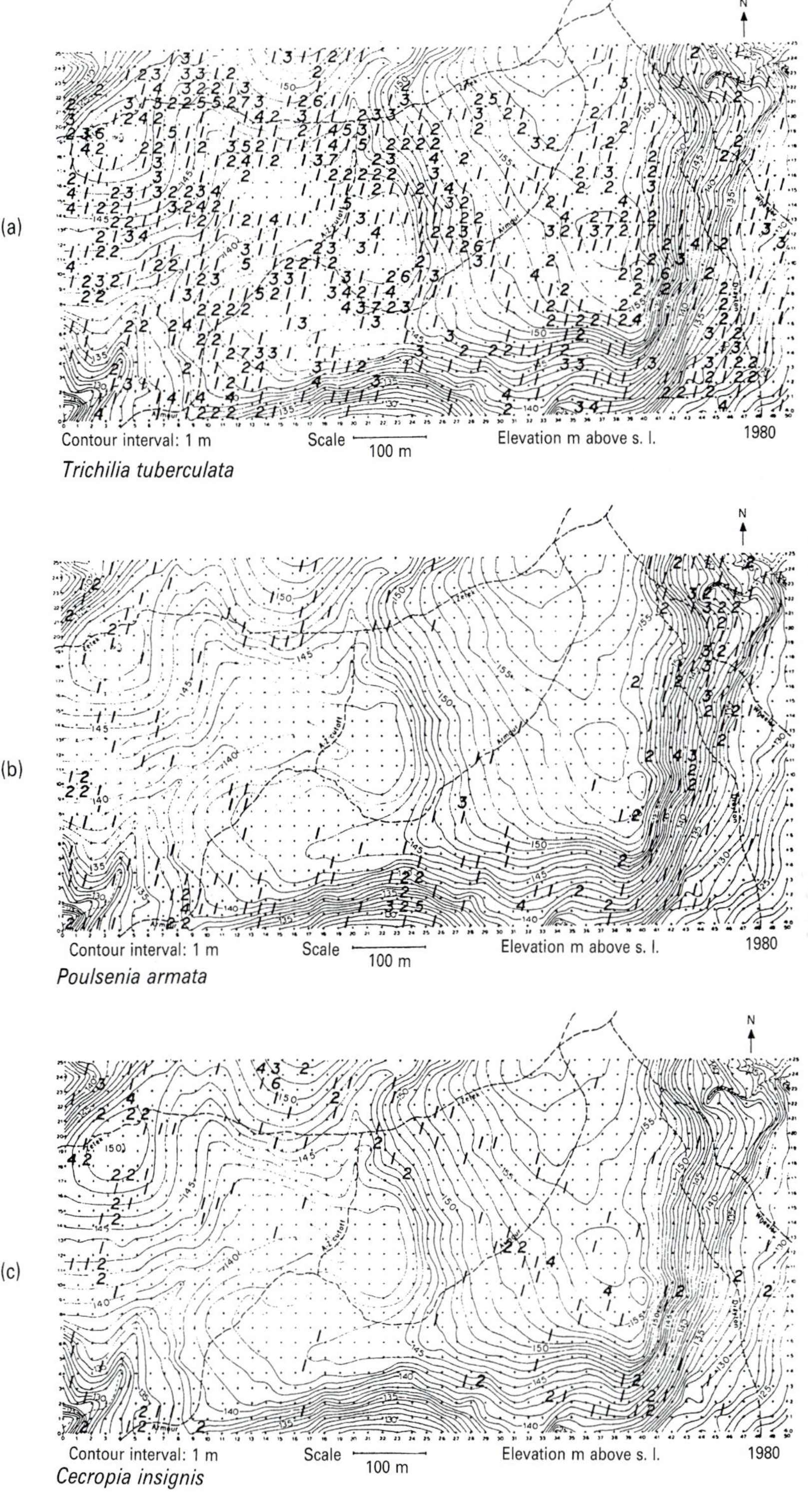
(a)
Contour interval: 1 m
Scale
100 m
Elevation m above s. l.
1980
Trichilia tuberculata
(b)
Contour interval: 1 m
Scale
100 m
Elevation m above s. l.
1980
Poulsenia armata
(c)
Contour interval: 1 m
Scale
100 m
Elevation m above s. l.
1980
Cecropia insignis

For example, at the edge of swiftly flowing rocky rivers and on their shingle banks, firmly rooted in crevices, is found the synusia of rheophytes (Fig. 2.30). On alluvial valley floors is found another set of species. For example, in Sarawak the occasionally flooded levées of well-watered, friable, deep, fertile soil support a very tall forest which includes the group of *Shorea* species producing illipe nuts, plus the Borneo ironwood (*Eusideroxylon zwageri*). An illustration of species that are commonest on ridge crests is the grey-crowned dipterocarp *Shorea curtisii* (Fig. 2.3) and its associate the stemless, prickly, giant bertam palm (*Eugeissona tristis*) in Malaya (Fig. 2.31). However, not all species are strongly linked to topography; Fig. 2.32 gives examples from Panama.

Soil

Most rain forest soils are low in plant nutrients and physical factors have a more important role than fertility in determining species ranges. Soil physical factors probably determined which species were more abundant over either granite or shale in a survey of 26 628 trees on 676 ha in Ulu Kelantan, Malaya.[65] At Korup, Cameroon, there large emergent species of Leguminosae/Caesalpinoideae (*Microberlinia bisulcata*, *Tetraberlinia bifoliolata*, *T. moreliana*) form groves *c.* 600 m across which have been shown to be associated with low concentrations in the wet season of extractable soil phosphorus.[66]

Presence of seedlings

Another more subtle level of variation in tree species composition is dependent on the presence of seedlings on the forest floor at the time a gap develops. Different species establish their seedlings better on different microsites (section 7.3). Species also differ in how frequently they produce fruit and hence replenish their seedling bank (section 3.3); for example, the dipterocarp Borneo camphorwood (*Dryobalonops aromatica*), which dominated certain Malayan forests, probably owed its abundance to the 'reproductive pressure' of more frequent fruiting than other dipterocarps. At this lowest level of variation, which has been named gap-phase replacement, chance plays a strong role: trees must flower, the flowers must set fertile fruit, dispersal must operate, seeds germinate, seedlings establish and survive damage by falling debris or from herbivores, and then a canopy gap must develop over them within a few years before they die in the dimly illuminated forest undergrowth.

As new studies are made to investigate the fundamental problem of the nature of variation from place to place in tropical rain forest, they confirm the model of a hierarchical series of kinds of variation, and illuminate the part of the hierarchy the particular study happens to relate to. Nowadays, attention has increasingly swung away from this problem to address others, which form the subject of the next few chapters.

2.6. WHAT ARE TROPICAL RAIN FORESTS?—CHAPTER SUMMARY

1. The wetter tropical climates have closed forests, collectively described as tropical moist forests. Tropical rain forests occur where there is only a short dry season or none. Monsoon (tropical seasonal) forests occur where the dry season is stronger.

2. Distinct forest formations occur in different habitats (Table 2.1), which are usually sharply bounded. They differ mainly in structure and physiognomy and the same formations occur in similar sites throughout the humid tropics.

3. Forests consist of a mosaic of gap-phase, building-phase, and mature-phase forest (Figs. 2.20, 2.24). These phases are arbitrary subdivisions of a continuous forest growth cycle. The spatial pattern of structural phases represents processes of change with time.

4. There are two contrasting ecological species groups whose key features are that climax species can germinate and establish seedlings below a canopy, whereas pioneer species require full light. Climax species grow up in small gaps and pioneer species in big gaps.

5. There are differences in species composition at all scales, between the three tropical areas and within them. Tree species numbers per hectare on small plots vary from *c.* 20 to over 300. Africa is poorest. These factors have historical explanations (Chapter 6).

6. Tropical rain forest communities show variation at a range of scales which can be arranged in a rough hierarchy. Availability of flora comes first, then differences between formations and that resulting from massive disturbance such as cyclones. Within formations variation is more continuous and less obvious and is related to topography (riverine forests and rheophytes are especially distinctive), to soil; as well as to seedling arrival and success at a particular spot, for which there is a strong element of chance.

3

Plant life

> The traveller walks out of an abandoned resthouse on to what was a lawn 18 months before, to find self-sown saplings of *Tectona grandis* [teak] 5 m high. They have long internodes, and leaves the size of dinner plates. He feels like Alice in Wonderland. Or he notices the weeds growing on abandoned farm land, including such objects as *Vernonia conferta*, a composite with vegetative parts like a rosette of *Taraxacum officinale* [dandelion] 3 m across, on a stalk up to 4 m high...[67]

The exuberance of plant life in the humid tropics continues to dazzle scientists from the more sober temperate biomes. More 'purple passages' have been penned on lowland evergreen rain forest than any other vegetation type. Trees have a great diversity of form and size and some have the uniquely tropical attributes of huge buttresses or trunk-borne flowers. They support a wealth of climbers and epiphytes: one stands and marvels. Bamboos (Fig. 3.1), palms (Figs. 2.31, 3.4, 6.3), cyclanths, pandans[68] (Fig. 3.2), stranglers (Fig. 3.3–3.5, 10.4), and banyans add peculiar extra dimensions. As acquaintance deepens some genera are found to have numerous species growing together, distinguished in the forest by details of bole, bark, buttress, and leaf. Dipterocarpaceae in western Malesia is the extreme example. Of *Shorea* M. Jacobs has written:[69]

> complete fugues could be composed from the leaves ..., varying in size, thickness, venation and hairs; with

Fig. 3.1. *Schizostachyum grande* showing the characteristic clumped growth of bamboos. Malaya. This elegant species becomes abundant in seriously disturbed lowland evergreen rain forest in the mountains at 600–800 m elevation.

Fig. 3.2. *Sararanga sinuosa*, showing the typical open branching and strap-like leaves of the Pandanaceae, dendroid monocots of the Old World tropics. Solomon Islands.

Fig. 3.3. (Below left) Strangling fig, showing multiple descending and anastomosing roots wholly obscuring the trunk of the host tree. North Sumatra.

Fig. 3.4. (Below right) Strangling fig with fan palms, *Licuala ramsayi*. Mission Beach, Queensland. © Ted Mead/Woodfall Wild Images.

Fig. 3.5. The development of a strangling fig (From Corner 1940 as fig. 2.25 in Whitmore 1984*a*). Strangling figs occur in all parts of the tropics. In Asia, *Wightia* and *Metrosideros* and in the New World, *Clusia* also have strangling species.

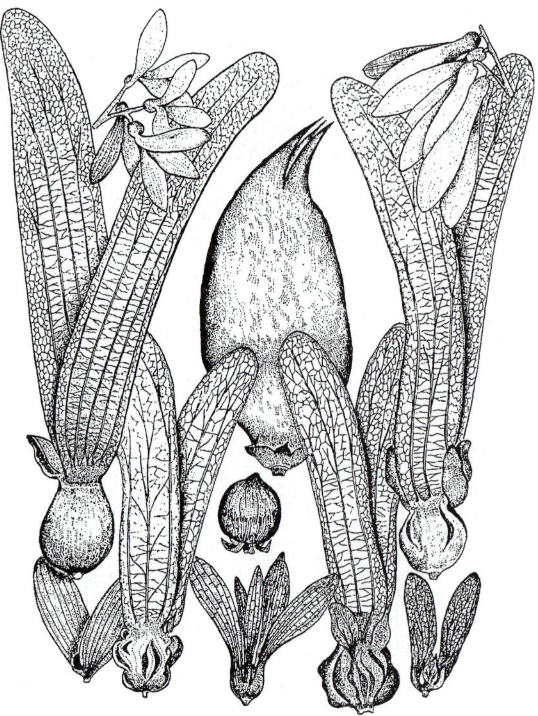

Fig. 3.6. Fruits of a miscellany of Dipterocarpaceae. (FAO 1985.)

The family is named from the two-winged fruits of *Dipterocarpus* (Greek di—double, pteron—wing, karpos—fruit). The wings are the flower sepals which become elongated and may be 2, 3, or 5 in number or absent.

Fig. 3.7. The monopodial undergrowth treelet *Agrostistachys longifolia*, here seen with an erect tuft of pale green young leaves. Singapore.

Fig. 3.8. (Below left) *Eucalyptus deglupta*, twelfth tallest tree species in the world, here *c.* 60 m tall. New Britain.

This is the only rain forest species of *Eucalyptus*. It is a pioneer of riverine alluvium from Sulawesi east to New Britain and is now commonly grown in plantations throughout the humid tropics.

Fig. 3.9. (Below right) *Eucalyptus deglupta* has very attractive bark, sloughing as long papery scales to reveal pale green fresh surfaces that darken through shades of green to become warm reddish brown.

Table 3.1 The World's tallest trees

Height (m)	Species (family)	Local name	Origin
143	*Eucalyptus regnans* (Myrtaceae)	Blue gum	Victoria, Australia
115	*Sequoia sempervirens* (Coniferae)	Californian redwood	California, USA
107	*Eucalyptus regnans* (Myrtaceae)	Blue gum	Victoria, Australia
96	*Sequoiadendron giganteum* (Coniferae)	Wellingtonia	California, USA
89	***Araucaria hunsteinii*** **(Coniferae)**	Klinki pine	New Guinea
87	***Koompassia excelsa*** **(Leguminosae)**	Menggaris, tualang	Sabah, Malaysia†
85	*Abies nobilis* (Coniferae)	Noble fir	Washington, USA
76	***Dryobalanops lanceolata*** **(Dipterocarpaceae)**	Kapur	Sarawak, Malaysia
76	*Eucalyptus grandis* (Myrtaceae)	Flooded gum	New South Wales, Australia
75	***Shorea superba*** **(Dipterocarpaceae)**	Balau	Sabah, Malaysia
75	*Agathis australis* (Coniferae)	Kauri	New Zealand
71	***Eucalyptus deglupta*** **(Myrtaceae)**	Kamerere	New Britain
70	***Agathis dammara*** **(Coniferae)**		Sulawesi, Indonesia
70	***Ceiba pentandra*** **var. *caribaea*** **(Bombacaceae)**	Kapok	Africa

Various sources, including Richards P.W. (1996)
Rain forest species shown in bold
†Relict tree on BAL Estate near Tawau

countermelodies of fruits and stipules; a single fermenta [flourish] for a particularly large flower; and a tremolo for an aberrant calyx. And from time to time the recurrent melody of the flowers' exquisite scent.

Just so, Fig. 3.6. If the kaleidoscope of striking impressions is analysed we see that trees are the predominant life form, in all sizes from unbranched pygmies with a single apical tuft of leaves (Fig. 3.7), which never get taller than 1–2 m, to emergent giants, which include some of the world's tallest (Table 3.1; Figs. 3.8, 3.10). Shrubs (i.e. woody plants with several main stems) are rare. Forest floor herbs are patchy, and much of the surface is bare except for a, usually thin, layer of leaf litter. On landslips and along rivers giant herbs often form thickets, gingers and Marantaceae everywhere, bananas in Asia (Fig. 3.11), *Heliconia* in America and Melanesia, bamboos in slightly seasonal climates.

Fig. 3.10. A giant *Shorea curtisii*. Note the two men up the tree. Brunei. P.S. Ashton gives instructions (1957).

Fig. 3.11. Extensive clumps of the wild banana *Musa truncata* colonizing a bared roadside, with two tree ferns, *Cyathea contaminans*. 1500 m elevation, Malaya with P.J. Grubb (1977).

Dependent on the free-living autotrophic forest plants are a few heterotrophs. There are many Loranthaceae (mistletoes), as hemiparasitic epiphytes. Santalaceae, which includes the sandal-wood (*Santalum*), is a family of small hemiparasitic trees. Full parasites include *Balanophora* (Fig. 3.12) and *Rafflesia* (Fig. 3.13), which in *R. arnoldii*, has the world's largest flower, 1 m in diameter. Saprophytes are uncommon and of only a few families, e.g. Burmanniaceae (Fig. 3.14).

Fig. 3.12. *Balanophora latisepala* showing a male inflorescence. Lower montane rain forest, Malaya. Balanophoras are heterotrophs, they have no leaves or chlorophyll and are parasitic on tree roots.

Fig. 3.14. *Thismia aseroë*. Tiny saprophytes of the rain forest floor which live on dead plant parts and have only the flowers above ground. Lower montane rain forest, Malaya.

Fig. 3.13. *Rafflesia pricei*. Note fly for scale. Sabah. Rafflesias are parasites which live inside the stems of the woody climbing vines *Tetrastigma* from which only the flowers emerge. These are blotched red and white, smell of carrion, and are pollinated by flies.

3.1. CLIMBERS AND EPIPHYTES

There are two different and distinct sorts of climber. Big woody climbers are those that hang freely from trees, for example *Gnetum* (Figs. 3.15, 3.16) with its hooped stem, gnarled *Bauhinia*, and in the Old World, especially Asia, the spiny climbing palms with scaly fruits called rattans. Big woody climbers have their leaves up in the top of the canopy, and some have crowns as big as any tree. They are strongly light-demanding and grow up in canopy gaps besides being abundant on forest fringes. The other kind of climber adheres to the tree trunks by specialized roots and may completely hide the bark beneath (Fig. 3.17). These are called bole climbers. They are especially common in very humid climates and uncommon or rare in forests that experience a strong dry season. Bole climbers are usually mixed with epiphytes and with a group of species called hemi-epiphytes which commence life as bole climbers but

Fig. 3.15. A big woody climber, *Gnetum* sp, showing the distinctively hooped stem of this genus. Malaya.

Free-hanging big woody climbers like this are a characteristic life-form of many lowland rain forest formations.

Fig. 3.16. Cauliflory: *Gnetum cuspidatum* with its seeds borne on the stem (note swollen, hooped nodes). Malaya.

Fig. 3.17. (Right) Bole totally clothed by aroid bole climbers, (probably *Scindapsus* sp.). Big woody climber to right. Sumatra.

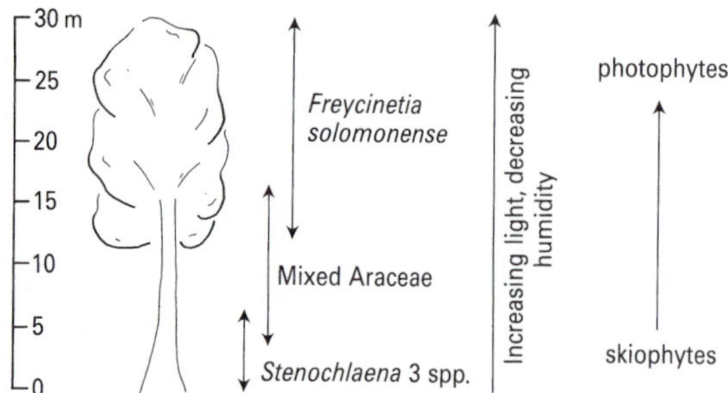

Fig. 3.18. Bole climbers occur in zones. Lowland rain forest, Kolombangara, Solomon Islands. (Whitmore 1974.)

Fig. 3.19. *Platycerium*, the stag's horn fern, an epiphyte of tree crowns; on *Campnosperma auriculatum*. Singapore.

This fern has two sorts of fronds, one sort form a nest in which humus collects, the others, the fertile fronds, are pendent and much divided.

Fig. 3.20. *Drynaria sparsisora*, another nest-forming epiphytic fern of tree crowns, forms gigantic masses, seen here in the crown of a *Toona australis* emergent tree flowering after it has shed its leaves. Queensland.

Fig. 3.21. Two epiphytic orchids *Maxillaria eburnea* (front) and *Diothonea imbricata* (behind) from the montane rain forests of Guyana. (Lindley's *Sertum Orchidacearum* 1838, plate 40.)
These are photophytic epiphytes of tree crowns. Crown epiphytes live in a periodically hot, dry microclimate and have many xeromorphic features. In these orchids the leaves are thick and leathery and the stems have swollen water-storing leaf bases called pseudobulbs.

become epiphytes by dying off behind; by this curious procedure they slowly climb the trunk of the supporting tree. Bole climbers and epiphytes are zoned (Fig. 3.18), being specialized to the different forest interior microclimates (Fig. 7.1). In the very humid, dark lower layers shade-dwellers, or skiophytes, occur. Some, for example, the filmy ferns, are poikilohydric: they have thin leaves which have the capacity to rehydrate without damage after desiccation. On the upper boles, in the crowns and in canopy gaps photophytes occur (Figs. 3.19–3.23). These are homoiohydric; they resist desiccation by various adaptations. The leaves are leathery with thick cuticles, there may be water-storage organs, e.g. the pseudobulbs of some orchids; and bromeliads, abundant in the neotropics, have leaf bases arranged to enclose a space or tank which collects water. Many photophytes have crassulacean acid metabolism (CAM). Some epiphytes trap nutrients by collecting falling detritus amongst their leaf bases. Humus then develops, and roots push out into it, for example, the stag's horn and bird's nest ferns (*Platycerium*, *Asplenium*: Figs. 3.19, 3.23).

Fig. 3.22. *Taeniophyllum* is a bizarre genus of Old World epiphytic orchids in which there are no leaves and photosynthesis takes place in green, flattened roots. The name means 'tapeworm leaf'. Malaya.

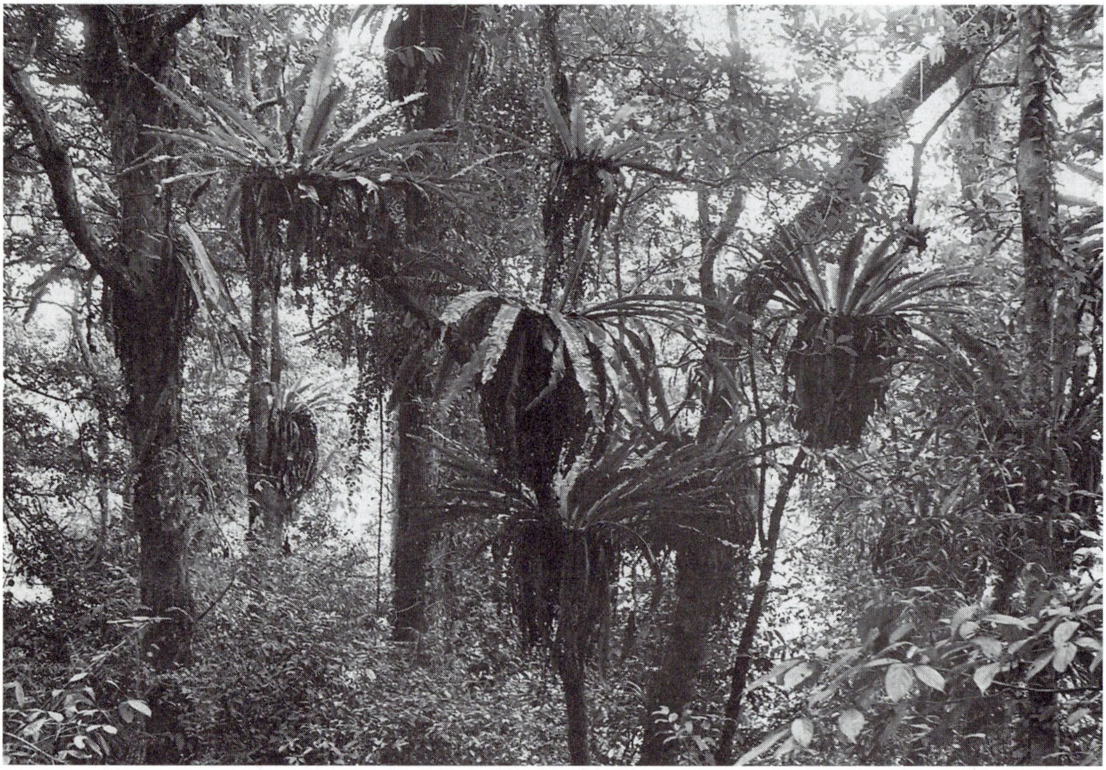

Fig. 3.23. *Asplenium nidus*, bird's nest fern, Old World tropics. Humus collects in the nest-forming frond bases. Java.

Other epiphytes may colonize and 'aerial gardens' form. The numbers of epiphyte species increase with wetness of climate. Counts in forests in northwestern tropical America have found 9–24 species in dry forests and 238–368 species in very wet forests, respectively 2 per cent and 23–24 per cent of the local flora.[70]

On old leaves a film of a special class of epiphytes, called epiphylls, often grows. These are mostly cyanobacteria, green algae, bryophytes, lichens or small filmy ferns. They must impair leaf function, and have been shown to do so in coffee plantations.

3.2. TREES

Crowns

Crowns have a single leading apical shoot or many, representing monopodial and sympodial construction, respectively. Most trees have monopodial crowns when they are young; many change at maturity. The plant with just an apical tuft of leaves (Fig. 3.7) has the extreme monopodial construction, a form best exhibited by palms (Fig. 6.3). There are no branches, just the terminal leaves. In palms the leaf bases conceal the delicate, usually edible, single apical bud and are often spiny or fibrous. Very few palms can resist frost and the family is essentially tropical. If the bud goes, so does the stem: 'like a foolhardy gambler the palm stakes all on a single card'. Many palms grow by a tufting habit, as a result of basal sympodial branching, and this is common amongst Monocotyledons as a whole, as found also, for example, in gingers, bamboos, and bananas[71] (Figs. 3.1, 3.11). The apical leaf tuft may trap falling detritus, and this has been claimed to provide a source of humus and nutrients for the tree. Other monopodial crowns have lateral branches, and these are often whorled. Where the lateral branches are in tiers, the crown has a pagoda-like appearance (Fig. 3.24). In the monopodial juveniles of forest giants the lateral branches are short-lived and eventually fall off. They behave like giant pinnate leaves. In these big trees at maturity, permanent lateral limbs develop (Figs. 2.26, 3.25, 3.29), and on them the juvenile tree form may be repeated—so-called 'reiteration'. The metamorphosis from juvenile to mature form occurs in big trees when the canopy top is reached,[72] presumably triggered by the change in microclimate (Fig. 7.1). Knot-free timber is produced below the sympodial crown. Trees that are tall in virgin forest are

Fig. 3.24. *Fagraea crenulata* has a tiered, pagoda-like crown. Singapore.

Tropical trees have a huge diversity of crown form whose construction has been shown to result from the interaction of three basic processes. Crown form is a valuable aid to species identification, together with bark, buttress, and bole: there is no need to rely on flowers or fruit.

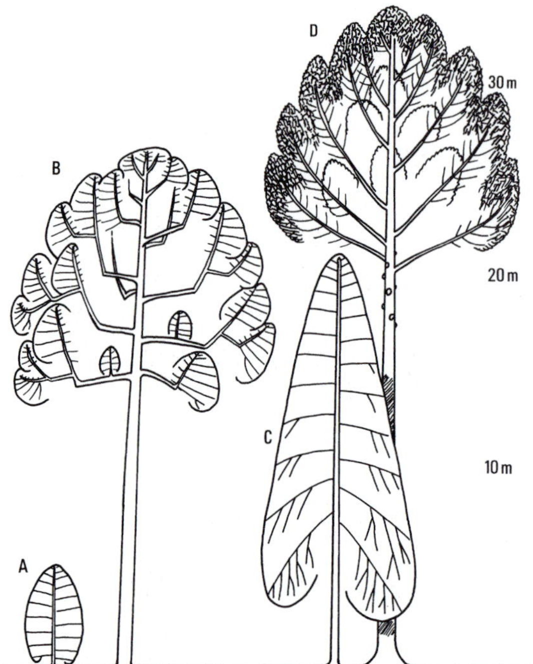

Fig. 3.25. Crown construction of two dipterocarps. *Shorea mecistopteryx* (A, B) and *Dryobalanops aromatica* (C, D). (Hallé and Ng 1981, in Whitmore 1984*a*, fig. 2.9.)
The juvenile tree (A, C) has monopodial crown structure. At maturity the crown is metamorphosed into a sympodial structure (B, D) which has numerous small subcrowns. In *S. mecistopteryx* these are reiterations of the juvenile tree.

usually much shorter where grown as specimens, in arboreta, or in forest regrown on open land.

Tree crowns in fact always have a precise construction[73] determined by the interaction of three main factors: apical versus lateral growth, as just discussed; radially symmetrical versus bilaterally symmetrical lateral meristems; and intermittent versus continuous growth. The construction is most easily observed in juveniles before metamorphosis. Twenty-three different crown 'models' have been recognized. For example, the papaya (*Carica papaya*) has apical growth with lateral inflorescences, is radially symmetrical, and grows continuously. There is little correlation between crown architecture analysed this way and taxonomy; for example, Euphorbiaceae have numerous models but Annonaceae and Ebenaceae all have monopodial, radially branched crowns of intermittent growth. Nor is there any correlation with ecology. The adaptive significance of architectural models is unresolved.

Leaves

The first impression is of gloomy, dull green, uniform foliage. This is quickly dispelled by closer inspection. There is actually a great diversity in leaf size, shape, nerve-pattern, margin, texture, and colour. Lamina sizes amongst lowland evergreen rain forest trees are mainly but not entirely notophyll and mesophyll (Fig. 3.26), but as was described in Chapter 2 microphylls are predominant in heath and upper

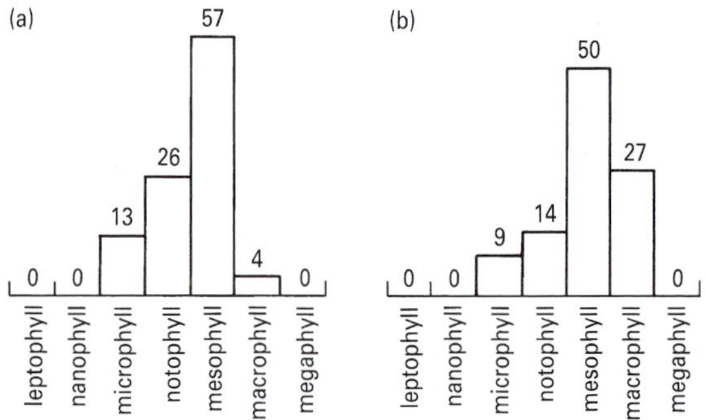

Fig. 3.26. Leaf size spectrum. (a) Lower montane rain forest (elevation 1710 m). (b) Lowland rain forest (520 m). Amazonian slope of the Andes in Ecuador. (After Grubb *et al.* 1963, fig. 5.)
Characteristically, the lowland forest has larger leaves. The leaf size spectrum has proved a useful tool in the analysis of rain forest physiognomy. Here the percentage of species with leaves of different size is shown (sometimes number of individuals is used). The spectrum was proposed by Raunkiaer (1934) and modified by Webb (1959) who introduced notophylls (2025–4500 mm^2) for part of the mesophyll class. They should be consulted for further details. Leaf area can be estimated with sufficient accuracy as 2/3 lamina length (to base of drip tip) × breadth.

Fig. 3.27. The giant palmate leaf of *Trevesia cheirantha*, 'Ghost's Footprint'. Malaya.

Fig. 3.28. Drip tips of *Ficus beccarii*. Malaya.

montane forests. In lowland rain forests, species with larger leaves are found in the lower part of the canopy where it is most humid and cool and with less water stress. Pinnately and palmately compound leaves are frequent in lowland rain forest (Figs. 2.5, 3.27). The biggest leaves are found amongst giant monocotyledonous herbs, the aroids (Figs. 1.8, 5.6), gingers and their relatives (Zingiberales), and palms.

Leaves commonly possess drip tips (Fig. 3.28). Recent research has demonstrated these really do hasten drainage of water films, which probably therefore retards the growth of epiphylls, as well as reducing loss of soluble nutrients by leaching.

A flash of colour glimpsed up in the canopy, as seen from the forest floor, or a coloured crown seen looking out over a hillside, is likely to be young leaves not flowers. Before full expansion the leaves of many species are tinged reddish, though they may be white and in a few rare cases are blue. This very striking characteristic of tropical rain forests is commonest in shade-tolerant species. It may be a defence against herbivores.[74] The young leaves have not yet become tough and inedible (cf. p. 82). They do not yet have chlorophyll and their delayed greening may make them less nutritious, and also may save the plant potential predation of valuable resources. Even so, young leaves are the site of nearly all herbivory. Unlike temperate deciduous forests, there are no spectacular autumn colours in the tropical rain forest. In fact, brightly coloured senescent leaves are sufficiently uncommon to be a useful aid to identification, for example, *Elaeocarpus* leaves wither a clear red.

Boles

The interior of rain forest is often likened to a cathedral with giant columns reaching up to a

Fig. 3.29. Cylindrical bole of the valuable timber species *Agathis robusta* ssp. *nesophila*. Highlands of Papua New Guinea.
The smooth bark with distant papery scales has scattered whitish microlichens. The big permanent limbs of the mature crown are visible, so are scars left by the temporary branches of the juvenile tree.

Fig. 3.30. No one can fail to recognize a mature *Pertusadina* tree because of its totally bizarre latticed trunk. Malaya.

leafy roof (Figs. 2.24, 3.29). Boles in fact commonly taper slightly, for which foresters make allowance when constructing volume tables. They may be fluted, and occasionally take on a bizarre form (Fig. 3.30).

Buttresses

Buttresses (Figs. 3.31, 3.32) are prominent in some forest formations. They differ in shape, size, and thickness between species and families; for example, they are particularly marked in Bombacaceae. Buttresses are tension structures, resonating when struck with an axe, and are mainly found on uphill sides of trees and counterbalancing asymmetric or epiphyte laden crowns. An analysis of the stresses to which a tree is subject and the strength of its wood shows buttresses are indeed of structural importance, helping to support the tree.[75]

Fig. 3.31. Steep plank buttresses of a big kapok tree, *Ceiba pentandra*. Ghana.
Kapok reaches 70 m tall in Africa; this variety (var. *caribaea*) is the tallest rain forest tree in the continent and also occurs in South America (Fig. 1.6) together with the other three species of *Ceiba*. It is likely that *C. pentandra* was dispersed on sea currents to Africa where it now grows wild from Senegal to the Great Rift Valley and Angola. The useful fibre kapok from inside the fruit is derived from a small spineless variety (var. *pentandra*) which is cultivated throughout the tropics (Fig. 10.2).

Fig. 3.32. Bombacaceae are characterized by particularly huge buttresses, which attain Brobdingnagian size in *Huberodendron duckei*. Brazilian Amazon.

Fig. 3.33. Flying buttresses and stilt roots of *Uapaca guineensis*. Ghana.

Flying buttresses occur in some families (Fig. 3.33), and are common in swamp forests. Stilt roots are another feature. They are very characteristic of pandans, and are also found in palms, some mangroves (Fig. 2.14), and a few other trees.

Bark

Bark is extremely diverse.[76] Colour varies from coal black in ebony (*Diospyros*) and some Myristicaceae to white in *Tristaniopsis*,[77] through bright rufous brown (*Eugenia*). *Calycophyllum* of the neotropics has bark like burnished copper. The bark surface may be smooth, scaly, fissured, scrolled, or dippled (Figs. 1.4, 2.26, 2.31, 3.9, 3.29). Smooth barks have a long persistent surface and these often house lichens which may be specific, e.g. *Disopyros* commonly has a dark green microlichen. The bark may contain a latex, resin, or coloured sap.

The botanist, confused by Jacobs' cacophony of fallen leaves, uses the crown, bole, buttresses, and bark to help him with identification. Flowers are seldom present and are hard to see or to examine high in the canopy but the characters of the living tree are so diverse they provide ample clues for identification. The most useful forest Floras are based on these 'field' characters and use them entirely for their keys.[78]

Roots

These are, for obvious reasons, less well known than the parts above the ground. Some rain forest species have a deep tap root. Others have 'sinkers', roots that descend vertically from spreading roots or buttresses. Most of the root biomass is in the top 0.3 m or so of the soil and there is sometimes a concentration or root mat at the surface. The relatively few deeper roots may be important in gathering nutrients from decomposing rock or which have leached down; species with such roots have potentially an important role in low-input sustainable agriculture based on tree and crop mixtures (p. 163). Deep roots are important in tapping water in dry weather (see p. 17). Fine roots are difficult to study; it is hard to extract them from the soil and when extracted to see which are alive. Roots up to 2 mm in diameter form 20–50 per cent of the total root biomass[79] and their believed rapid turnover is probably a significant part of ecosystem nutrient cycles (Chapter 8).

It has recently been realized that, as in other forests, in the rain forest nearly all trees have mycorrhiza. These are mainly endotrophic, vesicular arbuscular in type, but a few groups, some very important (Dipterocarpaceae, Fagaceae, *Eucalyptus*, Leguminosae/Caesalpinoideae), have ectotrophic mycorrhiza. Nitrogen-fixing nodules

occur in most Leguminosae (the bacterium *Rhizobium*), and *Alnus* and *Casuarina* (the fungus *Frankia*).

3.3. SEASONAL RHYTHMS

Shoot and leaf growth[80]

Tropical rain forests are evergreen and the climate is benign to plant life, always warm, and moist for all or most of the time. One might expect growth to be continuous. A few trees and other plants have indeed been found to grow all the time, as measured by production of nodes, each bearing one or a few leaves. The pioneer tree *Macaranga tanarius* is an example (Fig. 3.34). Internode elongation rate, another measure of growth, is constant in this species but in trees of other species, despite continuous production of nodes at a steady rate, elongation rate, and hence height growth rate, increases through time (Fig. 3.35). The exact measure used for growth matters; even with *Macaranga tanarius* we do not know if meristem activity is continuous. Palms exhibit continuous growth and the hoop-like leaf scars left on the trunk show elongation is also steady, though stem diameter waxes and wanes with growth conditions (Fig. 3.36).

Most rain forest trees, however, exhibit intermittent shoot growth (Fig. 3.37). A number of nodes form with their associated flush of new leaves, sometimes brightly coloured at first, and often as pendent, hanging tassels; followed by a resting period. The intermittent growth of the shoot tips is seldom reflected by growth rings in the wood, and where it is these are not annual and often not annular either. Rain forest trees, unlike those of seasonal climates, cannot be aged by counting wood rings (section 7.7).

Intermittent growth inhibits the build-up of fungi or insects which attack young leaves. In Malaya young leaves of *Hevea* are often attacked by two fungi, a mite, and a thrip any of which occasionally causes defoliation, even though the tree produces leaves in flushes.

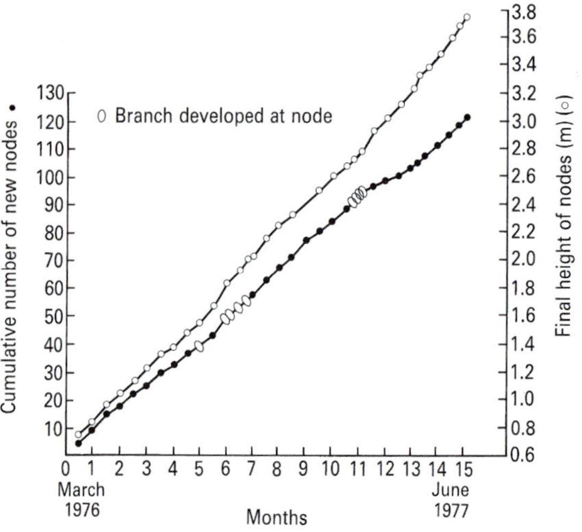

Fig. 3.34. In *Macaranga tanarius* height growth is steady. New nodes form continuously and regularly and every internode elongates by the same amount. (After Ng 1984, fig. 3).

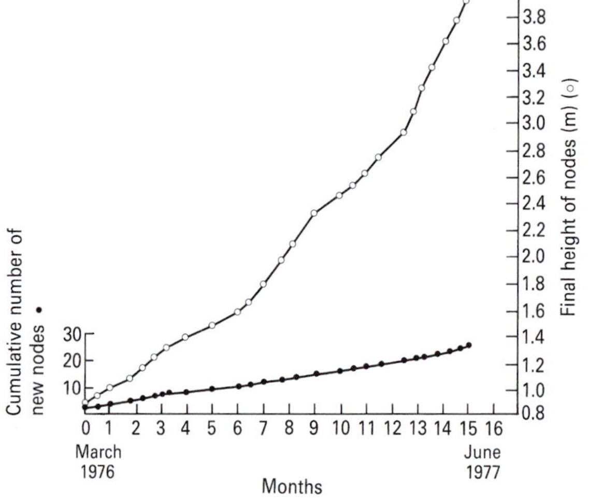

Fig. 3.35. In *Oroxylum indicum* height growth increases through time. New nodes form continuously and regularly (cf. Fig. 3.34) but the internodes continue to elongate. Growth rate measured by node production is linear but measured by height it accelerates. (After Ng 1984, fig. 5.)

Fig. 3.36. This old flowering *Corypha umbraculifera* (talipot palm) in the strongly seasonal climate of east Java has experienced good and bad growing periods.

In a study of 81 species of dipterocarps at Kepong, Malaya, 70 produced leaves in flushes and in the nearby Ulu Gombak forest 38 out of 41 species flushed.

Leaf fall[81]

The sharp distinction between deciduous and evergreen species of temperate climates does not exist in the humid tropics.

Trees with continuously growing shoots suffer continuous leaf fall. Leaf life span varies from about 3 to 15 months. A young palm may accumulate leaves and develop an oblong crown (Fig. 6.3), but at maturity for every new leaf unfurled, an old one drops off.

Amongst the majority of species that grow in flushes we may distinguish three patterns of leaf fall. Where leaves are shed well before bud-break the crown will be bare for a period, perhaps only a few days. Such trees are termed deciduous. In the humid tropics such species flower on the bare crown, for example, *Bombax*, *Firmiana* (Fig. 5.1), *Pterocymbium*, and *Toona* (Fig. 3.20), whose bird pollinators are probably assisted by the lack of leaves.

In other species leaf fall occurs at about the same time as bud-break (Fig. 3.38), but may be a few days earlier or later, depending on particular weather conditions or on the local climate. For example, the rain tree (*Samanea saman*) is evergreen in Singapore, but is bare for a few days in central America where it is native. In Singapore it briefly has two co-existent leaf flushes and in America it is briefly deciduous. Species with this behaviour are called leaf exchangers. The gardener whose task is to sweep up the fallen leaves only has to do so once or twice a year.

Finally, leaf fall may occur well after bud break. Species with this pattern are evergreen. Leaves live about 7 to 15 months and leaves of several flushes sometimes occur together. In young trees of the dipterocarp *Neobalanocarpus heimii* up to seven leaf generations have been found to be present simultaneously.

Synchronization

Leaf life spans on isolated relict and planted specimen trees observed in Singapore at the tip of the Malay Peninsula were found to be 12 months in four species and 6 months in another three species.[82] One might adduce a climatic trigger for leaf fall except that 17 further species had lifespans with no relation to the year, and were more or less constant in each species but with individual trees out of phase. This suggests there is a genetically determined life span; leaves become old and less efficient as they become

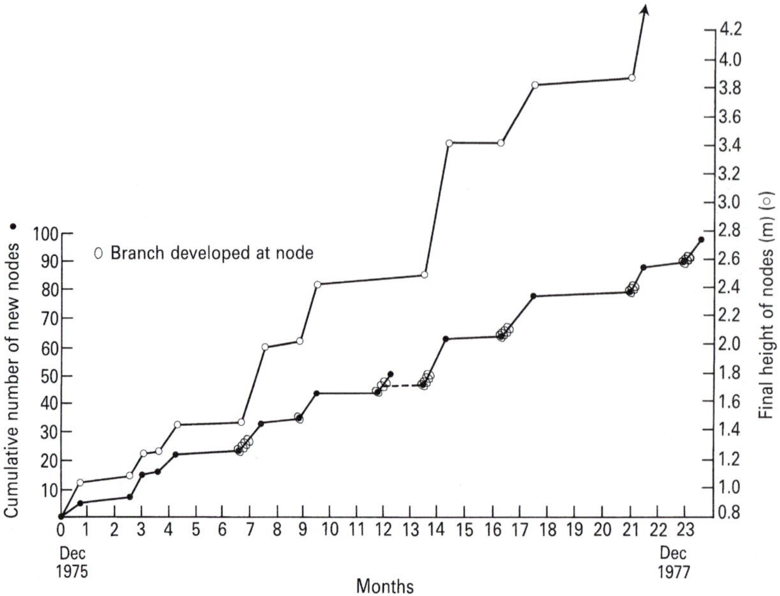

Fig. 3.37. In *Harpullia confusa* height growth is in flushes. New nodes form intermittently with resting periods between. The internodes continue to elongate after initial formation. (After Ng 1984, fig. 10.)

more shaded and accumulate epiphylls and are eventually dropped.

These specimen trees all behaved as individuals. A few species showed lack of synchrony within the crown between different limbs—mango (*Mangifera*) is a good example—and this has been called 'manifold growth'. It was not until a study was made of trees in their natural habitat that it was realized that even in the almost aseasonal climate of Malaya (Fig. 2.1) there is an annual rhythm in phenology. The study covered 9 years at the Ulu Gombak[83] forest and 61 trees of 46 species were observed monthly from a platform in the crown of an emergent. There were two leaf flushes per year, in February to April, and in September to October. Most leaf change occurred then, though different individuals of a species were often not in exact synchrony. The main peak of leaf growth came after the driest time of year, the second lesser peak began before but extended into the wettest time of year. Most trees flowered annually, mainly at the early leaf flush, but some at the later one. The Dipterocarpaceae however only flowered twice, in years which also saw heaviest flowering of the other trees. It is well known in Malaya that durians and other forest fruits are on sale twice a year, and there are sporadic bumper harvests.

In more strongly seasonal tropical climates than Malaya there are more numerous deciduous species. Crowns are bare in the dry season, but flowering and leaf flushing show complex patterns that are not all easily relatable to climatic seasonality. An interesting study in the tropical seasonal forest at Guanacaste, Costa Rica,[84] found that flushing and flowering were controlled by tree water status. On dry sites leaves were lost early in the dry season. In most species rehydration and bud break occurred following heavy rain, but in a few leaf loss allowed rehydration and was followed by bud break, sometimes with flowering, without rain having fallen. On wet sites the species remained ever-

Fig. 3.38. *Dyera costulata* (jelutong) is a leaf-exchanging species. On this tree a flush of new leaves is seen. The new leaves are expanding immediately after the old ones were shed. Note *Platycerium* crown epiphytes. Malaya.

green or just exchanged their leaves. Later, experimental irrigation of clusters of trees confirmed these deductions.[85] In the semi-evergreen rain forest of Barro Colorado Island (BCI), Panama, only a small minority of tree and woody climber species tested had leaf fall delayed by maintenance of moist soil by irrigation during the four-month dry season.[86]

Computer modelling supports the hypothesis that flushes of shoot growth, leafing and flowering result from genetically controlled endogenous periodicity being entrained to seasonal changes in water stress, progressively more strongly the stronger the dry season.[87] Thus, in Malaya endogenous controls dominate, at BCI soil moisture buffers most species against the moderate seasonal drought, whilst in Costa Rica the strong drought is important. It is seen that the interaction of internal rhythmic processes and external drought can have various consequences. However, one pattern, common in monsoon-climate Asia, still remains enigmatic. In the dry deciduous forests of Thailand, for example, new leaves are produced more or less synchronously *before* the first rains break the annual 4–5 month drought. How can shoot activity resume in water-stressed twigs? Further studies are needed.[88]

Flowering patterns

Climax species in rain forests mostly flower once a year and at about the same time. Some figs (*Ficus*) become fertile very irregularly in contradiction to this generalization, and with important consequences for their dispersers (section 4.3). There have been suggestions, which remain unsubstantiated, that small trees within the canopy flower more continually. By contrast, pioneer species, and hence secondary forests, are in continual flower, some species several times a year and others all the time (e.g. in Asia *Adinandra dumosa*, *Macaranga heynei*, and *Dillenia suffruticosa* and in Africa *Vernonia conferta* and *Trema orientalis*).

Monocarpy

A few species, mostly monocotyledonous plants such as some palms and bamboos, exhibit monocarpic or 'big bang' flowering. Usually the plants grow for many years and then over a whole district simultaneously flower, fruit, and die. This is best developed in the seasonal tropics, for example, many bamboos in continental Asia, including China, whose copious fruitfall is followed by plagues of rats, and then a dearth of materials for house construction until fresh clumps grow up. One of botany's mysteries is how subdivisions of a single bamboo clump grown apart time themselves to flower simultaneously. One striking example concerns *Arundinaria falcata* which flowered 35 years after introduction from India in Algeria, France, Luxembourg, and Ireland.[89] Most spectacular of all monocarpic species is the talipot palm (*Corypha umbraculifera*), which grows the largest inflorescence in the

Fig. 3.39. *Corypha umbraculifera*, the giant talipot palm, is monocarpic. It flowers once and then dies. Malaya.

An early European traveller wrote back from India in 1681 'one single leaf (is) so broad that it will cover some fifteen to twenty men, and keep them dry when it rains'.

world (Figs. 3.36, 3.39): a gigantic apical candelabrum, which then develops hundreds of thousands of fruits before the palm dies. A whole avenue of talipot at the Peradeniya Botanic Garden, Sri Lanka flowered and then died simultaneously.

There are fewer monocarpic woody dicotyledonous plants. The shrub *Strobilanthes* with several species through the Asian rain forests is one. *Tachigali*, with 24 species through the neotropics, *Harmsiopanax ingens* in the mountains of New Guinea,[90] *Cerberiopsis* (3 spp., New Caledonia) and some *Spathelia* spp. (tropical America) are the only trees so far known which have this habit.

Mass flowering in western Malesia

Every year in Malaya and Borneo there is a single period of heavy flowering and fruiting. As described above, careful studies over 9 years at

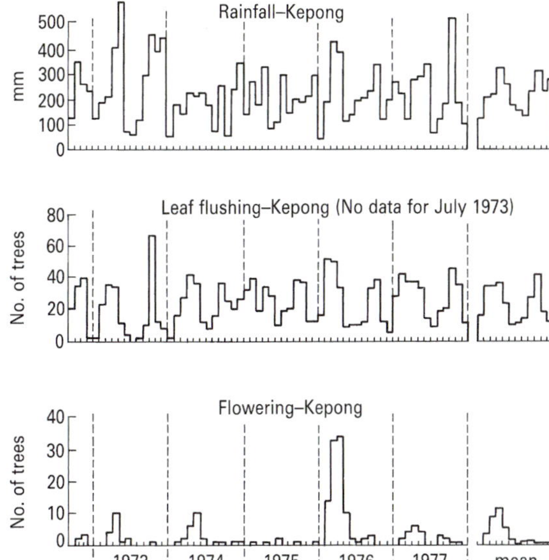

Fig. 3.40. Leaf flushing and flowering of dipterocarps, and rainfall. Kepong, Malaya. (After Ng, 1984, fig. 12.)

Ulu Gombak, Malaya, showed there is a second lesser episode and that production of new leaves also occurs mainly at these times. These two annual events coincide with two slightly drier periods, a few weeks of little rain. Once or twice a decade the early dry spell is unusually strong and is followed by heavier than usual flowering and fruiting, and in these years there is gregarious heavy flowering and fruiting of Dipterocarpaceae (Fig. 3.40).

Different species of dipterocarp flower after each other over a period of several weeks but fruiting is simultaneous, at about the time heavy rains begin. The fruits are highly favoured by pigs and mass fruiting is believed to satiate these marauders as well as seed-attacking beetles. There is strong selection pressure against any tree out of phase, so mass fruiting is a self-reinforcing phenomenon. However, mass fruiting in this family is confined to the west Malesian rain forest species, and we have no explanation as to why the gregarious dipterocarp *Anisoptera thurifera* of New Guinea flowers and fruits every year.

There are no other tropical rain forest plants that show such massive occasional fertility. It is this property that makes dipterocarps and the rain forests they dominate unique (see section 7.6).

Triggers for flowering

Chilling of already formed young flower buds, triggering them to complete their development, has long been known to be the flowering stimulus for some species.[91] For example, *Zephyranthes rosea*, the rain lily, flowers after the stimulus of rain and associated cooling; some orchids and trees (e.g. the pigeon orchid *Dendrobium crumenatum*, a common epiphyte of wayside trees of Malaya, and the angsana tree *Pterocarpus indicus*) burst into bloom some days after a midday storm causes sudden chilling.

Dipterocarpaceae may show some such trigger. As mentioned above, mass flowerings coincide with exceptionally dry weather. Up to about half of all dipterocarp trees burst into flower simultaneously. Later, the corollas carpet the ground like fallen, tinted confetti and are followed after 3–4 months by a heavy fruit fall. There has been much speculation as to the exact nature of the trigger. Drought, or an increase in hours of bright sunshine, or an unusually big difference between night and day temperatures because of the cloudless weather occurring at that season, have all been suggested. The problem has still not been finally resolved.[92]

3.4. PLANT LIFE—CHAPTER SUMMARY

1. There is a great richness of life forms in tropical rain forests. The trees provide support for climbers and epiphytes. There are species of all life forms adapted to the different forest interior microclimates; the broadest classification is into photophytes and skiophytes.

2. The first impression of a uniform phalanx of trees is quickly dispelled. There is enormous variety in crown, bole, buttresses, bark, roots, and leaves. Flowers are ephemeral and species

are usually identified from these characters of the living tree.

3. Only a few species grow continuously. For most, shoot and leaf growth is intermittent, in flushes. Young leaves are commonly highly coloured. Leaf fall follows various patterns, and often occurs after a new flush has grown, but there may be a bare period and this is more pronounced in seasonal climates.

4. Forests show synchronized leaf flushing, and usually bear flower and fruit annually. This is in response to climatic cues interacting with internal plant physiology, and becomes increasingly marked the more seasonal the climate. West Malesian rain forest Dipterocarpaceae are unique in their gregarious flowering and fruiting only a few times a decade.

5. Pioneer species flower more continually.

6. A few species, especially in seasonal forests, are monocarpic, flowering then dying.

7. Triggers to flowering include low night temperature and the chilling of bud initials early in their development.

4

Rain forest animals[93]

Animal life is, on the whole far more abundant and varied within the tropics than in any other part of the globe, and a great number of peculiar forms are found there which never extend into temperate regions. Endless eccentricities of form and extreme richness of colour are its most prominent features, and these are manifested in the highest degree in those equatorial lands where vegetation acquires its greatest beauty and fullest development.

A.R. Wallace *Tropical nature and other essays*, 1878.

4.1. RICHNESS AND DIVERSITY OF ANIMALS

The casual visitor entering a rain forest is overwhelmed by the lush diversity of the plants but disappointed by the apparent absence of animals, except for the rising and falling resonant trill of cicadas, and very soon too by the attentions of biting insects, and in the Eastern tropics of blood-sucking leeches. The observant naturalist however, detects the signs of bigger animals. In Borneo for instance he may notice an orang utan's nocturnal sleeping platform of branches roughly thrown together high in the canopy; scratches on a tree trunk where a big cat has sharpened its claws or where a sun bear has tried to break into a bees' nest; part of the forest floor swept clean of leaf litter as the display ground of an argus pheasant; the muddy hollows or disturbed soil surface of wallowing or rootling pigs. To see the animals and not just their signs he will walk quietly through the forest at dawn or dusk when birds are most active, or at night when many mammals and frogs are awake and can be seen in the light of a head-lamp.

Tropical rain forests are indeed fabulously rich in animal life as A.R. Wallace and the other nineteenth century naturalist explorers reported. Just how many species the world's rain forests contain is still, a century after their pioneering studies, only a matter of rough conjecture. For mammals, birds, and other larger animals there are roughly twice as many species in tropical regions as temperate ones (Fig. 4.1). These groups are fairly well studied, insects and other invertebrates much less so, and if the same proportion between temperate and tropical species holds for them then the tropics as a whole may hold a total of 2–3 million animal species; the great majority of these are undescribed insects, with their greatest abundance in rain forests.

The richness of various animal groups in different patches of rain forest is shown in Table 4.1. Most of the hyper-rich sites, all of them world records, are in the upper Amazon close to the Andes. These include birds at Limoncocha, Cocha Cashu (Table 4.1) and Tambopata, each with about 500 species in under 50 km^2, amphibians (frogs and toads) in

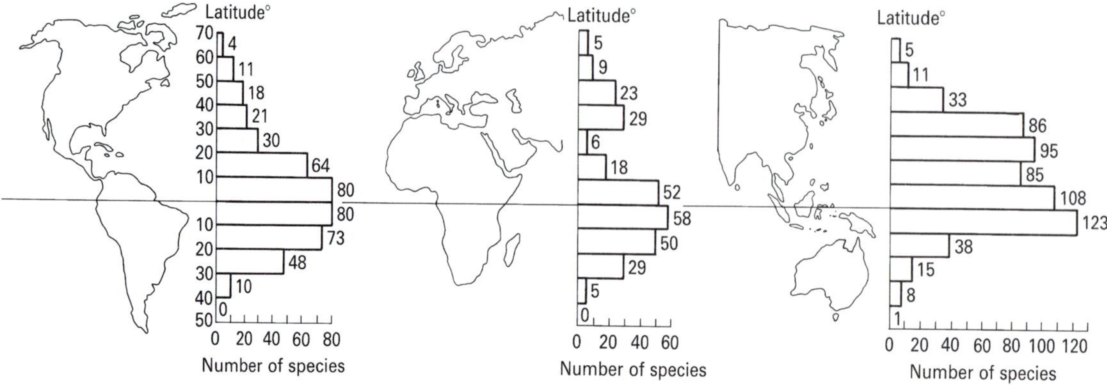

Fig. 4.1. Decrease away from the tropics in species numbers of swallowtail butterflies exemplifies a very common biogeographic pattern. (Collins and Morris 1985, fig. 2, in Longman and Jenik 1987.)

Table 4.1 Species richness of various groups in some well-studied lowland rain forests

	Area (km^2)	Number of species: Mammals (primates) [non-flying]	Birds	Reptiles	Amphibians	Moths	Vascular plants
Neotropics:							
Panama							
Barro Colorado Is.	15	97(5)[39]	443	81	52	*	1320
Costa Rica							
La Selva	*c.* 15	*c.* 117(4)[52]	410	86	43	*c.* 4000	1668
Ecuador							
Limoncocha	15	*	480	*	*	*	*
Santa Cecilia	3	*	*	92	93	*	*
Peru							
Cocha Cashu	> 50	*(13)[*c.* 70]	554	54	77	*	1370
Africa:							
Gabon							
Makokou	2000	199(14)	342	63	38	*	*
Far East:							
Malaysia							
Pasoh	8	89(5)	212	> 20	25	*	*
Papua New Guinea							
Gogol	10	27(10)	162	34	23	*	*

Various sources, including Gentry (1990), McDade *et al.* (1994)
* No data available

Ecuador (Table 4.1), and a single tree at Tambopata, Peru, with 43 ant species in 26 genera—about the same number as the whole British Isles or Canada.

The question of how many species live in the world's tropical rain forests is currently under debate, as part of the concern for their adequate conservation to be discussed in Chapter 10. The huge figure of 2–3 million species, or two-thirds of all forms of life on earth, may even, perhaps, be a substantial underestimate. T.L. Erwin[94] has argued that the global total count may be nearer 30 million species, based on his analysis of the insects living in the crown of *Luehea seemannii* in a scrubby tropical seasonal forest in Panama. Nineteen trees, sampled over three seasons, yielded 955 species of beetles not counting weevils. They were extracted by fogging the tree crowns with knockdown insecticides. Erwin's study breaks away from earlier loose arguments about species numbers by providing a new focused approach, and although one might disagree with the steps in his extrapolation he shows the way objective estimates can be obtained.[95]

Taxonomic description, mainly of vertebrates, has proceeded further than ecological studies, and investigations on the ecology of animals in rain forests have developed more recently than of plants. In this chapter something is said about the differences in vertebrate communities from place to place, followed by discussion of the main means whereby so many species are able to co-exist, and then on what factors may limit animal numbers. In Chapter 5 we go on to describe interactions between animals and plants, cohabiting components of the rain forest ecosystem. Invertebrates are important as decomposers of dead organic matter on the forest floor and this subject is discussed further in section 8.2.[96]

Differences and similarities can be seen between the faunas of the three tropical regions. Let us consider primates as an example. In tropical America and parts of Africa up to 14 sympatric primate species co-exist, but there are nowhere more than 9 in the Asian forests. Primate biomass is highest in African rain forests where it may reach over 2 tonne km^{-2}. In Africa, and also in Asia, there are numerous species specializing in eating leaves (folivores) which they are able to digest using commensal bacteria that inhabit modified, extended parts of the gut. America has a larger number of frugivore–insectivore species than elsewhere, none which are so dependent on leaves, and some species which are much smaller than are found elsewhere.[97] In both Africa and America mixed-species primate troops occur, but none have been found in Asia. The proportions of primates to other fruit- and leaf-eating canopy-dwellers varies from place to place. New Guinea has more frugivorous birds than elsewhere because there is only one competing mammal, a tree-kangaroo, and no primates or squirrels.

South America is sometimes called the bird continent. It has a particularly rich bird fauna, followed by Asia, with Africa trailing behind. In the Old World tropics hornbills (Fig. 4.2) form a guild of specialist canopy-top frugivores. This niche is filled in America by toucans, slightly smaller in size but with similar beak construction, which evolved convergently to feed on the same sorts of fruit (section 5.2). Tropical America is also much the richest region in bats,

Fig. 4.2. Rhinoceros hornbill (*Buceros rhinoceros*). Hornbills are specialist frugivores of Old World tropical forests.

and the Amazon has about half all known species of freshwater fish.

For many rain forest vertebrates fruits are an important component of the diet. This differs from the situation in tropical savannas where herbivores predominate and occur at higher biomass. For example, at the Cocha Cashu rain forest in Peru, mammal and bird biomass was *c.* 2 tonne km^{-2}, whereas it can reach 12 tonne km^{-2} in African savanna. Fruits are variously supplemented by other foods. For example, frugivorous birds augment their diet with insects, which are richer in protein, especially during the nesting and moulting season.

Besides differences between the three tropical regions there are other differences within them. One major pattern is that within the African and American rain forests there are areas of especially high species richness, set like islands in a sea of relative poverty. This is shown well by African birds and is paralleled in America by many groups, including birds and butterflies (Figs. 6.16,6.17). The upper Amazon near the Andes in Colombia, Ecuador and Peru is exceptionally rich. No such patchiness has been detected in Asia where the major pattern is set by Wallace's Line, one of the sharpest zoogeographical boundaries in the world and which delimits the continental Asian faunas from the Australasian (Fig. 6.5). These patterns are now realized to have explanations based on Earth history as will be discussed in Chapter 6.

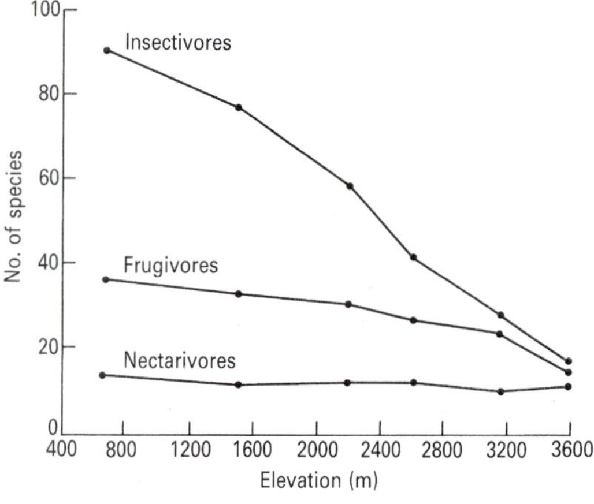

Fig. 4.3. Reduction in bird species numbers of three feeding guilds with increasing elevation. Peruvian Andes. (Data of Terborgh in Bourlière 1983.)

Finally, in all parts of the tropics there is a general diminution in species number with increasing elevation, from lowland to montane forest (Fig. 4.3).

4.2. MODES OF COEXISTENCE

Studies have now been made in many parts of the tropics to elucidate how so many animals co-exist in the same forest. They do so by specialization. Rain forest is lofty and provides a diverse three-dimensional living space and a huge variety of foods.

Specialization in time and space

Some animals are active by day and others by night, and many species live mainly in a single layer of the canopy. Figure 4.4 shows this vividly for the mammals in a rain forest in Borneo. The canopy layers occupied by the different birds at La Selva, Costa Rica, are shown in Table 4.2 and by nine sympatric squirrels at M'Passa, Gabon, in Fig. 4.5. Invertebrates have also developed particular niches in time and space. D.H. Murphy has described how at the Bukit Timah forest, Singapore, there are morning and evening rush hours when cryptozoans move from the soil to feed in the canopy and back, awaited at the foot of the trees by a predatory scorpion, frog, and spider, while during the day these dawn and dusk predators are replaced by others, including a flying lizard. Murphy has also discovered that insects do not fly at random through the forest canopy but follow particular, preferred flight paths, and that along these are concentrated the web-making spiders which prey on them.

The water-filled tanks of epiphytic bromeliads are a particularly special niche in the neotropics (Fig. 4.6). They are veritable aerial aquaria. In Jamaica 68 species have been found to inhabit them, including mosquito larvae. In fact,

Fig. 4.4. Space- and time-partitioning of non-flying mammals in the lowland rain forest of Sabah. (Data of MacKinnon in Whitmore 1984*a*, fig. 3.1.)

Table 4.2 Bird zonation at La Selva forest, Costa Rica

Above canopy	Vultures, hawks, swifts
Canopy top	Toucans, cotingas, parrots, cacique birds
15–25 m	Woodpeckers, woodhewers, large trogons, jacamars, puffbirds
Understorey	Most hummingbirds, antibirds, manakins, flycatchers, tanagers
Forest floor	Tinamous, great curassow, ground doves, wrens

Data of Slud in Bourlière (1983)

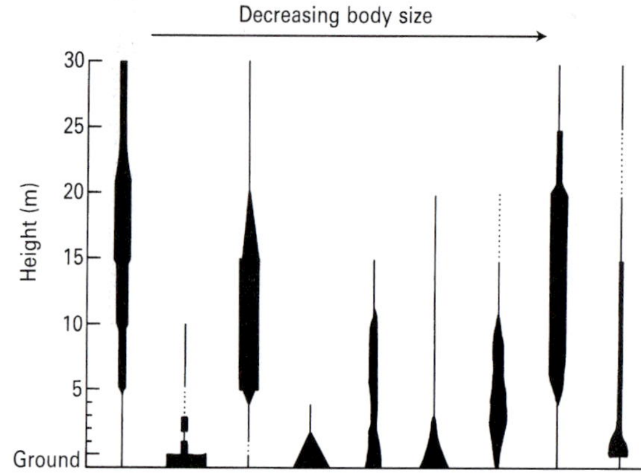

Fig. 4.5. Canopy height preferences of nine sympatric squirrel species in lowland rain forest, Gabon. (Data of Emmons in Bourlière 1983.)

Note how similar heights in the canopy are occupied by squirrels of very different body weight.

Fig. 4.6. Epiphytic bromeliads are a very distinctive feature of the neotropics. Amazonian Ecuador. Water collects in tanks at the overlapping leaf bases and is absorbed by the plant. This pool provides home for many animalcules, including mosquito larvae.

Fig. 4.7. *Nepenthes sanguinea*, Malaya. These pitcher plants are a feature of the Eastern tropics. Small invertebrates, which fall into the pitcher, drown and are digested by the plant.

malaria-control in the New World, unlike the Old, can never be by the elimination of stagnant water because of the presence of these water-tank epiphytes.

The pitcher plants *Nepenthes* of Madagascar, Seychelles, and the Eastern tropics provide another special niche for animals (Fig. 4.7). In Malaya 55 insect species have been recorded as inhabitants of the pitchers, two-thirds of which live and breed there. Amongst the more surprising occasional inhabitants of both bromeliad tanks and *Nepenthes* pitchers are species of freshwater crab.

Food preferences

Many cases have now been analysed which show how the animal inhabitants have evolved to share out the food resources of a forest.

Neotropical bats are by far the most complex assemblages anywhere in the world of sympatric mammals, with 35–50 species co-existing in a few square kilometres of forest. Amongst the 35 sympatric species of Barro Colorado Island, Panama, nine food-defined guilds can be distinguished (Table 4.3), each with from one to nine species. Within each guild there is specialization mainly for size of food particle, which is itself proportional to body weight.

In northeast Gabon in Central Africa there are five sympatric nocturnal species of lorises (prosimian primates), a potto (*Perodictus potto*), three bush-babies (*Galago alleni*, *G. demidovii*, *Euoticus elegantulus*), and an angwantibo (*Arctocebus calabarensis*). These avoid competition by partitioning both food and space. In the canopy *G. demidovii* is mainly insectivorous, the potto mainly eats fruits, and *Euticus* feeds on plant gums. The two other

Table 4.3 The nine feeding guilds found amongst the 35 bat species at Barro Colorado Island, Panama

1	Frugivores	In canopy
2	"	Near the ground
3	"	As scavengers
4	Omnivores	Feeding on nectar, pollen, fruit, and insects
5	Sanguivores	The vampires, feeding on blood
6	Carnivores	(Which glean)
7	Piscivores	(Fish eaters)
8	Insectivores	Slow-flying hawking
9	"	Fast-flying hawking

Data of Barrocorso in Bourlière (1983)

species inhabit the undergrowth being mainly frugivorous (*G. alleni*) and insectivorous (the angwantibo), respectively. There are additional differences in hunting technique, with specialization on slow- or fast-moving insects.

Amongst sympatric primates diet is always related to body size. The smallest species are insectivores, but insects cannot supply enough nutrition for medium-sized animals in relation to the hunting effort, and fruit is the most important sort of food. For the largest species, leaves and seeds are most important and they also eat some fruit. This is well exemplified at Cocha Cashu described below.

There are also some instances of intricate co-existence between insects and their plant foods.[98] One is *Uraina fulgens*, a day-flying moth of Central America, which is well known because it migrates in dense clouds over long distances. Research has shown that these moths are in search of a fresh food source. Their caterpillars feed on the leaves of the forest climber *Omphalea*. After a few generations the plant counterattacks by producing a toxic alkaloid, DMDP, structurally similar to fructose and a powerful insect-feeding deterrent, that interferes with sugar metabolism. These discoveries, starting from pure curiosity to explain why a moth migrates, have led to investigations into the pharmaceutical potential of DMDP for the treatment of cancer, obesity, diabetes and AIDS.

A second example is the South American leaf-cutter ant *Atta*, whose species tend to avoid taking to their fungus-gardens the leaves of species that are fungicidal.[99]

Breeding sites

Another way species may live together is demonstrated by the extremely rich frog and toad community of a forest at Santa Cecilia in Amazonian Ecuador, now unfortunately destroyed. There the 74 species, all living within 3 km^2, were mainly opportunistic feeders and differed mainly by a wide diversity of breeding sites. Ten modes of egg-laying behaviour were distinguished, including deposition in water, on vegetation, in tree cavities, in a depression in the soil made by the male, in foam nests, and on the back of the female. There was also some temporal division, only ten or fewer species being found to breed simultaneously.

4.3. CARRYING CAPACITY OF THE FOREST

Ultimately, by whatever manner the animals have evolved to co-exist within a particular forest, their numbers are limited by the amount of food available.

Data have been gathered from several rain forests which show how there are periodic lean periods when animals may go hungry. The Cocha Cashu area of the Manu National Park in Amazonian Peru may be used by way of example to show how animals subdivide the resources of the forest and cope with hungry periods.[100] At Cocha Cashu the dominant mammals and birds are frugivores, contributing about three-quarters of the total vertebrate biomass. Amongst the mammals, primates then rodents and peccary are commonest. Fruit production was estimated to be 2 tonnes ha^{-1} year^{-1}, but with a 3-month lean period, May to July, when abundance fell to between 2 and 5 per cent of its peak quantities. During the lean period frugivores are forced either to migrate or

to alter their diet. Five of the eleven primate species were studied in detail. These were found to shift to different foods and what they ate depended on their size. The two largest species, the capuchin monkeys (*Cebus albifrons, C. apella*; 3000–5000 g) concentrate on palm nuts which they are able to crack with their strong jaws, but only the larger species is really strong enough. The squirrel monkey (*Saimiri sciureus*; 800–1200 g) eats figs, but sometimes goes a week with no food or just insects, using up more energy than it gains. The two tiniest species, the tamarins (*Saguinus fuscicollis, S. imperator*; 400–500 g) were observed to spend 90 per cent of their feeding time sipping nectar. Only about 1 per cent of the 2000 tree species in the area fruit during the lean period, and these are essential to sustain the frugivorous animals.

There is evidence that palms and figs also play an important role as famine foods in other South American forests. Such plants have been called keystone species because of their vital function. They have obvious importance in forest conservation, and we shall discuss them further in section 11.3.

Figs as keystone species

> The bustling activity at a large fig is one of the unforgettable spectacles of the [Cocha Cashu] forest. Monkey troops arrive from all directions as if guided by some mysterious perception. We have seen over 100 monkeys of five species and 20 to 30 species of birds feeding simultaneously in a single *Ficus perforata*. How is it possible that so many animals independently discover a tree almost the first day the fruit ripens? We believe they are summoned by the shrill din of the myriads of parakeets (*Brotogerus* spp.) that quickly converge on the scene. These birds are fig specialists as much as *Saimiri*, and the sound of them in numbers is almost certain indication of a fruiting tree.[101]

Certain of the big tree fig species of western Malesia, mainly stranglers and banyans (*Ficus* sect. *Urostigma*; Figs. 3.3, 3.5) are just as attractive and important for birds and mammals. Fifty hornbills of four species have been observed feeding simultaneously in a single fig tree at Gunung Mulu, Sarawak. At Kuala Lompat, Malaya, 60 bird species feed mainly on 38 *Ficus* species in an area of 2 km^2. A study[102] at Kutai, east Kalimantan, found that, as at Cocha Cashu, certain figs were the keystone resource for many mammals and birds, and provide the baseline food because they bear fruit throughout the year. Fig fruits are supplemented by fruits of climbers of the Annonaceae and by some Meliaceae and Myristicaceae. At both Kuala Lompat and East Kutai individual fig trees, and indeed species, fruited sporadically, but collectively these figs were always fertile.

There is no record of figs as keystone species in the African rain forests. In a Gabonese forest it was found that figs were rare, and mainly fed on by wide-ranging bats. Moreover, they fruited sparsely and sporadically. Monkeys and large birds relied on two Myristicaceae and an Annonaceae for keystone famine fruits.[103]

Carrying capacity and conservation

Some animal species exist at low population densities either because of the intense competition for food or living space or from the constant attentions of predators. This is true of many invertebrates, and also of lower canopy birds, whose total biomass is not dissimilar from that in a temperate forest but is divided between numerous rare species instead of fewer abundant ones. Rain forest birds of the lower canopy characteristically breed seasonally when insect food is particularly abundant, lay only one or two eggs, and have a low rate of success in raising fledglings. Although each species occurs at a low density of only a few pairs per hundred hectares of forest, subadults can be seen wandering around trying to find living space. Adults are long-lived. It seems likely that a spell of lean years could lead to local extinction of rare species, followed later by re-invasion from adjacent forest when better conditions return. We have little data to substantiate this speculation, and long-term monitoring of populations is required. It is a reason for concern because as

rain forests become fragmented into isolated, small, relict patches, there may not be enough space for the survival of those species that exist near their lower sustainable limit of density (see p. 227).

4.4. RAIN FOREST ANIMALS—CHAPTER SUMMARY

1. Rain forests are very rich in animal species. Numbers diminish with elevation and away from the tropics. There are examples between the three regions of convergent evolution, for example, amongst specialist frugivorous birds. There are also differences, for example, neotropical bat diversity has no parallel elsewhere.

2. American and African rain forests have patches of species richness and of relative poverty. In the Eastern tropics the main zoogeographical pattern is associated with Wallace's Line which separates Asian and Australasian faunas.

3. These many animals co-exist in the same forest by sharing the resources in numerous ways. This is shown by considering related groups, for example, primates or bats. Many exhibit preference for a particular layer in the forest canopy, and are active either by day or by night (Fig. 4.4). There is strong specialization for particular foods. There may be differences in time and place of reproduction.

4. For many rain forest vertebrates fruits are an important food. The amount of fruit varies through the year and the number of animals the forest can support is determined by the amount produced during the lean period. A few so-called keystone species fruit all the year round and provide the baseline food. In America and Asia certain figs are important keystone species.

5. Many animals occur at very low population density and probably suffer local extinction during exceptionally lean periods, but re-invade when conditions improve. Such species are likely to disappear if rain forests are reduced by man to small relict patches.

5

Interconnections between plants and animals—the web of life

Animals are dependent on plants for food, consumers on producers, heterotrophs on autotrophs, and forests also provide diverse living space for their animal denizens, as was outlined in Chapter 4. But tropical rain forests are more than just a restaurant and a home. In this chapter we explore aspects of the complex web of interconnections within rain forests which scientists have begun to disentangle over recent years. Plants make use of animals as pollinators (very few are wind-pollinated in contrast to temperate forests), and many use them also as dispersers (though wind dispersal also does occur). It has recently been fashionable to invoke the term co-evolution, in a one-to-one relationship of specialization, as a driving force for species evolution. Some symbiotic relationships do bear this stamp, but in most cases the forces of natural selection are now realized to be more diffuse.

5.1. ANIMALS AS POLLINATORS[104]

The same combinations of flower and inflorescence features which attract particular pollinators, so-called character 'syndromes', occur in all parts of the tropics. The plant attracts its animal visitors with both nectar[105] and pollen.

Bird flowers are robust and harsh in colour—reds, oranges, yellows, and greens; the nectar is watery and copious. *Firmiana* (Fig. 5.1) is a good example. This syndrome, as others, is recognized by the local birds when a plant is cultivated outside its place of origin. Bird flowers are held away from the leaves, or are borne when the crown is deciduous.

Fig. 5.1. *Firmiana malayana*. Malaya.
The robust inflorescence borne on the bare crown, with vivid orange flowers is typical of species adapted to bird-pollination.

Fig. 5.2. *Kostermansia malayana* like many other Bombacaceae has flowers that are nocturnal and ramiflorous. They are pollinated by bats which brush pollen on to their body while seeking at the petal bases for nectar. (See also Fig. 5.3.) Malaya.

Fig. 5.3. *Pachira macrocarpa* of tropical Africa, and also in the Bombacaceae, exemplifies the other kind of bat-pollinated flower (cf. Fig. 5.2). Mengla Botanic Garden, Yunnan, China.

Fig. 5.4. *Parkia speciosa* showing the pendulous inflorescences which hang free, where the pollinators can easily reach them without collision with the crown. Planted for its garlic-like seeds. Malaya.

All species of this pantropical genus (Leguminosae, subfamily Mimosoideae) have drumstick-like inflorescences held free of the foliage. This species is bat-pollinated. In others, pollination has been recorded by bees, butterflies, birds, or in Africa by the potto, a nocturnal primate.

Bat pollination involves two different syndromes. Bat flowers open at night, have viscid nectar, often a musty or sour scent, and are often dull cream in colour. They are either deeply cup-shaped, as in *Durio* and its relative *Kostermansia* (Fig. 5.2) and pollen is conveyed on the bat's head, or else they have a mass of exposed stamens, like a shaving brush, which deposits pollen on to the bat's breast; examples are *Parkia* and *Pachira* (Fig. 5.3). Bats, like birds, cannot easily penetrate foliage and bat flowers are either borne outside the crown, sometimes on long processes—so-called penduliflory shown well by bananas (*Musa*), *Oroxylum*, and *Parkia* (Fig. 5.4), or are borne behind the leafy twigs on the limbs (ramiflory; e.g. *Kostermansia*, Fig. 5.2) or on the trunk (cauliflory, Fig. 1.2).

There is sometimes overlap between bat and bird pollination. *Mucuna*, *Spathodea*, and

Fig. 5.5. *Ixora lobbii* showing the long corolla tube characteristic of moth-pollinated flowers. Malaya.

perhaps *Erythrina* are visited (though perhaps not pollinated) by both groups.

Moth flowers, like bat flowers, are mostly nocturnal. They are commonly pale in colour, sweetly scented, and have a corolla tube which is often very long and down which the moth inserts its proboscis to reach nectar produced at the base. *Randia* and other Rubiaceae are good examples (Fig. 5.5).

Beetle flowers, many of them highly fragrant, are as characteristic of the tropics as bee flowers are of temperate and semi-arid lands. Beetles pollinate by clumsy scrambling, eating pollen as they go, and many beetle flowers are open dishes or bowls usually set amongst the leaves.

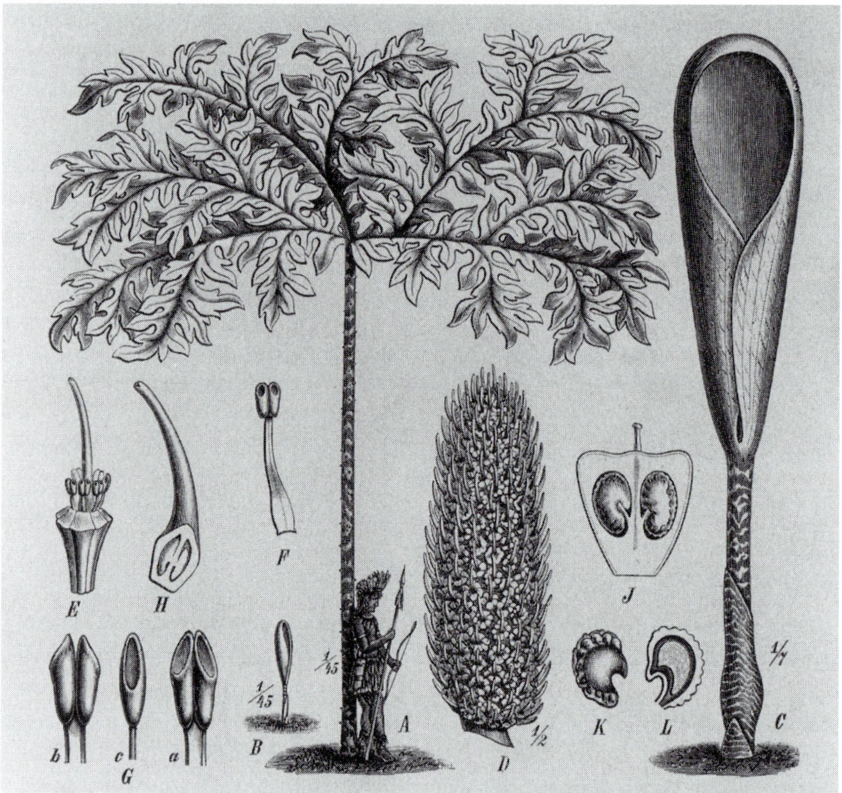

Fig. 5.6. The giant South American aroid *Dracontium gigas* with a single multipalmate leaf (A) 4 m tall, and inflorescence (C) 2 m tall. (A. Engler and K. Prantl, *Pflanzenfamilien* II (3) 1889.)
The Old World counterpart is *Amorphophallus* some of whose species are also gigantic.

Annonaceae, Magnoliaceae, Myristicaceae, and the palm family provide many examples.[106]

Bee flowers, so common in temperate climates, are also extremely common in the humid tropics and are perhaps the major flower-type. Bee flowers are diurnal, often zygomorphic, brightly coloured, sometimes with lines of colour (which may only be visible in ultraviolet light to which bees are sensitive), and are provided with footholds. Orchids, and Leguminosae subfamily Papilionatae are prime examples. Some orchids have evolved exceedingly complex mechanisms, which in some species the intended pollinator has learned to bypass by chewing through the corolla base to reach the nectar.

Foul smelling, mottled, brown and carrion-coloured flowers attract beetles and flies by deception and often trap them; *Amorphophallus* and *Dracontium* (Fig. 5.6), *Aristolochia* and *Rafflesia* (Fig. 3.13) are examples.

A recent discovery is of sequential flowering in species which share the same pollinators. The likelihood of successful cross-pollination is enhanced, firstly by conspecific plants all blooming simultaneously, and additionally, by blooming at a different time from other species. Four of the five *Heliconia* species at La Selva, Costa Rica, share their humming-bird pollinators in this manner and with other food-plants (Fig. 5.7). Six closely related *Shorea* in Malaya similarly share the same thrip pollinators.[107] Also at Pasoh, Malaya, ten unrelated species, pollinated by the same carpenter bees (*Xylocopa*), have been discovered to have staggered flowering times (Fig. 5.8).

Deeper intricacies, to promote cross- rather than self-pollination, continue to be discovered. For example, *Oroxylum* in Malaya produces *c.* 18 ml nectar per night but in bursts of 0.05 ml each.[108] This stimulates its bat pollinator to travel from plant to plant, periodically returning, a mode of foraging that has been evocatively termed 'trap-lining' by analogy to a hunter who sets and monitors a line of snares.

A distinction has been discovered between the three nectarivorous bats of Malaya which matches them to the phenology of the species they depend on.[109] Two bats feed on the nectar of plants that flower continually, banana and *Sonneratia* (a mangrove genus). These bats, *Macroglossus sobrinus* and *M. minimus*, respectively, roost singly or in small groups. The third

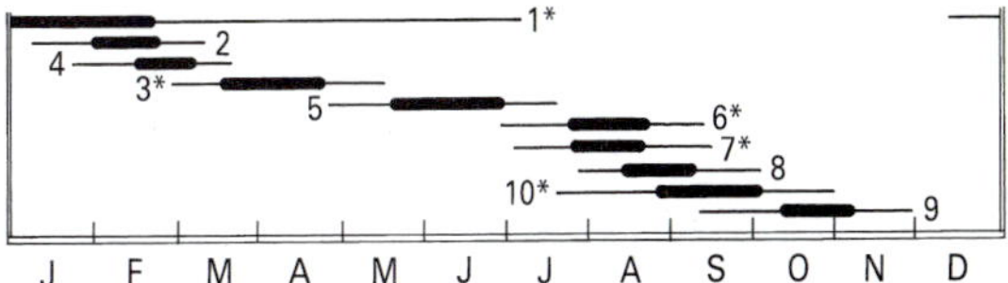

Fig. 5.7. Sequential flowering of the undergrowth species fed on and pollinated by hermit humming birds. Lowland rain forest, La Selva, Costa Rica. The five *Heliconia* are starred. (From Stiles 1977.)
1, *Heliconia pogonatha*; 2, *Passiflora vitifolia*; 3, *H. wagneriana*; 4, *Jacobinia aurea*; 5, *Costus ruber*; 6, *Heliconia* sp. 18; 7, *Heliconia* sp. 16; 8, *Aphelandra sinclairiana*; 9, *Costus malortieanus*; 10, *Heliconia* sp. 3.

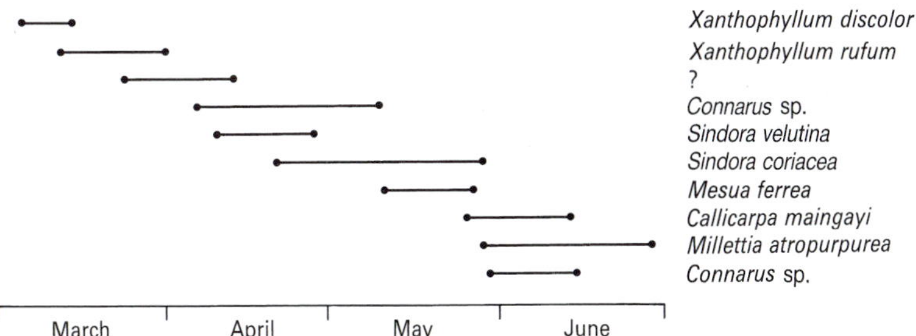

Fig. 5.8. Sequential flowering of ten species all pollinated by the same carpenter bees (*Xylocopa*). Malaya. (Appanah 1985.)

bat, *Eonycteris spelaea* feeds on tree species that only flower once or twice a year, including the important fruit trees durian and petai (*Durio, Parkia* Fig. 5.4). *Eonycteris* inhabits caves and lives in huge colonies, which emerge at dusk, like a plume of smoke, from crevices in the rock, and may have to travel many kilometres to feed. Their gregarious living is believed to enable them to communicate to one another the location of patchy food sources. Destruction of forest within the bats' range reduces the food supply, made up of many species each essential for part of the year.

5.2. ANIMALS AS DISPERSERS

Two kinds of bird fruits can be recognized.[111] Non-specialist frugivores feed on fruits with watery, sugary flesh and small seeds, which provide only part of their diet, mostly carbohydrates. These fruits are common in secondary forest, for example, Melastomataceae and *Trema*, and the birds that feed on them often congregate in flocks. By contrast, fruits evolved for dispersal by specialist frugivorous birds provide a higher quality diet, rich in fats and proteins; the seed is large; many are drupes. Lauraceae contain many examples in all three rain forest regions, plus many palms and Burseraceae in Asia and America. Other specialist bird fruits are brightly coloured, dehiscent, and arillate. This syndrome is found in many families. It is well exemplified by the nutmeg family, Myristicaceae, in which a yellowish pericarp splits open to expose a vivid red edible aril, partly concealing a big black seed (Fig. 5.9). Both types of bird fruit occur in Elaeocarpaceae: *Elaeocarpus* has an oily drupe, *Sloanea* fruits are dehiscent and arillate.

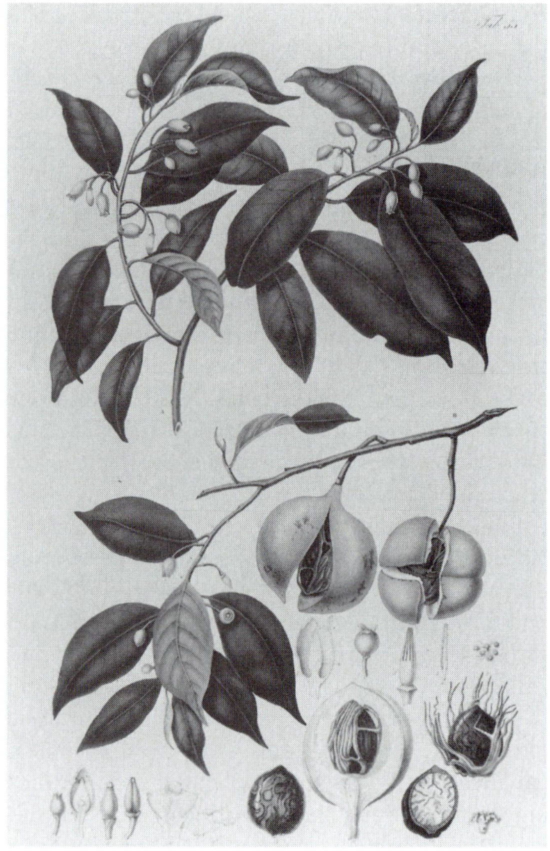

Fig. 5.9. *Myristica fragrans*, the nutmeg, has fruits adapted for dispersal by specialist frugivorous birds (Blume 1835, plate 55.)

The fawn fruits dehisce to expose a single, big black seed, partially enclosed in a finely divided, scarlet aril.

The Dutch East India Company tried in the seventeenth century to control the spice trade by confining nutmeg cultivation to just a few islands of the Moluccas, but were thwarted by nutmeg-dispersing pigeons.

At Makokou, Gabon,[112] fruit type was found to be more closely related to mode of foraging and energy needs than to animal taxonomy. This forest had 39 species of fruit consumers (seven large canopy birds, eight small and two large rodents, nine squirrels, seven ruminants and six monkeys). Analysis of 122 species of fruits eaten by these showed, as prime distinctions, that the birds and monkeys favoured brightly coloured fruit with succulent pulp or arillate seeds, whereas the ruminants and large rodents favoured large, indehiscent, fibrous fruits.

There are no well-defined character syndromes for mammal-dispersed fruits (Fig. 5.10).

Fig. 5.10. The siamang (*Hylobates syndactylus*), a large ape of Malaya and Sumatra (weight 9–12 kg), seen here feeding on *Aglaia*, includes much fruit in its diet.

Civets, mongooses, and bears are probably attracted by a strong smell. Colours are often dull and muted. Durian (*Durio zibethinus*), most famous fruit of the East, certainly has a distinctive pungent aroma. Tigers are notorious for their passion for durians. All these animals are mainly nocturnal. The close study of five primate species at Cocha Cashu, Peru (section 4.3), found that, although all fed on fruits, the only common feature the fruits possessed was yellow to orange colour.[113]

It has recently come to be realized that fish are major dispersers of fruits of the riverine forests of Amazonia.[114] They feed during the high water season in the riparian, periodic swamp forests. During the low water season they retreat to lagoons and live largely on accumulated fat. No particular features characterize fish-dispersed fruits. Two species of rubber tree, *Hevea brasiliensis* and *H. spruceana*, have seeds that are a major food source for certain Amazonian fish. They occur in seasonal swamp

forest (and in the case of *H. brasiliensis* also on dry land), and have big fruit capsules that dehisce explosively to disperse the 20- to 40-mm-diameter seeds for 10 m or more. Some seeds fall on dry land. Fish congregate below the trees to catch the rest. Seeds that escape predation float until the floodwaters recede, at which time they reach the soil surface and germinate.

The relationship of particular animal dispersers to particular fruit-bearing plant species is less close than with pollination. Temporal partitioning has however been discovered. For example, in Trinidad there occur 19 bushy epiphytic *Miconia* species, Melastomataceae, which bear brightly coloured sugary fruits, typical of the family. Each species only bears fruit for a short period and these periods do not overlap, which enhances the likelihood of dispersal for any particular species.[115] It is, however, difficult to determine in this and other examples, how much selection pressure there has been to evolve non-overlapping fruiting periods because, with nineteen species each of which only fruits for a short while, there is a strong probability periods will not overlap purely by chance and, as described above (section 3.3), flowering and fruiting are under the control of a plexus of internal factors interacting with climate.[116] In the Makokou study discussed above, climatic factors were paramount.

5.3. PLANT WEBS, MOBILE LINKS, AND KEYSTONE SPECIES

A rain forest may be universally green and monotonous to our eyes but a herbivore sees it as a poisonous place with patches of alkaloid, saponin, or siliceous spicules, and seeks out what it can safely eat.[117] Plants have evolved toxins that deter herbivores, and related groups are constrained by their genetic constitution to manufacture similar chemicals. Thus, for example, Apocynaceae are rich in alkaloids,[118] especially in the milky latex that nearly all possess. Drug companies exploit these similarities when screening forest plants for potential new medicines. Instead of an expensive blind search they concentrate on those families most likely to be of interest.

Herbivores have evolved adaptations that enable them to detoxify particular chemicals. Thus there have arisen guilds of insects all specialized to feed on a particular group of related, and therefore chemically similar, plants. Such a group may be called a plant web.[119] In the forest many parallel insect guilds co-exist on these different plant webs, without overlap of feeding sites.[120] In some cases the insects utilize the chemicals they ingest. Birdwing butterflies and some other Papilionidae feed on the herbaceous climbers of family Aristolochiaceae, which provide them with a distasteful chemical and render them unpalatable to predators. The butterflies can therefore afford to congregate near the plant, and this facilitates mating. They are mimicked by other butterflies and also by day-flying moths.

Two complex plant webs have been disentangled in the neotropical rain forests, one involving Solanaceae and the other Passifloraceae.[121]

There are *c.* 500 species of Passifloraceae in the New World tropics, as wiry climbers and small trees; *Passiflora* is the biggest genus. The main insect feeders are Heliconiine butterflies, along with flea beetles, coreid bugs, and diopterine moths as less important herbivores. At any one locality there are between ten and fifteen passiflores of various habits and occupying different habitats. These are pollinated and dispersed by generalist bees, birds, bats, and moths. The pollinators and dispersers cannot obtain their food solely from Passifloraceae because flowering is sporadic. Instead they feed on species of different plant webs at different times of year, and thus may be called mobile links because they bridge otherwise unconnected components of the forest ecosystem.

If a forest is degraded partly to pasture, which is happening extensively in parts of the neotropics (Chapter 10), some of the local passiflore species disappear along with their habitat. Likewise, if an area becomes entirely climax forest, there is also a loss of diversity in the

plant webs and a parallel loss in the dependent insect guild. Moreover, Heliconiine butterflies are the principal model for major mimicry complexes so loss of any of these butterflies diminishes the numbers of the (edible) mimics.

The plant web and associated insect guild on neotropical Solanaceae is similar. In this case Ithomiine butterflies are the principal feeders and mimetic model.

Another dimension of complexity is that the mobile link species which are essential to the reproductive success of the plants of several separate plant webs, are themselves commonly totally dependent on a few keystone species for providing their food at times when nothing else is available. The importance of some figs as keystone species for birds and mammals was described in section 4.3. The continuously flowering pioneer forest-fringe shrub *Melastoma malabathricum* is today a keystone species at Pasoh, Malaya, for carpenter bees which are heavily dependent on it during lean periods. When the primary forest burst into heavy flower in late 1981 these bees were observed to migrate from the secondary scrub with which the remaining primary forest is now surrounded on to the new ephemeral rain forest food source.[122]

In America, Euglossine bees are important mobile links. Thirty to fifty species may occur together and any one bee species may pollinate a dozen plant species, including orchids and aroids. Trap-lining (p. 77) is common. Orchids provide pheromones to male Euglossines which are essential for courtship and hence mating. The loss or absence of orchids thus causes the ecosystem to start to unravel. The failures to set fruit of plantation-grown Brazil nut trees is a good example of these intricate interactions.[123]

The Brazil nut (*Bertholletia excelsa*) is pollinated by Euglossines. However, these bees need some other source of pollen for the 11 months when it is not in bloom, and they also need epiphytic orchids to complete their own life cycle. Wild Brazil nut trees occur scattered through the forest. When concentrated in isolated monospecific plantations or remaining as relict trees after forest destruction they do not have the linkages essential for their success. The huge, thickly woody fruits weigh several kilograms and fall heavily to the ground. They contain 10–20 strongly armoured seeds, the Brazil nuts of commerce. The final link in the Brazil nut life cycle is the agouti (*Dasyprocta*), a giant rodent, which gnaws open the outer carapace and stores the nuts in caches scattered through the forest. Agoutis forget some of their nut hoards, which is how the scattered distribution of Brazil nut trees arises.

If any one plant-web becomes particularly abundant, insects of its feeding guild will find it easy to locate in the forest and may increase in numbers. 'Pest pressure' is likely always to keep any particular plant-web at a low density, below a threshold at which it becomes too 'apparent' to its herbivores. This is one mechanism that maintains high species diversity in many ecosystems, including tropical rain forests. One particular case of the reduction of tree seedling population density is discussed later in section 7.3.

5.4. CO-EVOLUTION

Extreme richness in species of plants and animals is one of the best known features of tropical rain forests. How it has arisen is one of the questions that scientists find perpetually fascinating about these ecosystems. It is a question that has therefore sparked off a plethora of discussion and speculation.

Co-evolution, i.e. close relationships between individual plant and animal species, has been invoked as a major motor of species evolution.[124] An arms race has been postulated in which an insect species succeeds in detoxifying a protective chemical, feeds well, and multiplies until the plant responds to this pest pressure by evolving a new toxin. This is chemical warfare and its extreme form has sometimes been called the Red Queen Hypothesis—because 'it takes all the running you can do to stay in the same place'.[125] But such stepwise co-evolution seems, on inspection and reflection, to be improbable.

In fact, pest pressure from the plant's point of view is general, and toxic chemicals are developed against all its herbivores. 'Arms races' are unlikely to proceed as relentless and progressive bouts of defence and counter-defence. They are likely to be much more erratic and to lead to the development of groups of mutually dependent plants and animals, the plant-webs and their associated herbivore guilds which were discussed above.

Speculation that particular ecological species groups of trees are more likely to have their leaves chemically defended against herbivores than others has also not survived scrutiny. Chemicals that are possibly toxic are indeed commonly present, but it now seems likely that mechanical toughness rather than chemical composition is the most important deterrent to insect herbivory. Most loss of leaf area in fact occurs before full expansion of the lamina has taken place, because young leaves, which are soft and not fully formed, are especially susceptible.

A study at Barro Colorado Island, Panama, demonstrates these general conclusions.[126] There, 70 per cent of the 46 species investigated had young leaves damaged more than mature ones. Even though their content of phenolic compounds was 2.3 times higher, they were less tough and fibrous. In the same study mature leaves of the pioneer species were grazed six times more rapidly than those of climax species, and were also found to be less tough and fibrous.

Symbiosis

There are various different rain forest examples of symbiotic associations between ants and plants.[127] About 90 species of *Hydnophytum, Myrmecodia* (Fig. 5.11) and relatives, family Rubiaceae, centred on New Guinea, have swollen tubers, like irregular knobbly hand grenades, with chambered, hollow interiors that are ant-inhabited. The excretions of the ants and the humus that they accumulate nourish the plant.[128] Other epiphytes may grow on the humus. In the New World tropics certain tree-

Fig. 5.11. The ant-plant *Myrmecodia tuberosa* with its swollen stem cut open to show the internal chambers in which the ant *Iridomyrmex cordatus* lives. Note covered ant-runs on tree trunk. New Ireland.

living ant species collect seeds of various epiphytes and plant them on their earthen nests to form aerial 'ant gardens'.[129] They collect animal faeces to nourish the plants which reward them by production of starch grains or sugary secretions from extrafloral nectaries. Ant gardens are particularly abundant in heath forest.

Certain climbing rattan palms are associated with ants. Some *Korthalsia* species have a swollen modified woody ligule, or ochrea, within whose concave interior the ants dwell. Should one accidentally brush against an ant-*Korthalsia* it emits a curious whispering, rustling, rattling noise, as the ants all rush out and beat their mandibles on the hollow box against the attack. In other rattans, species of *Calamus* and *Daemonorops*, ants make nests amongst the bases of interlocking plates of spines on the leaf sheaths (Fig. 5.12), but no benefit to the palm has been demonstrated.

The West African rain forest treelet *Barteria fistulosa* is defended against large browsing herbivores by the ant *Pachysima aethiops* whose sting can penetrate elephant hide and which numbs lesser mammals, including humans, for several days.

The genera of pioneer trees *Cecropia* (New World; Fig. 7.6) and *Macaranga* (Old World, mainly Malesia; Figs. 7.7, 7.8) both have numer-

Fig. 5.12. The climbing rattan palm *Daemonorops crinita* of Borneo and Sumatra, showing the interlocking combs of spines in whose bases ants make their nests. (Blume 1835, plate 136.)

Fig. 5.13. *Macaranga triloba* showing the holes made in the internodes by cohabiting ants. Note also the recurved stipules on whose margins starch grains form which the ants feed on. West Malesia.

ous species that harbour ants, related groups of *Azteca* and *Crematogaster*, respectively. In *Cecropia* the trunk, branches and twigs are all hollow and prone to inhabitation; in *Macaranga* just the apical few internodes of the twigs (Fig. 5.13), where the ant keeps scale insects (mainly *Coccus penangensis*) which tap the tree's sap. The trees produce starch grains on the edges of their recurved stipules on which the ants feed. Experimental removal of ants has led to these *Macaranga* spp. being attacked by leaf-eating insects. Furthermore, these ant macarangas are usually free of epiphytes and climbers because the ants bite off any part of a foreign plant that comes into contact with their host (33 per cent of uninhabited *Macaranga triloba* in Malaya had climbers on them compared to only 5 per cent of ant-inhabited ones).[130] The ant-*Macaranga* story has a further twist. Plants of several *Macaranga* species have been found with their leaves badly eaten despite the presence of ants. Very careful searching has shown that the culprits are tiny caterpillars of the Lycaenid butterfly *Arhopala*, with never more than a few on any one tree. The caterpillars produce a sugary secretion when touched by an ant, so are

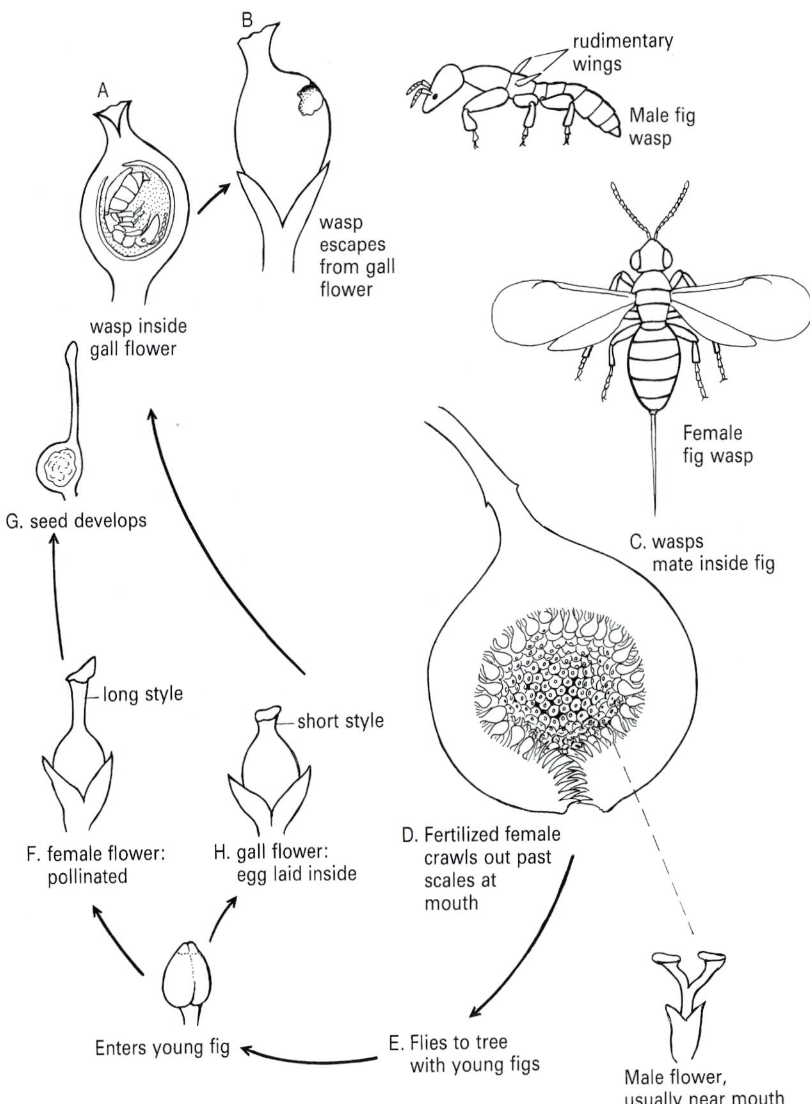

Fig. 5.14. The pollination of fig flowers is one of the most intimate plant-animal relationships known. (Partly after Corner 1988, figs. 162, 164.)

Special fig wasps, a different species for every species of fig, develop inside gall glowers *(A)*; and bite their way out *(B)*; they mate inside the fig *(C)*; the male dies but the female crawls out of the mouth of the fig, getting pollen on her body as she passes the male flowers *(D)*; she flies to another tree, probably attracted by the scent of young figs *(E)*; the young fig either contains female flowers with long styles and these are pollinated *(F)* so develop a seed *(G)*, or the fig may contain sterile gall flowers with short styles down which the female inserts her ovipositor and lays an egg *(H)*; this hatches, the grub feeds on the gall flower ovary and develops into either a male or female wasp, which then repeats the cycle.

tolerated. They are coloured like the stipules or leaf nerves of the *Macaranga* and in this way escape predation. The butterfly has evolved to parasitize the ant-*Macaranga* symbiosis.[131]

Other equally complex mutualisms have also been discovered that do not involve ants. One such is between Passifloraceae and Heliconiine butterflies.[132] The butterfly lays its eggs on young shoot tips and the caterpillars feed there. The mother butterfly seeks out unoccupied plants but the plants have evolved either yellow extrafloral nectaries that mimic eggs, or orange-yellow modified stipules that imitate young caterpillars. Heliconiine butterflies have evolved an acute sense of vision and an ability to learn. It can be observed that they counter these defences by probing potential laying sites with their antennae and forelegs to see if they are really occupied. The Passifloraceae have exceptionally polymorphic leaves, variable both within and between species; it is claimed this enables them to escape detection by the butterflies. What lends credence to the claim for parallel evolution with considerable mutual influence between butterflies and passiflores, is that the more complex avoidance and detection systems occur in both the insect and the plant in those species which taxonomy suggests are more highly evolved, whereas the most primitive species have the simplest adaptations.

A particularly close symbiosis is that between figs and the wasps that pollinate them. Every species of fig has its own wasp, and where a fig tree is cultivated outside the range of its pollinator fertile seeds fail to develop. This intricate microcosm of evolution is described in Fig. 5.14. As a further complication some wasps have a species-specific parasite (Fig. 5.15).

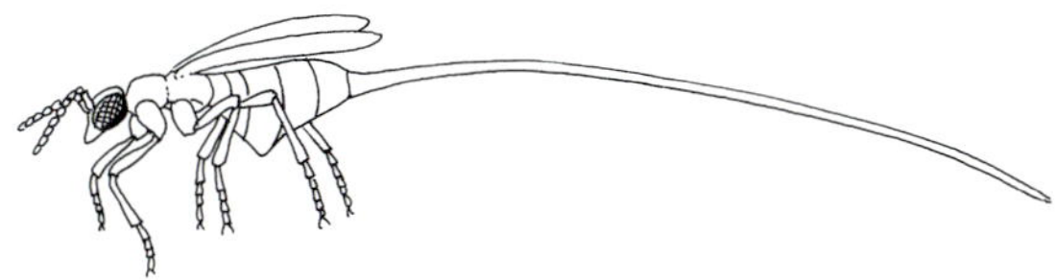

Fig. 5.15. The fig to wasp relationship is rendered even more complex by parasitic insects called Inquilines which lay their eggs inside the developing fig wasp grub by pushing their very long oviposter through the wall of half-grown fig fruits right into a gall flower. (Corner 1988, fig. 164.)

5.5. IMPLICATIONS FOR FOREST CONSERVATION

Tropical rain forests are becoming fragmented into small relict patches or simplified in structure and composition by exploitation for timber, as will be discussed further in Chapters 10 and 11. Interconnections between plants and animals in the eco-system, which are only now beginning to be studied, are essential for its functioning, and are prone to disruption. The ecosystem may then slowly collapse. Trees live for perhaps a century or more (Fig. 7.35), so are likely to persist for decades after potential pollinators and dispersers have gone. At the Sungai Menyala forest in Malaya, which consists of only a few relict hectares as an island in a sea of rubber plantation, fleshy mammal-dispersed fruits of *Chrysophyllum roxburghii* nowadays lie in piles around the parent tree.[133] That forest, which to the causal observer still looks fine, is in fact breaking down. However, as was discussed above, plant species are not always dependent on only a single animal species for pollination or dispersal; where relict forest patches occur which are big enough to retain some suitable animals and where hunting has not exterminated them, the ecosystem may continue to function even though it has an incomplete species assemblage. It is a matter of high priority for biologists today (section 11.4) to explore and monitor the changes that take place in forest fragments, with a view to manipulation if necessary to conserve species richness, because in a century or so from now much of the world's tropical rain forest will exist as fragments. Can these hold the full species complement? As yet we do not know.

5.6. PLANT BREEDING SYSTEMS[134]

Plants have evolved the various syndromes of flower and inflorescence characters described above, to attract particular animal pollinators which then convey pollen from flower to flower. Where pollination occurs between different plants the amount of genetic recombination is high. These are so-called outbreeding species. Genetic recombination is less in self-pollinated species, so-called inbreeders, in which flowers on the same plant are involved, and is zero in apomicts, viz. those species whose reproduction bypasses the sexual process (although many apomictic species do also reproduce sexually on rare occasions).

The question has been debated whether the prevalence of one particular breeding system is the driving force that accounts for the enormous richness in plant species of tropical rain forests. A. Fedorov observed that many rain forest species occur as scattered individuals, and speculated that cross-pollination is low or negligible with the result that inbreeding or apomixis prevails. As mutations occur over the course of time the isolated populations evolve into different species. P.S. Ashton countered Fedorov with the opposite view. He believed cross-pollination prevails to give high gene flow between individuals. Each generation is genetically diverse as a result of outbreeding and is operated on by selection, with resulting strong specialization for different niches in the forest ecosystem.[135]

There are in tropical rain forests some genera with many sympatric species and these could have arisen from either mechanism. One of the best known species-rich families is Dipterocarpaceae, and in any one west Malesian lowland rain forest up to twenty or so species mainly of *Dipterocarpus*, *Hopea*, and *Shorea* commonly co-exist. Other Malesian examples are *Polyalthia* (Annonaceae), *Calophyllum* and *Garcinia* (Guttiferae), and *Syzygium* (Myrtaceae). Further instances are *Eschweilera*, *Inga*, and the forest floor palms *Geonoma* of the neotropics, and in West Africa *Diospyros* and *Drypetes*.

Various studies on breeding systems in rain forest trees now cast light on the questions raised by Fedorov and Ashton. An investigation of dipterocarp breeding systems in Malaya[136] demonstrated two apomictic species, five probable apomicts, sixteen with slight or only tentative evidence, and seven with negative evidence for apomixis. In addition ten species were cross-pollinated and a further two probably so. Apomixis is well-known in cultivated rain forest fruit trees, namly in citrus, clove (*Syzygium aromaticum*), mangosteen (*Garcinia mangostana*; Fig. 5.16), duku, langsat (*Lansium domesticum*), and mango (*Mangifera*), and probably occurs in wild *Garcinia* and *Syzygium*, both of which genera have numerous sympatric forest species.

By contrast dioecism, the occurrence of male and female flowers on different plants, ensures that cross-pollination must occur. It is found in many rain forest species, e.g. many Ebenaceae, Euphorbiaceae, and Myristicaceae. Meliaceae and Sapindaceae whose flowers look hermaphrodite are in fact functionally dioecious. At La Selva, Costa Rica, 23 per cent of 333 tree species were dioecious, and 26 per cent of the 711 tree species with stems 0.1 m in diameter or over were dioecious on an 8.8 ha study area in Sarawak. But in the flora as a whole, when herbs and climbers are included, dioecy is less common: only 17 per cent of 507 species at La Selva and 9 per cent of 1320 species at Barro Colorado Island.[137]

Apomicts are obligate inbreeders. Dioecious species are obligate outbreeders. However, in a species that is outbreeding, most gene flow is in fact restricted to short distances. This has been demonstrated in rain forest in a study of isoenzymes and of morphological variation in the leaves of the small tree *Xerospermum intermedium*, and isoenzymes of the emergent tree *Shorea leprosula* at Pasoh, Malaya; it was found that the further trees are apart the more they differ.[138] It appears that most pollination occurs between near neighbours and most seeds fall near the parent tree.

The conclusion from all these studies is that rain forests, like other ecosystems, have a diver-

Fig. 5.16. *Garcinia mangostana* (mangosteen) of Malesia has extremely uniform fruits; the species reproduces apomictically. Malayan wayside fruit stall.

sity of breeding systems and therefore no single one is the driving force of species evolution. All kinds of breeding systems contribute to species richness, none offers an exclusive mechanism. The patterns of genetic variation in the forest confirm, however, that there is restricted gene flow in the few species studied.

5.7. INTERCONNECTIONS BETWEEN PLANTS AND ANIMALS—THE WEB OF LIFE—CHAPTER SUMMARY

1. Numerous animals co-exist in tropical rain forests by subdividing its food resources. There has been loose co-evolution between plants and animals.

2. Flowers have developed the same syndromes of features for particular groups of pollinators in all parts of the tropics, and we can distinguish bat, bird, beetle flowers, etc.

3. There are also syndromes of fruit characters, likewise pantropical, each adaptive to particular dispersers.

4. Some plant species that have the same pollinators or dispersers co-exist by flowering or fruiting at different times.

5. Related plants have similar defensive chemicals and form plant webs, each one fed on by a particular guild of insects evolved to detoxify those particular compounds. 'Pest pressure' is likely to prevent members of any one plant web becoming particularly abundant. Several plant webs may share the same pollinators or dispersers: these have been called 'mobile links'. At lean seasons these depend for food on plants that are fertile when little else is, the so-called 'keystone species'. Both mobile links and keystone species are essential for the maintenance of full ecosystem diversity.

6. Closer co-evolution is less common. Ants and rain forest plants have evolved several kinds of mutualism. New World Heliconiine butterflies and Passifloraceae have co-evolved and the simpler adaptations in both partners are in the more primitive species. Figs and their pollinating wasps show one-to-one species relationships (Fig. 5.14).

7. All kinds of plant breeding systems found in other biomes exist side-by-side in tropical rain forests. No single breeding system can be invoked as the mechanism that has generated high species richness.

8. Rain forests are becoming fragmented, or changed by exploitation. It is likely that there is a loss of complexity and diversity and the web

of interconnections is unravelling. For example, Amazonian fish need swamp forest fruits; durians are pollinated by bats that need other pollen sources for the 11 months when durians are not in flower; guilds of Heliconiine butterflies and their mimics become impoverished if their food-plant Passifloraceae begin to disappear because of loss of habitat. Because increasingly more rain forests are being altered, biologists urgently need to monitor these changes with a view, if necessary, to manipulation in order to conserve biodiversity.

6

Tropical rain forests through time

The great contrast between the two divisions of the archipelago is nowhere so abruptly exhibited as on passage from the island of Bali to that of Lombock, where the two regions are in closest proximity. In Bali we have barbets, fruit-thrushes, and woodpeckers; on passing over to Lombock these are seen no more, but we have an abundance of cockatoos, honey-suckers, and brush-turkeys which are equally unknown in Bali or any island further west. The strait is here fifteen miles wide, so that we may pass in two hours from one great division of the earth to another, differing as essentially in their animal life as Europe does from America.

A. R. Wallace *The Malay archipelago*, (3rd edn.), 1872.

Patterns of distribution of plants, animals, and vegetation have long fascinated biologists. Their detection is part of the search for generalizations about Nature, the reduction to simplicity of initially bemusing variety. The scientist then goes on to seek for the causes of the patterns he has detected. Over the past few decades two major causes have been discovered for many of the present-day patterns seen amongst rain forest plants and animals. These are continental drift, and past fluctuations of climate; both have contributed to a recent revolution in our understanding of tropical rain forest biogeography.

6.1. PALAEOGEOGRAPHY

The occurrence of separate continental plates that have drifted by ocean-floor spreading is an established fact.[139] Plate tectonics caused the supercontinent Pangaea to break up from about 180 million years ago (mid-Jurassic) onwards (Table 6.1, Fig. 6.1). The northern and southern halves, Laurasia and Gondwanaland, continued to fragment and today's arrangement of land and sea was not achieved until about late Tertiary, about 10 million years ago.[140] The break-up of Gondwanaland is of special significance for the tropical biogeographer because all three tropical regions contain fragments of it. Thus, continental drift has had powerful influences on patterns of distribution of families, genera, and species.

Gondwanan ranges

Similarities in flora between the three regions of tropical rain forest, described in general terms in Chapter 2, occur because all are mainly parts of old Gondwanaland. The major evolution of the flowering plants had occurred before Gondwanaland began to break up and has continued on the different fragments. Today 334

Table 6.1 Geological periods and epochs mentioned in the text

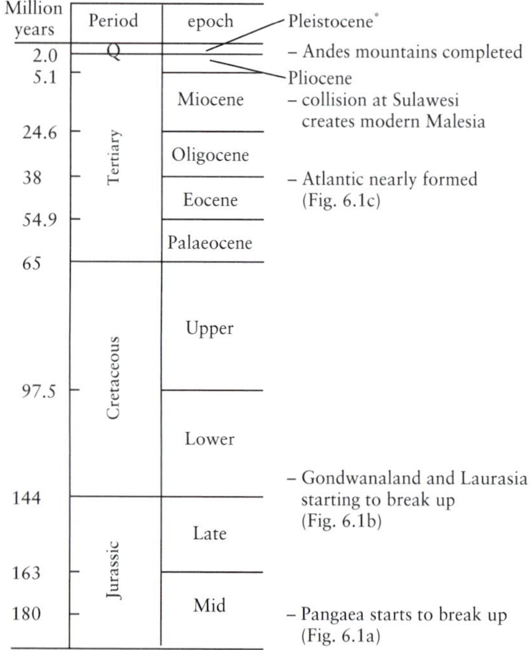

Q = Quaternary * Last 10^4 years of Quaternary is the Holocene

genera and 59 families of flowering plants are essentially pantropical. Looking more closely, other patterns can also be seen.

Campnosperma has a typical Gondwanan distribution. This genus of light-demanding lowland rain-forest trees occurs in Panama, Colombia, Brazil, Madagascar, Seychelles, and from India to Micronesia. Absence from continental Africa is believed to reflect massive extinctions there due to periods of strongly dry climate.

Disjunct ranges, with one part of a plant-group occurring on the Guyana Shield of northern South America and the other in Malesia, are also believed to reflect descent from a pan-Gondwanan ancestor, with extinction in Africa. Examples are Bonnetiaceae, a family of small trees (*Bonnetia* America, *Ploiarium* Asia), and Tetrameristaceae, with *Pentamerista*, recently discovered in Guyana, as a relative of the well-known peat swamp forest genus *Tetramerista* (punah) of western Malesia.

The geographical distribution of Dipterocarpaceae also reflects a Gondwanan ancestory. Sub-family Dipterocarpoideae contains the important big timber trees, pre-eminent in today's international timber trade and on which so much of the discussion in this book is focused. They range from Malesia westwards to India and Sri Lanka, with one outlying endemic genus in the Seychelles and fossils in Somalia and Uganda. Subfamily Monotoideae has three genera. Two are small trees of sub-Saharan Africa, mainly of seasonal forests. The third, discovered in the mid 1980s, is a big rain forest tree of Colombia in South America.[141] South America is also home to the third subfamily, Pakaramoideae, comprising one species, a shrub or small tree of the Guyana Shield rain forests, in Guyana and Venezuela, discovered in the 1970s.

There are other groups besides Dipterocarpaceae which span the southern Indian Ocean, occurring on fragments of old Gondwana. The fan palms *Borassus* and its allies are one example (Figs. 6.2, 6.3), *Nepenthes*, the pitcher plants (Fig. 4.7), found in Madagascar and Seychelles as well as Malesia, are another.

Another common pattern is a geographical range that spans the southern Atlantic Ocean, with representatives in both tropical America and tropical Africa.[142] Twelve families of flowering plants are essentially limited to these two regions, and in many cases are much more numerous in one; for example, Bromeliaceae and Cactaceae which are mainly American but have a few Old World species. One hundred and ten genera have the same amphi-Atlantic range, e.g. *Annona, Chlorophora, Ocotea*, and the palms *Elaeis* and *Raphia* (oil palm, raffia).

Closure of the eastern Tethys Ocean[143]

Gondwanaland and Laurasia were separated by the great Tethys Ocean. Tethys was closed by the northwards movement of parts of Gondwanaland (Fig. 6.1). First Africa and then

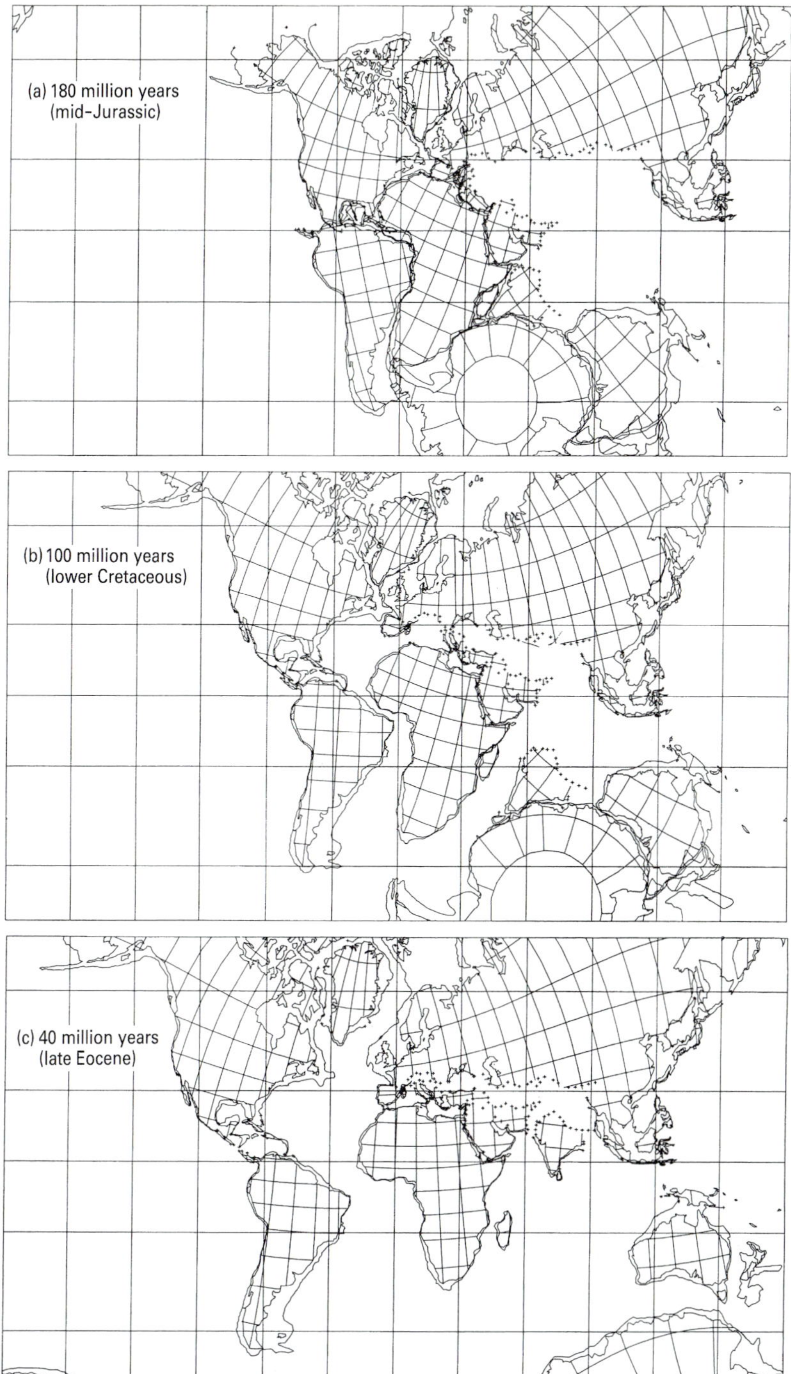

Fig. 6.1. Continental drift. (a) Mid-Jurassic: Laurasia (north) and Gondwanaland (south) in contact as Pangaea, separated in the east by the Tethys Ocean. (b) Lower Cretaceous: Gondwanaland breaking up and Laurasia starting to do so; Tethys Ocean being closed; Australia/New Guinea still joined to Antarctica. (c) Late Eocene: Atlantic Ocean nearly formed; Tethys Ocean much smaller; Africa and India have nearly collided with Laurasia; Australia/New Guinea has separated from Antarctica but the Malay archipelago has not yet formed. (Smith and Briden 1977; present-day coastlines and continental shelves shown, Mercator projection.)

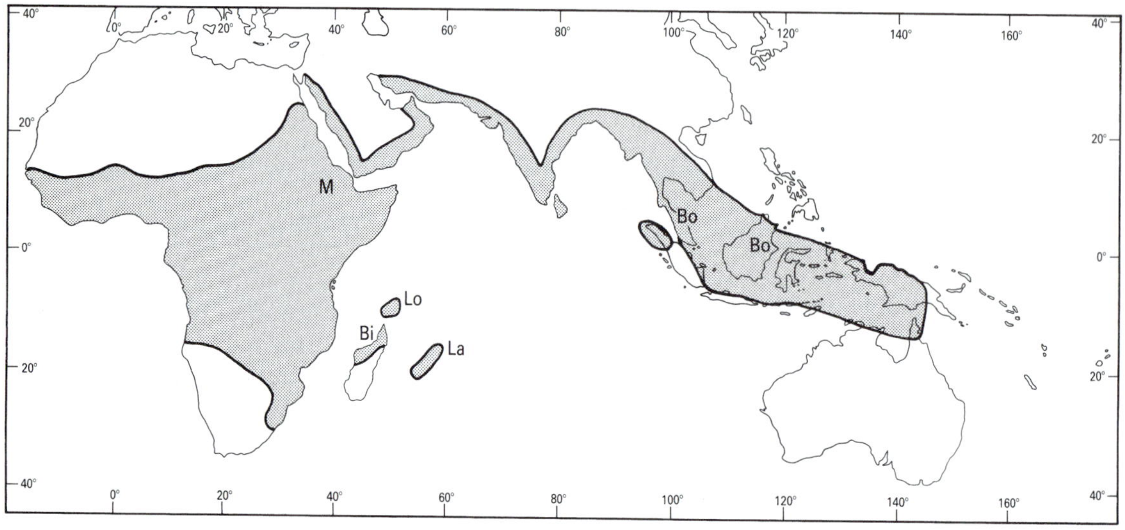

Fig. 6.2. (Above) The fan palms of tribe Borasseae originated on Gondwanaland. They continued to evolve as it broke up and occur today on its Old World fragments. Outline, *Borassus* and *Hyphaene* (10 spp. each), both widespread and probably with ranges extended by man. *Bi, Bismarkia* (1 sp.); *Bo, Borassodendron* (2 spp.); *La Latania* (3 spp); *Lo, Lodoicea* (1 sp., the double coconut; *M*), *Medemia* (1 sp., last seen 1964, probably extinct.) (After Uhl and Dransfield 1987.)

Fig. 6.3. (Left) *Borassus flabellifer*, the lontar palm, here seen planted in east Malaya and exhibiting the deep, oblong crown of youth.

India drifted north and collided with the southern margin of Laurasia. Further east the continental plate which comprised Antarctica/Australia/southern New Guinea moved northwards, broke in two leaving Antarctica behind, and, as a simplification, collided with the southeast extremity of Laurasia, at about 15 million years ago, the mid-Miocene; this created the Malay archipelago (Malesia) as it exists today. Both super-continents had their own set of plants and animals.

India rafted Gondwanan plants northwards and after collision these mixed with Laurasian ones to give the dual-origin flora of that region today. The collision within Malesia was at what is the modern island of Sulawesi. Western and eastern Malesia have very different animals, demarcated by a very sharp boundary, Wallace's Line (Figs. 6.4, 6.5). The strength of Wallace's Line as a boundary varies with the ease with which the animal group under consideration disperses over salt water. Present day zoogeography is a clear reflection of the plate tectonic history. Plants do not divide so sharply at Wallace's Line, but groups of northwestern and southern (Gondwanan) origin can be recognized (Figs. 6.6, 6.7), centred in western and eastern Malesia, respectively. The northwestern group has Laurasian plants, such as Magnoliaceae (Fig. 6.6b), as well as others that come from the Indian part of Gondwana, such as Dipterocarpaceae (Fig. 6.6a), *Ctenolophon*, *Durio*, *Eugeissona* and *Gonystylus*.[144]

Both Laurasian and Gondwanan floristic groups occur in the tropical rain forests of Queensland, northeast Australia. Recent research has shown that Laurasian elements were well represented there before the mid-Miocene collision and there was no sudden influx. This is because the evolution of the Malay archipelago was in fact more complex than a single collision.[145] Various shards progressively broke off Gondwana from the Jurassic onwards, drifted northwards, and became embedded in what is now continental Asia (Fig. 6.8). Then the easternmost limit of Laurasia, southwest Sulawesi, broke off from Borneo at about 40 million years ago taking its Laurasian flora with it, including *Ctenolophon*, *Durio* and *Gonystylus*.[146] This flora is thought to have subsequently colonized land further east, including further fragments rafted off Gondwana. More land progressively arose, both as island arcs and further fragments, and at about 15 million years collision occurred, facilitating further exchanges of plants and animals.

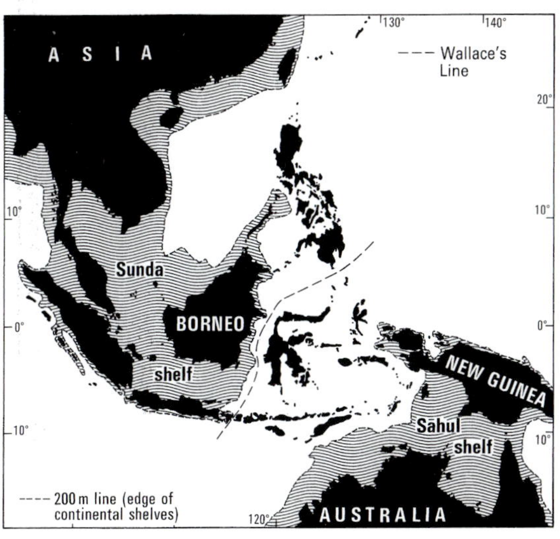

Fig. 6.4. Wallace's Line, which runs between Borneo and Sulawesi, marks the boundary between the Asian and Australasian faunas of Laurasian and Gondwanan ancestry, respectively. (After Whitmore 1984*a*, fig. 1.9.)

Fig. 6.5. Wallace's selection of animals to demonstrate the two distinct faunas that occur in Malesia. These are (left to right and top to bottom) (a) Borneo: the western tarsier *Tarsius bancanus*, one flying lemur *Cynocephalus variegatus* in flight and another one seated, the pentail treeshrew *Ptilocerus lowii*, the malay tapir *Tapirus indicus* and a couple of lesser mousedeer *Tragulus javanicus*; (b) New Guinea: a tree kangaroo *Dendrolagus inustis*, the fairy lory *Charmosyna papou*, the twelve-wired bird of paradise *Seleucides melanoleuca*, the common paradise kingfisher *Tanysiptera galatea*, and a crowned pigeon *Goura cristata (coronata)*. (Wallace 1876, Vol. 1.)

Fig. 6.6. (Right) Plant geography of Malesia derived from plate tectonics. (a) Dipterocarpaceae, and (b) Magnoliaceae are centred on the western, Laurasian, part of Malesia. They originate from the Indian part of Gondwanaland and from Laurasia respectively. (c) *Styphelia*, Epacridaceae, has an eastern, Gondwanan, centre of distribution. (Whitmore 1981, figs. 8.3, 8.4, 8.8.)

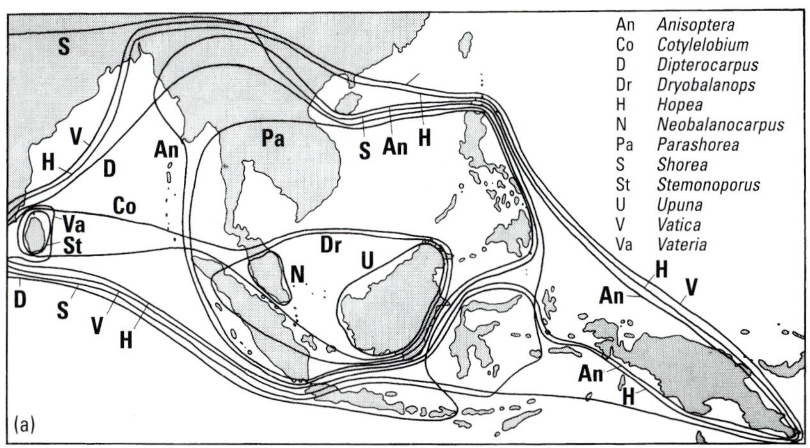
An Anisoptera
Co Cotylelobium
D Dipterocarpus
Dr Dryobalanops
H Hopea
N Neobalanocarpus
Pa Parashorea
S Shorea
St Stemonoporus
U Upuna
V Vatica
Va Vateria
(a)

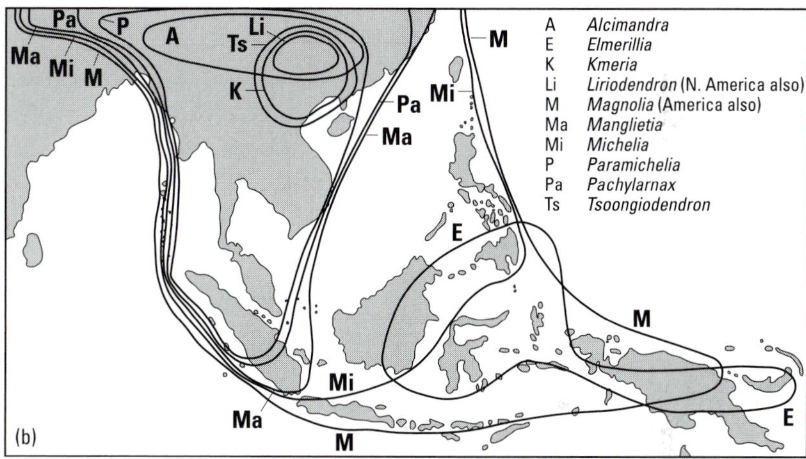
A Alcimandra
E Elmerillia
K Kmeria
Li Liriodendron (N. America also)
M Magnolia (America also)
Ma Manglietia
Mi Michelia
P Paramichelia
Pa Pachylarnax
Ts Tsoongiodendron
(b)

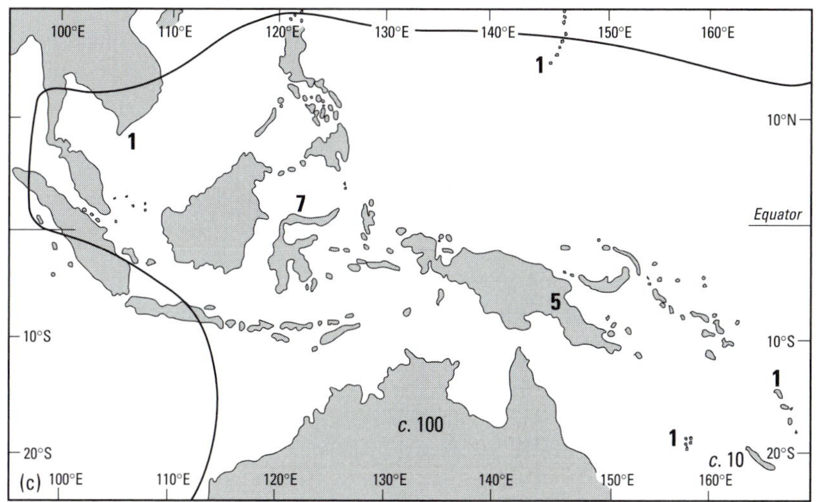
Equator
c. 100
c. 10
(c)

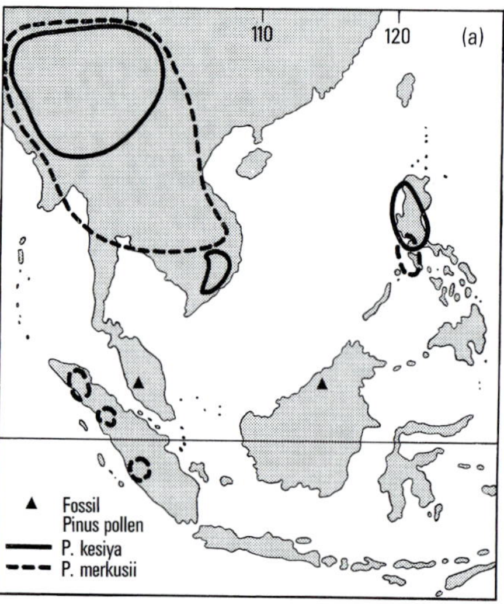

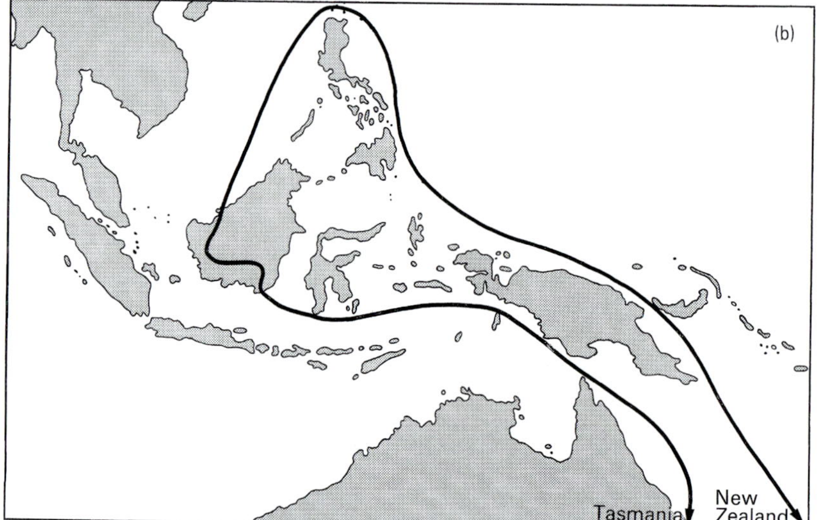

Fig. 6.7. In Malesia conifers have either a northern, Laurasian or a southern, Gondwanan, centre of distribution (cf. Fig. 6.6).
(a) *Pinus*, (b) *Phyllocladus*, celery pine. (Whitmore 1981, figs. 8.1, 8.6.)

The northern Andes

Another result of plate tectonics, which has left its mark on neotropical rain forest biogeography, is the creation of the Andes which continued till the end of the Pliocene, 2 million years ago. The Andes arose by the progressive uplift of the western edge of South America as it overrides the edge of the Pacific continental plate which is being pushed under, a process known as subduction. One consequence has been to divide in two the ranges of lowland rain forest species, so that today they are disjunct with one part in the Pacific coast rain forests and the other part in the Amazonian forests. In many groups there has been evolution since the range became broken so that the two parts are no longer identical. An example, the vegetable ivory palms, subfamily Phytelephantoideae, is shown in Fig. 6.9.

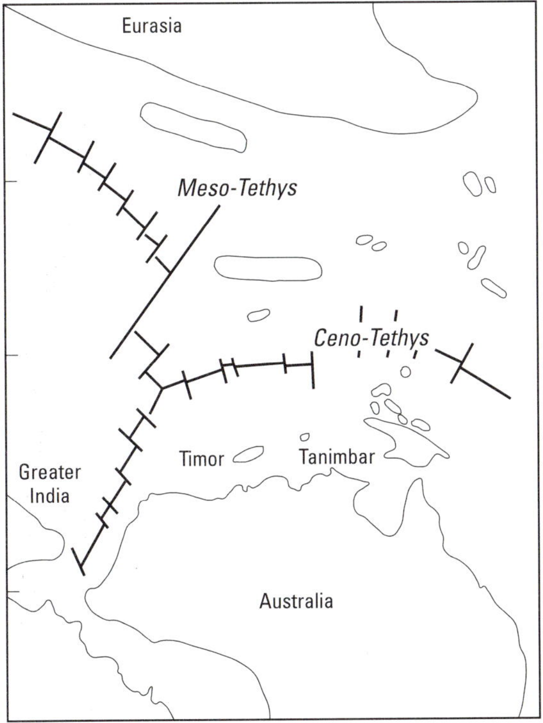

Fig. 6.8 Plate tectonics and the evolution of the Malay archipelago. From the late Jurassic (150 million years BP) onwards at the eastern end of the Tethys Ocean (cf. Fig. 6.1) fragments broke off northwest Australia and drifted northwards. This map shows the situation at the Early Cretaceous (100 million years BP). (After Metcalfe 1996, fig. 15b.)

The full biogeographical implications of these 'stepping stones', and earlier ones that also drifted northwards, remains to be worked out.

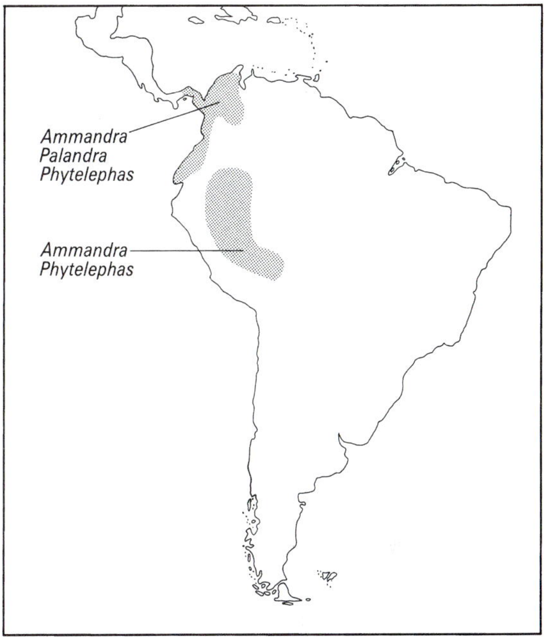

Fig. 6.9. The vegetable ivory palms, subfamily Phytelephantoideae, are confined to northwest South America where their range is disjunct. The uplift of the Andes, which continued to the end of the Pliocene 2 million years BP, split the range. Evolution continued. Today no species is found in the rain forests on both sides of the mountains. (After Uhl and Dransfield 1987.)

Many groups show this pattern, with one set of species centred on the Choco region of Colombia, the other in the Amazon, for example *Couepia*, *Licania*, and *Parinari* of the Chrysobalanaceae.

6.2. PALAEOCLIMATES

The climate of the tropics has been continually changing. The old idea of fixity is quite wrong; climatic changes have had profound influences on species ranges.

The Quaternary

Most knowledge about past climates is for the last 2 million years, the Quaternary period, during which there has been repeated alternation at high latitudes near the poles between Ice Ages or Glacial periods and Interglacials. During Glacial periods tropical climates were slightly cooler and drier, with lower and more seasonal rainfall. During these times rain forests became less extensive and seasonal forests expanded. Most of the Quaternary was like that; present-day climates are extreme and not typical of the period as a whole. Today we live at the height of an Interglacial.

The last Glacial maximum (the Wisconsin–Würm Glaciation) was about 18 000 years ago

(before present, BP). Since then the climate has got warmer, and during the last 10 000 years, the period known as the Holocene, was warmest 5000–8000 years ago. At the Glacial maxima sea levels were lower by as much as 180 m because much water was frozen in the greatly extended polar ice caps. In Malesia the Sunda shelf, extending from the Asian continent, and the Sahul shelf, reaching out from New Guinea, were both exposed (Fig. 6.4), so the separate islands west of Wallace's Line became joined as one land mass and those to the east as another. The islands of the Caribbean also became partially joined. Sea surface temperature was cooler than today, by 5 °C or more[147] at 18 000 BP in the tropics.

The evidence for past differences in climate in the tropics is two-fold, from the record of past vegetation deduced from palynology, and from direct physical traces.

Palynology[148]

There are montane peat deposits in former lakes in all three tropical regions. Pollen from nearby vegetation gets incorporated in peat as it forms and can be extracted for study. The picture revealed by pollen analysis of peat from a site at 2600 m near Bogota, Colombia, in the northern Andes (Fig. 6.10) is typical. The repeated oscillations in the relative abundance of the pollen of the three different species groups shown on Fig. 6.10 are a reflection of fluctuating climate. In cold epochs vegetation belts were depressed so that the site lay in paramo, high mountain moorland, in which grasses and the high mountain shrubs *Acaena* and *Polylepis* occur. By contrast, in Interglacials the tree line rose above the site and pollen of trees and shrubs predominated. Similar oscillations have been found in African and Malesian montane peat deposits, though none stretch so far back in time. Radiocarbon dating has shown that the oscillations were simultaneous at different places and also coincide with the alternating Glacials and Interglacials of temperate latitudes. The oscillations of the tree line in all three tropical regions, reconstructed from numerous pollen profiles that have now been analysed, are shown in Fig. 6.11. The reduction in temperature during Glacial periods was greater at high elevations than in the lowlands and consequently vegetation zones were compressed and depressed (Fig. 6.12).

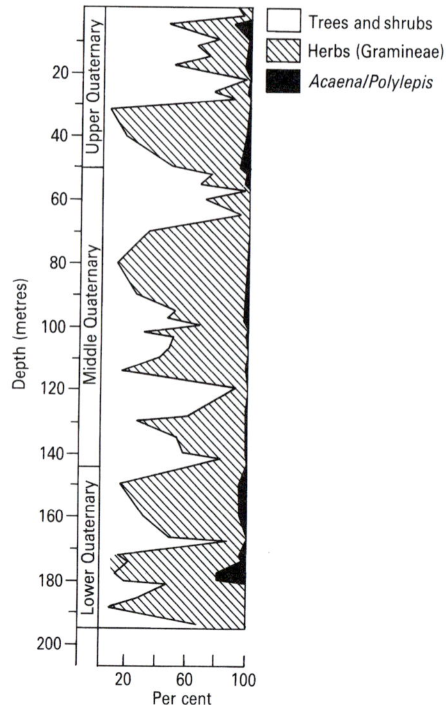

Fig. 6.10. Pollen diagram of a 200 m core of lake sediments and peat, extending for the whole Quaternary. Northern South America, at Bogota, Colombia, 2600 m altitude. (Data of van der Hammen in Flenley 1979, fig. 4.12.)

The repeated fluctuations in abundance of the three species groups is a reflection of continual climatic change.

There are fewer records of changes in vegetation from the lowlands because suitable peat is uncommon and, although pollen in mineral soil can be used instead, few suitable pollen deposits have been found. Two profiles from the lowlands of northern South America are reproduced in Fig. 6.13. These show repeated oscillations between dominance by savanna herbs (including grasses) and by trees, which reflect drier and

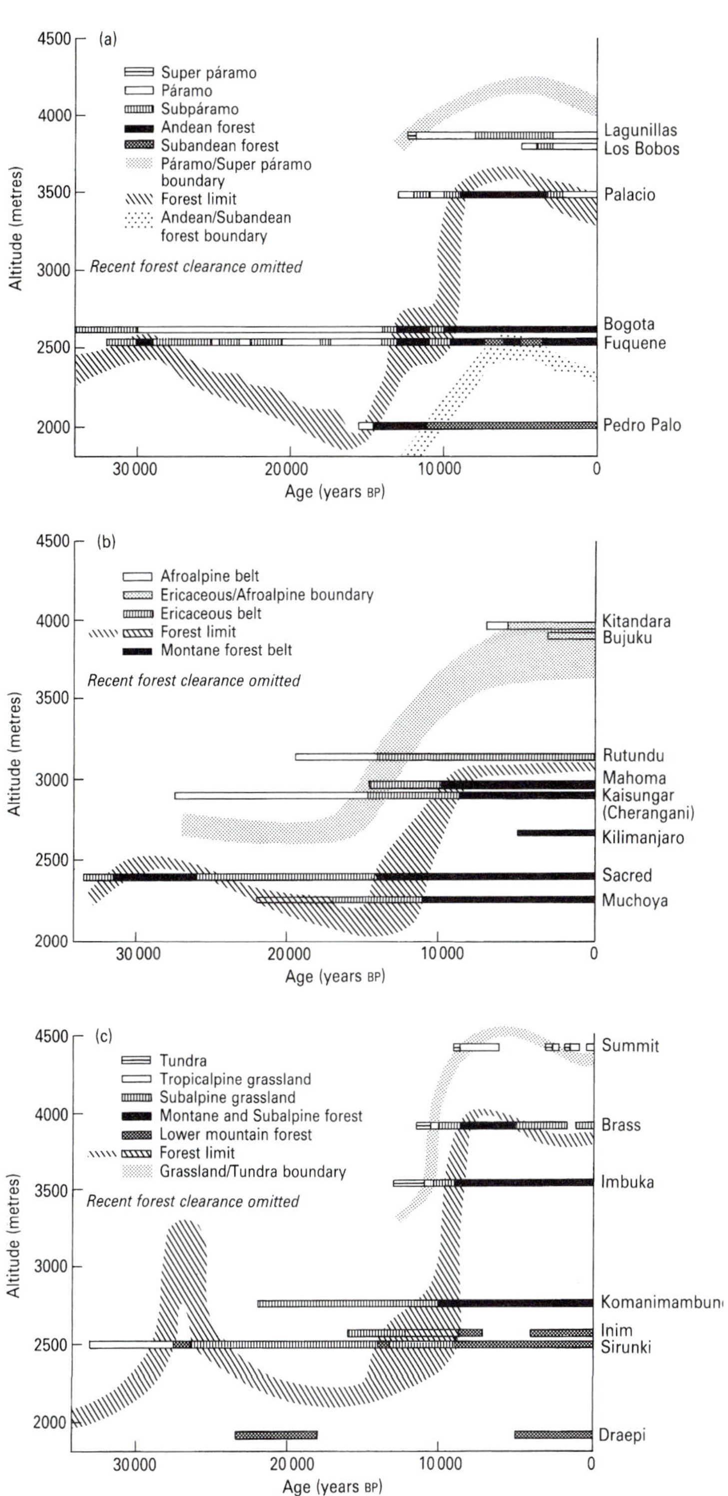

Fig. 6.11. Oscillations in the upper limit of forests during the late Quaternary in (a) the Colombian Andes, (b) the mountains of east Africa, and (c) the New Guinea highlands. (After Flenley 1979, figs. 4.26, 3.24, 5.23.)

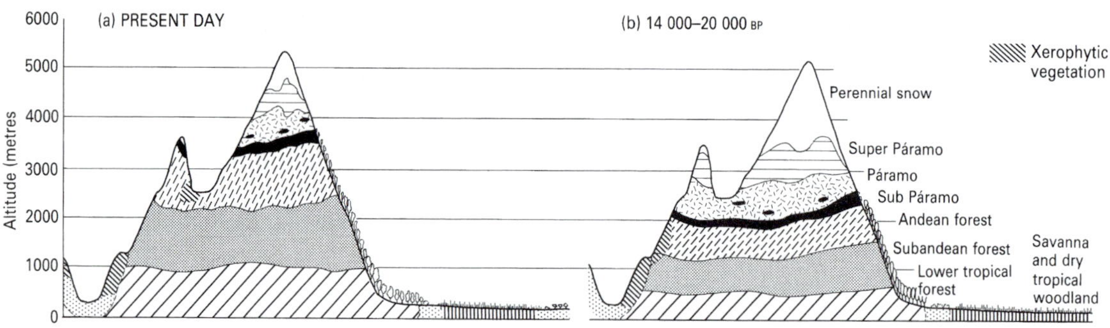

Fig. 6.12. Vegetation zones on the Andes near Bogota, Colombia. Compared to (a) the present day, the zones were (b) both depressed and compressed at the last Glacial maximum. (After van der Hammen in Flenley 1979, fig. 4.27.)

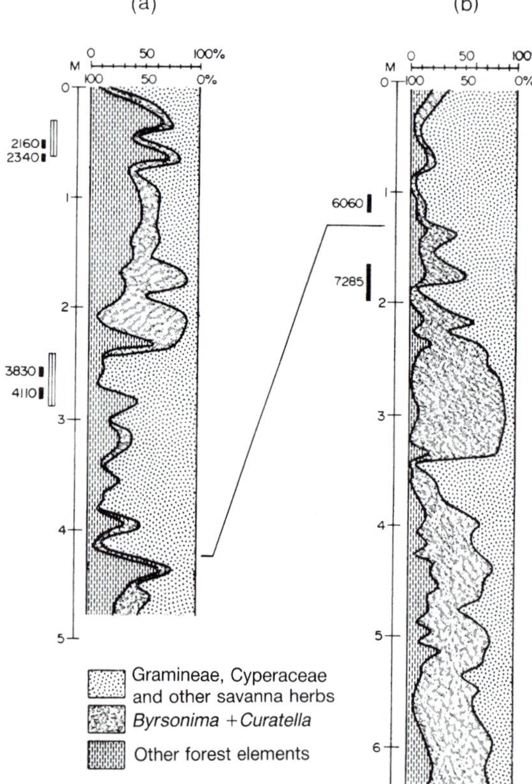

Fig. 6.13. Pollen diagrams from lowland northern South America (a) eastern Colombia and (b) Guyana. Radiocarbon dates are shown in years BP. (After Haffer in Whitmore and Prance 1987, fig. 1.8.) The repeated fluctuations in abundance of the three species groups shows an alternation between drier and wetter periods, with herbs and trees dominant, respectively. The recent sharp decline in trees may reflect the influence of burning by humans.

wetter conditions, respectively. Radiocarbon dates show that the oscillations correspond to Glacial and Interglacial periods. In central Malaya a mid-Pleistocene peat deposit below what is now the runway of Subang airport, Kuala Lumpur, was dominated by *Pinus* and grass pollen. Pine savanna is confined to strongly seasonal climates and is not found in Malaya today (Fig. 6.7a). At a site in lowland Amazonia pollen in riverine alluvial deposits shows repeated episodes of abundant grass at 4000, 2100, and 700 BP. This, however, probably reflects local variations in water level, hence in vegetation, and only indirectly indicates climatic fluctuation.

In the lowland rain forests of Queensland, pollen profiles have been described from peat deposited in several volcanic crater lakes. Repeated oscillation between dominance by rain forest and by wet sclerophyll forest with much *Eucalyptus* (Fig. 2.6) has occurred. The present-day boundary between these two forest formations is only a few kilometres from the sites. Although there have undoubtedly been changes in vegetation these might be only slight and may

not reflect more than small fluctuations in climate.

Geoscientific evidence

Previous seasonal climates in places that are at the present day perhumid can also be detected by physical signs. Soils containing true laterite, which only develop in seasonal climates with fluctuating soil water conditions, have been found below the South China Sea between Malaya and Borneo, part of the Sunda continental shelf. This sea-bed is also dissected by deep river valleys. On the Sahul shelf kunkar nodules, which similarly reflect a seasonal climate, have been found. The implication is that at Glacial maxima, when sea-level was lower and these continental shelves were exposed, the climate was indeed seasonal, as the pollen record indicates. Off the coast of South America sea-bed deposits also indicate past seasonal climate.

In northern South America sand dunes have been detected in the llanos savannas of Colombia, a region that today has a humid climate. The southern part of the Congo rain forest of Africa grows over sands that have affinity with the Kalahari desert to the south. In both these regions the sand is thought to have blown on the wind and deposited in places that were open and very dry.

Land-forms provide further clues of past climate. Braided streams choked with debris develop in climates in which rivers flood during periodical wet seasons. Steep slopes form, surrounded by gentle plains with gravel beds, as periodic heavy rains erode the hills. In Amazonia, aerial survey has identified areas under rain forest today with these geomorphological signs that they developed in a seasonal climate. By contrast, in perhumid climates that do not have marked seasonality, rivers develop meanders, and hills are rounded and with deeper soils that are formed by chemical weathering and less mechanical erosion.[149]

Today the very highest equatorial mountains of New Guinea, East Africa, and the Andes have a permanent snow and ice cap, above *c.* 4500 m. During Glacial maxima the snow-line was depressed by as much as 1500 m (Fig. 6.12). Signs of glaciation can be seen today below the present snow-line and on some mountains that at present have no snow cap; for example Mt. Kinabalu, 4101 m, in northern Borneo was glaciated down to *c.* 3000 m (Fig. 6.14).

The Tertiary

Less is known about Tertiary climates, i.e. before about 2 million BP, but they too fluctuated. Cores from below the South China Sea between Malaya and Borneo, which are used by oil companies for stratigraphical analysis, include a band which contains pollen of the

Fig. 6.14 The ice-smoothed rocky summit of Mt. Kinabalu, Sabah, 4101 m, which was covered by glaciers during the Quaternary ice ages.

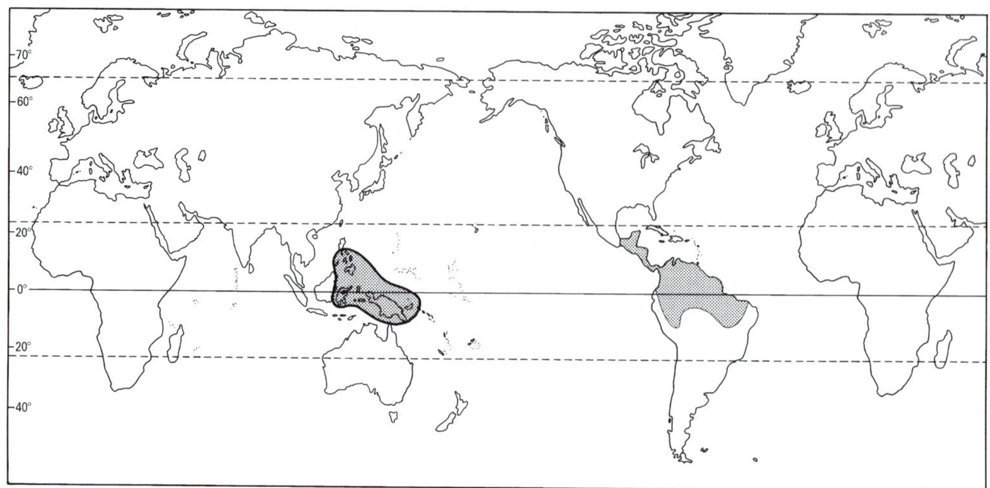

Fig. 6.15. The aroid genus *Spathiphyllum* occurs on both sides of the tropical Pacific Ocean in lowland rain forest.
Eighty-eight other genera show the same amphi-Pacific disjunction, including *Guettarda, Heliconia*, *Saurauia*, *Sloanea*, and *Trigonobalanus*. (Map 5, van Steenis 1962).

conifers *Abies*, *Keteleeria*, *Pinus*, and *Tsuga*, and of the xeromorphic shrub *Ephedra*, all of which are today confined to the much cooler, drier climates of continental Asia. This horizon spans 14 million years from the Oligocene into the early Miocene (about 35 to 21 million BP). Rain forests were more extensive than at any time in the Quaternary during the early Pliocene, parts of the Miocene, and especially the early Eocene; so these were all warm periods. Then, in the late Tertiary, fluctuations similar to those of the Quaternary occurred. The pollen record near Bogota discussed above was truly remarkable in extending back 3.5 million years, during which period it recorded 27 complete climatic oscillations.[150]

Tertiary climates have left their mark on present-day ranges of flowering plants in several ways. There are 89 genera of flowering plants represented on both borders of the Pacific Ocean in both tropical America and tropical Asia (Fig. 6.15).[151] These so-called 'amphi-Pacific tropical disjuncts' can hardly be explained by continental drift. They are in fact believed to have achieved their present range by migration north-wards around the Pacific rim via the Bering Straits; during warm periods of the Tertiary the climatic gradient polewards from the Equator was less steep than today. This is substantiated by Tertiary fossil floras in Alaska which contain many subtropical and tropical plants, e.g. *Alangium*, *Cinnamomum*, *Firmiana*, *Macaranga*, and *Saurauia*.

Africa has a much poorer flora than the other two rain forest regions.[152] This is believed to be because it was much more strongly desiccated during the Tertiary. We noted in Chapter 2 Africa's poverty in bamboos, ferns, and palms.

Australia too suffered strong Tertiary desiccation. At that time its mesic vegetation became mainly confined to the eastern seaboard. The strip of tropical rain forests found today in north Queensland is only 2–30 km wide and is of particular interest because it contains the relicts of the old mesic flora. This includes the ancestors from which many modern Australian species adapted to hot dry climates are believed to have evolved, e.g. *Brachychiton*, *Flindersia*, *Gardenia*, in some cases by explosive speciation, e.g. *Acacia*. New Caledonia is a shard of Gondwanaland which drifted away eastwards from northeast Australia starting in the Upper Cretaceous 82

million BP. Because it is an island its vegetation has suffered less from the drier Glacial climates so more of the old flora has survived.

The lands bordering the western Pacific have the greatest concentration of primitive flowering plants found anywhere, including Amborellaceae, Austrobaileyaceae, Calycanthaceae, Degeneriaceae, Eupomatiaceae, Himantandraceae, Lardizabalaceae, Magnoliaceae (Fig. 6.6*b*), Tetracentraceae, Trochodendraceae, and Winteraceae. It is most likely that they survive here as relicts. Winteraceae, for example, is a very ancient family with fossils back to the Barremian and was cosmopolitan in the mid-Cretaceous.

6.3. PLEISTOCENE REFUGIA

The pollen record gives direct evidence for fluctuations in vegetation which reflect fluctuations in climate, and there are also various different clues in the physical environment from which the same climatic fluctuations can be deduced. It has been discovered that rain forests have waxed and waned in extent during the Quaternary, and probably in the Tertiary too, and are not the ancient and immutable bastions where life originated which populist writings still sometimes suggest. In the present Interglacial they are as extensive as they have ever been, or nearly so. At Glacial maxima lowland rain forests are believed to have contracted and only to have persisted in places where conditions remained favourable for them, as patches surrounded by tropical seasonal forests, like islands set in a sea. In subsequent Interglacials, as perhumid conditions returned, the rain forests expanded out of these patches, which have come to be called Pleistocene refugia.

In the late 1960s it was shown that within Amazonia birds have areas of high species endemism and richness which are surrounded by relatively poorer areas. The same was soon demonstrated for lizards.[153] Subsequently many groups of animals have been shown to exhibit such patchiness, notably butterflies, and so have flowering plants, though for them only a few families were included in the analysis. The centres of concentration more or less coincide with each other (Fig. 6.16a). The existence of these overlapping centres of species richness and endemism is not seriously disputed, though for flowering plants some (e.g. the centre around Manaus city) have been shown to be merely reflections of high collecting intensity.[154] These loci overlap with areas that geoscientific evidence suggests retained rain forest during Pleistocene glaciations (Fig. 6.16b). The original interpretation was that rain forests have repeatedly fragmented to become islands in a sea of savanna and then expanded again, and that the overlapping centres of high species richness and endemism are the places that continuously remained rain forest.[155] This interpretation has been criticized. Firstly, it is more likely that rain forest was replaced by seasonal forest than by savanna. Secondly, it has been questioned whether the geoscientific evidence really does reflect past seasonal climates. The alternative postulation to counter the second criticism is that patchiness in species' ranges and richness is a consequence of diverse past fluctuations in the environment across the vast 3500-km width of the Amazon basin. These fluctuations include basin-wide lower temperature, rainfall and atmospheric carbon dioxide concentration, and increased seasonality during Glacial maxima, with in addition local forest disruption by mobile rivers (see section 7.8) and by flooding, wind and fire.[156] It is also now realized that species have evolved and migrated within the Amazon basin in an environment that has been fluctuating on a much longer time scale than the past two million years of the Pleistocene.

In the African rain forests four groups of loci of species richness and endemism are now recognized by students of mammals, birds, reptiles, amphibians, butterflies and flowering plants.[157] These are centred on Upper Guinea, on Cameroon and neighbouring Gabon and Congo, on the central Congo basin, and fourthly on its eastern rim (Fig. 6.17). These are

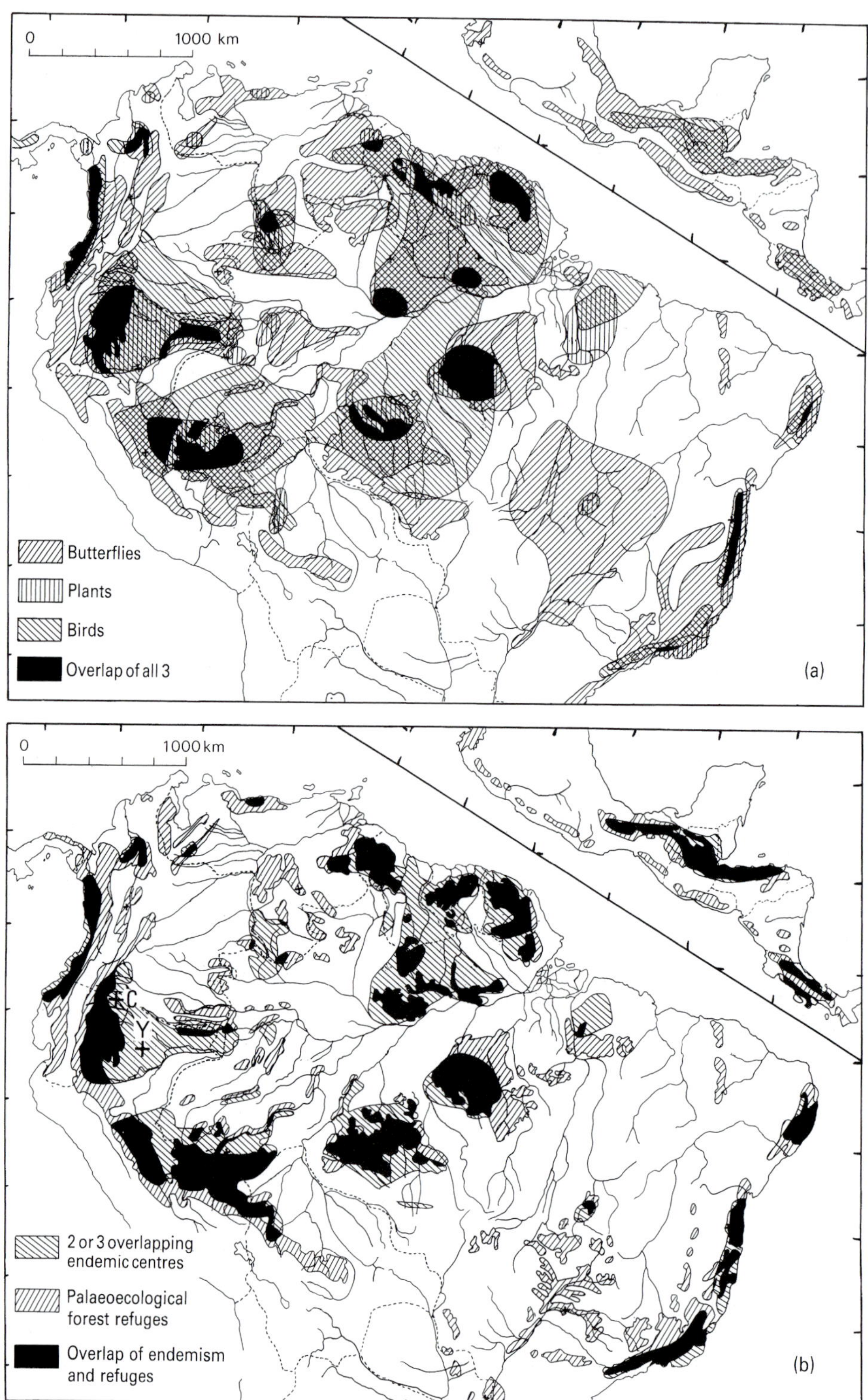
0
1000 km
Butterflies
Plants
Birds
Overlap of all 3
(a)
0
1000 km
C
Y
2 or 3 overlapping endemic centres
Palaeoecological forest refuges
Overlap of endemism and refuges
(b)

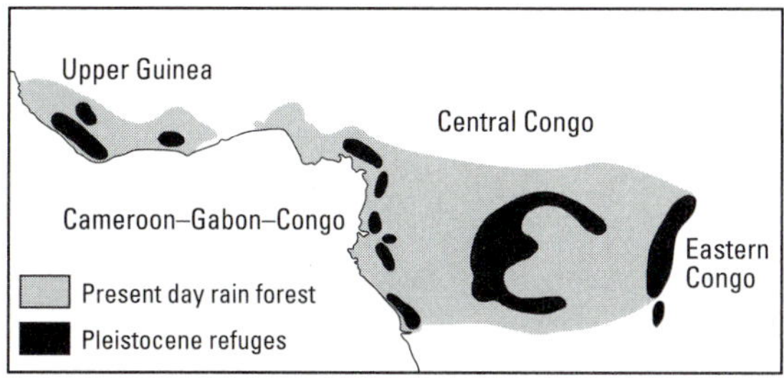

Fig. 6.17. Present-day extent of tropical rain forest in Africa. Areas of high species richness and endemism, which are postulated locations of Pleistocene forest refugia, also shown. (After Maley in Alexander *et al.* 1996, fig. 5.)

the areas to which it is believed rain forest retreated at the Glacial maxima. Tree species per hectare of over 0.1 m in diameter pick up this pattern. The Cameroon refuge had 138 species on the richest of several 0.64-ha sample plots at Korup and contrasted to less than 100 species per hectare for non-refugial sites in the African rain forests (Fig. 2.27).

The Pygmies who dwell within the central African rain forest fall into three groups with distinctive language and culture and these groups have genetic differences that suggest 10–20 000 years of isolation. It is believed that as the forest fragmented into three parts, so did the Pygmy population; later to spread out again with the Holocene climatic amelioration.[158]

Most parts of Malesia are today about as equally rich in species, including endemics, as the Pleistocene refugia of Africa and America. At the Glacial maxima the Sunda and Sahul continental shelves were exposed by falling sea-level. Rain forests were likely to have become confined to the more mountainous places where there was more, orographic, rain. The main development of seasonal forests in this region is likely to have been on the newly exposed low-lands, and when sea-level rose again at the next Interglacial these and the physical signs of seasonal climates (described above) were drowned. The parts of Malesia that are above sea-level today probably remained, largely perhumid and covered by rain forest, which explains their extreme species richness and their lack of geo-scientific evidence of seasonal past climates.

Present-day lowland rain forest communities consist of plant and animal species that have survived past climatic vicissitudes or have immigrated since the climate ameliorated. Thus many species co-exist today as a result of historical chance, not because they co-evolved together. Their communities are neither immutable nor finely tuned. This point is of great importance to the ideas scientists have expressed concerning plant–animal interactions which were discussed in Chapter 5, as well as to theoretical discussions on the nature of plant communities.

Those parts of the world's tropical rain forests that are most rich in species are those that the evidence shows have been the most stable, where species have evolved and continued to accumulate with the passage of time without episodes of extinction caused by unfavourable climatic periods. This is similar to the pattern seen in other forest biomes, for example in the warm temperate forests of

Fig. 6.16. (Left) Pleistocene refugia in the New World rain forests. (a) Centres of endemism of butterflies, plants and birds. (b) The overlap of these with palaeoecological forest refuges as deduced from palaeoclimates, topography, geomorphology, soils, and vegetation structure. (Whitmore and Prance 1987, figs. 7.4, 7.5.)
Note the extensive refugia in the western Amazon near the Andes. This is the region of the world's richest rain forests in terms of numbers of both animal and plant species (see Chapter 2). The locations of the two hyper-rich forest plots at Cuyabeno (Ecuador) C and Yanomono (Peru) Y are shown (see fig. 2.27). There are not yet sufficient sample plots spread across Amazonia to confirm that tree species numbers on small plots reflect the pattern of refugia detected from other evidence. The strong overlap between butterflies, plants and birds shown in (a), as well as the ranges of other groups of animals, are interpreted to reflect a wide range of past environmental disturbances stretching back through much of the Pleistocene and possibly into the Tertiary.

China[159] where there are species-rich patches with a concentration of relict conifers including the dawn redwood (*Metasequoia glyptostroboides*); and in the contrast between the species-rich Tertiary-relict Colchic forest at the eastern end of the Black Sea in Transcaucasian Russia with the rest of the European temperate broadleaf forest.

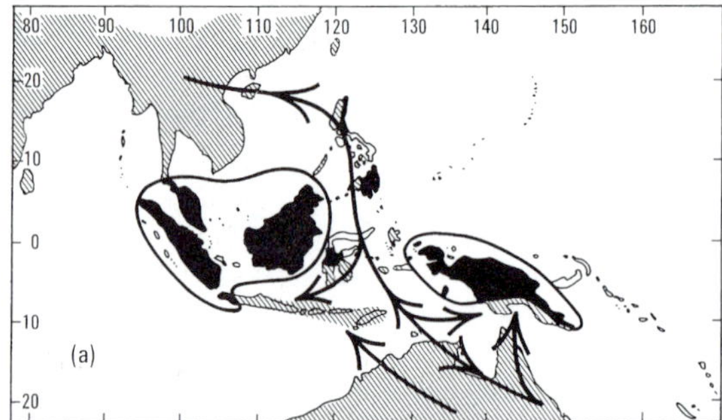

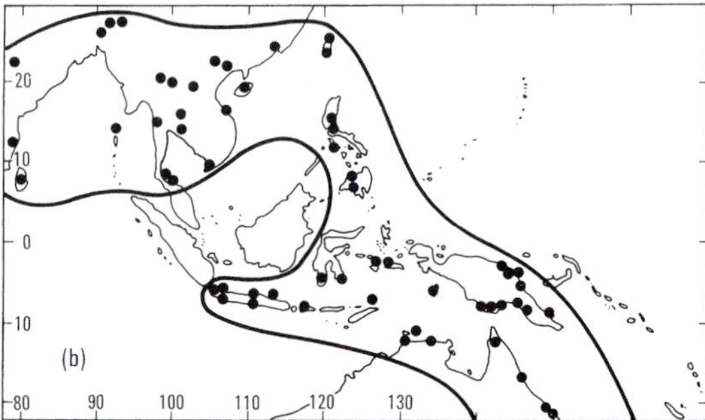

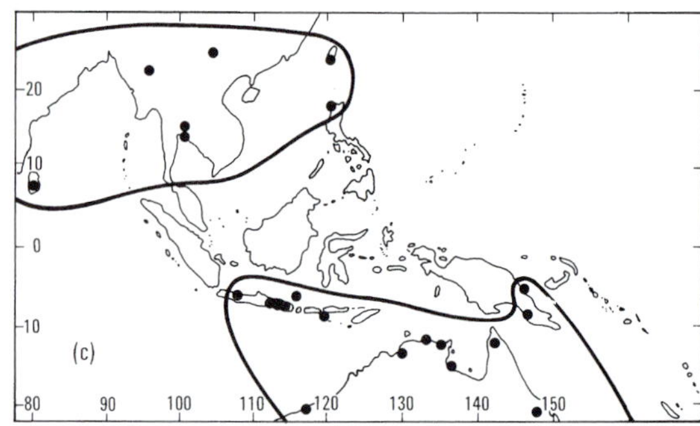

Fig. 6.18. Geographical ranges of two seasonal climate species in Malesia. (a) The two big cores of ever-wet rain forest as barriers for distribution of seasonal plants; the hatched areas either dry or consisting of a mosaic of ever-wet and dry areas. (b) *Pycnospora lutescens*, requiring some degree of dry season. (c) *Rhynchosia minima*, requiring a much stronger dry season; the disjunction is much more marked. After van Meeuwen, Nooteboom and van Steenis 1961 in Whitmore 1984*a*, fig. 14.4.)

6.4. SEASONAL CLIMATE ELEMENTS IN MALESIA

Seasonal climate plants in Malesia today have disjunct distributions, confined to continental Asia and Australia; their extents depend on how strongly seasonal a climate they require (Fig. 6.18), and on the two population areas being out of dispersal range of each other. In Glacial times when there was more seasonal forest, as described above, these ranges would have extended and become continuous or nearly so.

Elephant and gaur (*Elephas maximus*, *Bos gaurus*) today have races living in the rain forests of western Malesia though they are mainly animals of the seasonal forests of continental Asia. Elephant browse on forest fringe and gap plants and gaur are commensal with shifting cultivators, inhabiting early regrowth forest on fallowed fields. It seems likely that these animals adapted to live in rain forest when their preferred seasonal vegetation disappeared as the climate ameliorated and sea-level rose.[160] The same may apply to the Sumatran rhinoceros (*Didermocerus sumatrensis*).

6.5. HIGH MOUNTAIN PLANTS

A larger part of the land area of the humid tropics of Malesia is mountainous than is the case in Africa or America. European botanists living in Malesia, and perhaps nostalgic for home, have given much attention to the herbs and shrubs found above the tree line. There are two floristic elements, one northern, epitomized by *Primula* and *Ranunculus* (primrose and buttercup), the other southern, e.g. *Gunnera* and *Nertera*.[161] At the present day, populations on different mountains are outside dispersal range of each other and of source areas in the

Fig. 6.19. *Coprosma hookeri* at 4000 m on Mt. Kinabalu, Sabah.
Coprosma is a genus of south temperate latitudes which extends into the tropics on high mountains.

Himalaya and Australasia. This no longer presents the enigma it formerly did, now it is realized that all montane vegetation belts have been strongly and repeatedly depressed during Glacial maxima (cf. Fig. 6.12), with the consequence that populations would have been much closer. Mt. Kinabalu, the only really high mountain in Borneo, however, seems always to have been very isolated.[162] Nevertheless, it possesses a rich temperate flora (Fig. 6.19).

Patches of paramo moorland in the high Andes which are today widely separated islands would likewise have been closer together, if not continuous, at times in the past.

6.6. CONSERVATION

The central objective of rain forest conservation is to preserve adequate samples of the total diversity of species and ecosystems. Interactions between species and fluctuating climates on short, medium, and long time-scales are one motor of continuing evolution. The discovery that some rain forests have areas of high endemism provides criteria for the design of conservation areas. The biologist can state with conviction that it is necessary to include parts of these centres of endemism and of their boundaries, and also of surrounding species-poor forest, in order to conserve samples of the full diversity of the forest.

6.7. TROPICAL RAIN FORESTS THROUGH TIME—CHAPTER SUMMARY

1. In his search for generalizations the naturalist looks for repeating patterns of distribution and then for their interpretation. Two causes explain many patterns of tropical rain forest biogeography.

2. Continents have drifted. Some plants have ranges that arise from their occurrence on Gondwanaland and its fragments as these have moved apart (Fig. 6.1). Biogeography of the Eastern tropics (Figs. 6.5–6.7) reflects the convergence, then collision, of east Gondwanaland with Laurasia (Figs. 6.1, 6.8). Northern South American rain forest bigeography reflects the uplift of the Andes (Fig. 6.9).

3. Climatic history explains many other patterns. On the continental scale it explains some disjunct ranges, for example genera that occur in the lowland humid tropics on both sides of the Pacific (Fig. 6.15). Tropical climates have oscillated throughout the Quaternary, and the later Tertiary, between warm, wet periods, such as today, and cooler, drier, more seasonal periods. These correspond, respectively, to Interglacials and Glacials of high latitudes. Rain forests today are at or near their maximal Quaternary extent. During Glacials their area was reduced and tropical seasonal forests became more extensive.

4. Evidence for fluctuating climates comes from two sources. Firstly, pollen analysis shows vegetation zones have fluctuated. Most data are from mountains and show vegetation types have moved up and down as climate has changed (Fig. 6.11). In parts of the lowlands, rain forest and more seasonal vegetation have alternated (Fig. 6.13). Secondly, direct physical evidence from geomorphology and soils, shows some rain forests occur today in places that once had seasonal climates.

5. In the neotropics many animals and some plants have areas of high species richness and endemism. These more or less coincide for different groups, and are believed to be refugia to which rain forest retreated at Glacial maxima, Geoscientific evidence for refugia is also more or less coincident (Fig. 6.16). In Africa there is evidence of four such refugia (Fig. 6.17).

6. Disjunct distributions of seasonal climate plants in Malesia (Fig. 6.18) and widely separated populations of high mountain plants would have been continuous or more nearly so in Glacial periods.

7. These patterns of distribution provide a rational basis for the design of conservation areas.

7
Forest dynamics

In Chapter 6, tropical rain forests were examined on the long-time scale of tens of thousands of years and shown to have fluctuated in area. Delving further still, over millions of years, continental drift and evolution come into play. Present day fauna and flora bear witness to these long-term changes. On these very long time scales rain forests have been unstable and have changed profoundly in extent and composition, as indeed have all other biomes.

In the present chapter we look at temporal change at the other extreme, time scales of a few years to a few centuries, up to the life spans of one or a few generations of trees. Change is examined in the context of development and disintegration of the forest canopy, the forest growth cycle, introduced in section 2.3. Two spatial scales of canopy dynamics can be distinguished: patch disturbance, which involves one or a few trees, and community-wide disturbance. Patch disturbance is sometimes called 'forest gap-phase dynamics' and since about the mid-1970s has been one of the main interests of forest scientists in many parts of the world.[163] In the first part of this chapter the processes of regeneration and maintenance in rain forests are described. They have close parallels elsewhere. There seems to be a general model of forest dynamics which holds in many different biomes, albeit with local variants. Ecology as a science is always seeking for generalizations to make sense of the bewildering diversity of Nature, and it appears that the forest growth cycle and its ramifications is one such set of generalizations. This is a topic on which understanding has advanced rapidly and a good framework has been built. It has now reached a mature stage where the multitudinous continuing research is essentially filling in details, 'fleshing out the picture'.

The chapter then goes on to discuss silviculture: the ways that humans make use of forest dynamic processes to manipulate forest composition for their own benefit. It concludes with a consideration of community-wide disturbances to rain forests, now realized to be very widespread, and including human disturbance and recovery therefrom.

7.1 FOREST MICROCLIMATES

Species differ in the microclimate in which they successfully regenerate. It is important therefore to gain an understanding of the microclimates within a rain forest. These are mainly determined by size of canopy gap.

The microclimate above the forest canopy, which is similar to that in a large clearing, is substantially different from that near the floor below mature phase forest. Figure 7.1 shows readings taken in the canopy at various heights

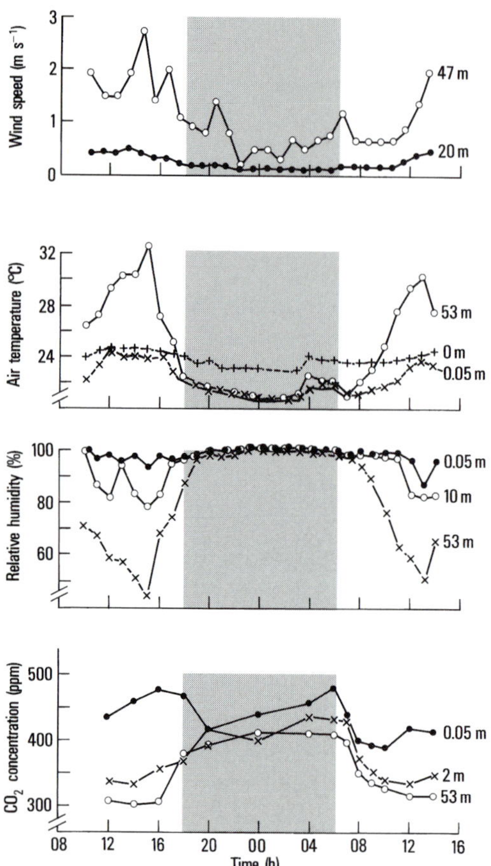

Fig. 7.1. The microclimate above the forest canopy strongly differs from that at ground level as is shown here by the daily march of wind speed, air temperature, relative humidity, and CO_2 concentration at various canopy levels in lowland rain forest at Pasoh, Malaya, 21–22 November 1973. Night time shaded. (After Aoki *et al.* 1978, in Whitmore 1984*a*, fig. 4.7.)

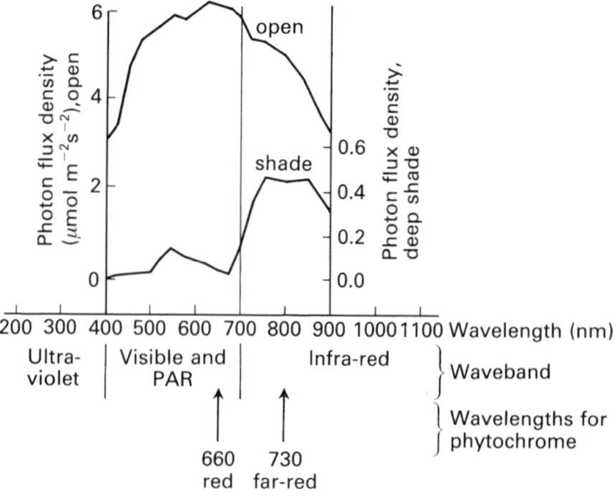

Fig. 7.2. The light in the open ('white light') and in deep forest canopy shade compared. There is a change in both quantity and quality. (Curves based on fig. 1 in MacLellan and Frankland 1985.)

above the ground. Outside, wind speeds during the day are higher, as is air temperature, while relative humidity is lower. Close to the ground within the forest, carbon dioxide content of the air remains high all the time but up in the canopy it drops during the day due to uptake by photosynthesis.

The light climate within a forest is complex. There are four components, skylight coming through canopy holes, direct sunlight, seen as sunflecks on the forest floor, light transmitted through leaves, and light reflected from leaves, trunks and other surfaces. The sky usually has some clouds and the first two components have approximately the same spectral composition. Light transmitted through, or reflected from, leaves is greenish because the orange to red wavelengths have been absorbed and utilized for photosynthesis (Fig. 7.2). The waveband 400 to 700 nm (which is approximately the visible spectrum) is utilized for photosynthesis and is known as photosynthetically active radiation or PAR. The forest floor only receives up to *c.* 2 per cent of the PAR incident on the forest canopy (Fig. 7.2), and 50–80 per cent of this is contained in sunflecks (Table 7.1). Plants living below closed forest probably rely on the sunfleck component of the light climate for photosynthesis, the other components are too small to be utilizable. Compared to the open, most sunflecks have low energy (Fig. 7.3). Very bright sunflecks are rare and may have more energy than shade-adapted leaves can utilize.

In addition to reduction in quantity of PAR within the forest canopy, PAR also changes in quality with a shift in the ratio of red to far-red wavelengths (660 and 730 nm) (Fig. 7.2). This is because the canopy preferentially absorbs the

Table 7.1 Forest floor light conditions

Location (latitude)	Mean per cent transmission PAR and range	Maximum % total PAR due to sunflecks
Queensland, Australia (28°15′S)	0.5 (0.4–1.1)	62
Oahu, Hawaii (21°30′N)	2.4 (1.5–3.8)	80
La Selva, Costa Rica (10°26′N)	*c.* 1.0–2.0	55–77
Singapore (1°20′N)	*c.* 2.0	*c.* 50

Mainly after Chazdon and Fetcher (1984) table 1. Very little photosynthetically active radiation (PAR) penetrates the canopy and most of what does is contained in sunflecks.

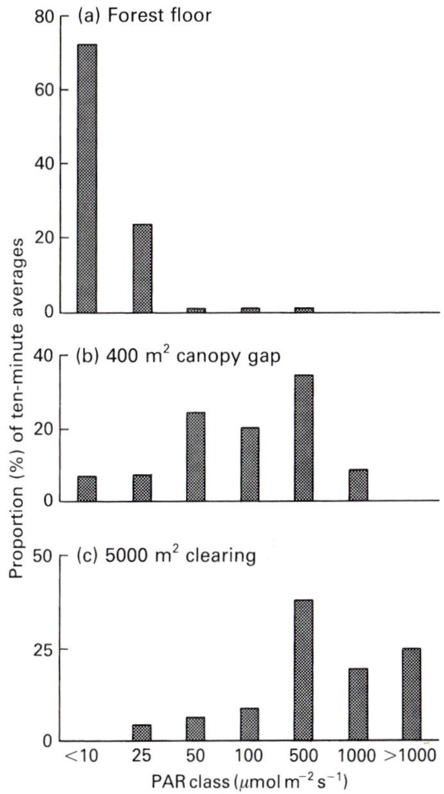

Fig. 7.3. Ten minute average values of photosynthetically active radiation (PAR) at La Selva forest, Costa Rica. Inside the forest (a) most values are small with a few big ones due to sunflecks. In a canopy gap (b) values are larger, but less than in a large clearing (c). (After Chazdon and Fetcher 1984, fig. 2.)

red waveband. At La Selva, Costa Rica, the change was from a median value of the red/far-red ratio of about 1.2 in the open to 0.99 in sunflecks and 0.42 in forest shade.

Canopy gaps have an aerial microclimate like that above the forest but the smaller the gap the less different it is from the forest interior (Table 7.2, Fig. 7.4). In particular, the amount of photosynthetically active radiation is reduced (Table 7.1) and its distribution through the day alters. Both the quantity and quality of light reaching the plant is known to be of profound importance in the mechanisms of gap-phase dynamics, as will be shown later in this chapter. Due to the increase in solar radiation, the soil of gaps is more prone to drying out in rainless spells than soil below the canopy[164] and the temperature of the top few millimetres of soil may become as hot as 35 °C compared to only *c.* 25 °C below the canopy (Table 7.2). The manner in which different species are adapted to different dimensions of microclimate, especially the light climate, are described below. The microclimate has turned out to be more complex than had been expected. Gaps were at first regarded as having a microclimate varying with their size, to be contrasted with closed-forest microclimate. But this is a simplication.[165] In microclimatic terms 'forests are not just Swiss cheese', gaps are neither homogeneous holes nor are they sharply bounded.[166] Within a gap the microclimate is most extreme in the centre and changes outwards to the physical gap edge and beyond (Figs. 7.5, 7.16). The larger the gap the more extreme the microclimate of its centre. There is likely to be a larger area of small-gap-centre microclimate around the periphery of large gaps than in the centres of small gaps.

Furthermore, there is much more variability between small gaps than large ones in microclimate. The amount of direct sunlight controls the

Table 7.2 Microclimates of open, canopy gap, and closed lowland rain forest compared. (Danum, Sabah)

	Open	Big gap	Small gap	Closed forest
Per cent sky visible[a]	74	31	10	6
Aerial microclimate:				
Temperature, max./min. (°C)[b]	30[c]/22.5	38/21.5	34.5/21.5	28.5/21
Relative humidity, min. (%)[d]	50	52	67	85
Photosynthetically active radiation (mol $m^{-2}day^{-1}$)[e]	35.0	19.2	4.9	0.5
Soil mean temperature (°C):				
10 mm max./min.	35.5/25	35/23	30.5/23	25/22.5
50 mm max./min.	29.5/25.3	31.5/24.5	25.5/24	24.5/23.5

Unpublished data of N. Brown and D. Kennedy
Measurements made in gap centres
The smaller the gap the more closely it resembles closed forest except that minimum air and soil temperatures are similar everywhere
[a] Assessed from a hemisphere photograph (Fig. 7.4)
[b] Mean of weekly shaded value for 2 weeks every month for 12 months
[c] Open maximum temperature kept low by river breezes
[d] Mean daily value, from measurements of 2-week periods for 12 months
[e] Daily total, recorded 2 weeks every month for 12 months

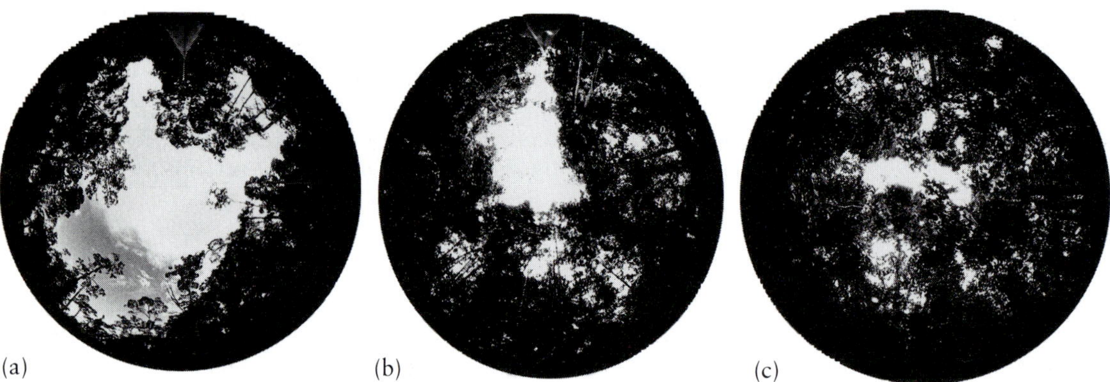

Fig. 7.4. Hemisphere photographs of (a) big gap, (b) small gap and (c) closed mature phase canopy, with 31, 10 and 6 per cent sky visible, respectively.
These are the sites in the Danum forest, Sabah, whose microclimates are shown in Table 7.2. Photographs taken with a 180° fish-eye lens. North at top. Sky percentage computed with an image analyzer; the human eye is very deceptive. Note cloud in (a).

maximum temperature, relative humidity and PAR (Table 7.2). Large gaps are open to direct sunlight for most of the day, for most or all of the year. Small gaps are only open for the much shorter periods when the sun is opposite them and the position of the gaps becomes important. Where (as at Danum, Sabah) afternoons are cloudy, small gaps west of the zenith let through

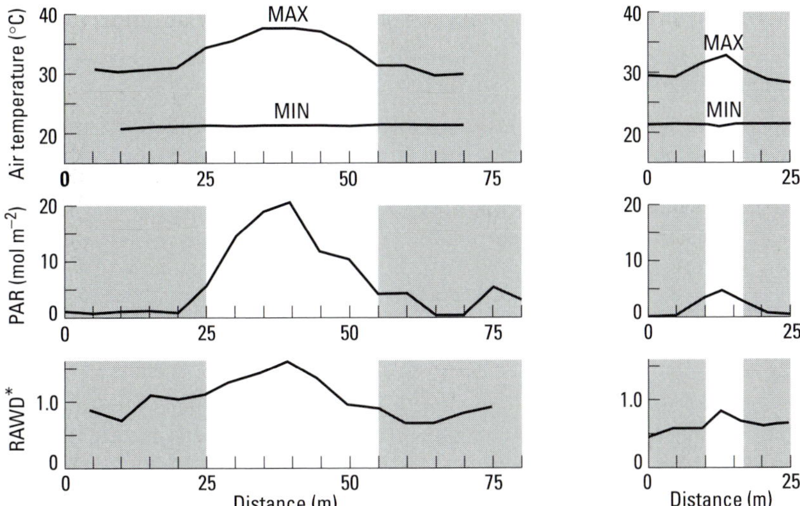

Fig. 7.5. Microclimate along north-south transects through a big and tiny gap (Danum, Sabah). Daily values. Measurements made over a year. (Brown 1993, fig. 12.)

*Relative atmospheric water demand.

The large gap is that also shown on Table 7.2 and Fig. 7.4. Note that there is no sharp change in microclimate at the physical edge of the gap and that its effect extends over 5 m into the forest.

less direct sunlight and have a less extreme microclimate than those to the east, which the sun shines through during the morning.

Thus, gap size is a poor surrogate measure of microclimate, most markedly over short periods. These complications were revealed by experiments designed to unravel the growth of seedlings of different tree species in different microclimates. This is an example of how a science advances. Here we started with the interesting, vital and apparently simple question of the extent to which seedlings of different tree species have different responses to different sizes of gap and discovered that this in fact was based on a very naive view of the forest floor microclimate.

7.2 PIONEER AND CLIMAX TREE SPECIES[167]

It is quickly discovered, even by the novice, that tree species differ in the amount of solar radiation required for their regeneration. At one extreme are species only ever found to grow up along roadsides, in old fields and in big canopy gaps. At the other extreme are species that grow up in tiny canopy gaps, or, in the ultimate case of undergrowth treelets, in no perceptible gap.

Ecologists and foresters continue to engage in vigorous debate as to whether species along this spectrum of light climates can be divided into clear, separate groups, as any glance at the research literature soon shows.

As described in section 2.3, some strong light-demanders require full light for both seed germination and seedling establishment. These are the pioneer species, set apart from all others by these two features.[168] By contrast, all other species have the capacity to germinate and establish below canopy shade. These may be called climax species. They are able to perpetuate in the same place, but are an extremely diverse group. Amongst climax species are some that, although they can germinate in deep shade, have seedlings that die unless they get full light within about a year, examples being some *Entandrophragma* (Fig. 7.10) of West Africa, *Swietenia macrophylla* of tropical America

Table 7.3 The main characters of pioneer and climax tree species in tropical rain forests

	Pioneer	Climax
Common alternative names	Light-demander, (shade-) intolerant, secondary	Shade-bearer, (shade-) tolerant, primary
Germination	Only in canopy gaps open to the sky which receive some full sunlight	Usually below canopy
Seedlings	Cannot survive below canopy in shade, never found there	Can survive below canopy, forming a 'seedling bank'
Seeds	Usually small, produced copiously and more or less continuously, and from early in life	Often large, not copious, often produced annually or less frequently and only on trees that have (almost) reached full height
Soil seed bank	Many species	Few species
Dispersal	By wind or animals, often for a considerable distance	By diverse means, including gravity, sometimes only a short distance
Dormancy	Capable of dormancy ('orthodox') commonly abundant in forest soil as a seed bank	Often with no capacity for dormancy ('recalcitrant'), seldom found in soil seed bank
Growth rate	Carbon fixation rate, unit leaf rate, and relative growth rates high	These rates lower
Compensation point	High	Low
Height growth	Fast	Often slow
Branching	Sparse, few orders	Often copious, often several orders
Growth periodicity	Indeterminate ('sylleptic'), no resting buds	Determinate ('proleptic'), with resting buds
Leaf life*	Short, one generation present, viz. high turn-over rate	Long, sometimes several generations present so slow turn-over rate
Herbivory	Leaves susceptible, soft, little chemical defence	Leaves sometimes less susceptible due to mechanical toughness or toxic chemicals
Wood	Usually pale, low density, not siliceous	Variable, pale to very dark, low to high density, sometimes siliceous
Ecological range	Wide	Sometimes narrow
Stand table	Negative	Positive
Longevity	Often short	Sometimes very long

From Swaine and Whitmore (1988); Whitmore (1991)
* Reich *et al.* (1991)
These sets of characters separate two ecological species groups, but exceptions can be found. Climax species have a wide range of characters such that, except for the first two listed, some are very similar to pioneers. See text for fuller details.

(p. 142), and some *Parashorea* and *Shorea* of Malesia. Apart from the absence of the two key characters that distinguish true pioneers, these species have many of their other features (Table 7.3). To the forester, for example, their rapid growth and pale, low density timber, shared with true pioneers, is important. Thus, although the pioneer versus climax distinction has critical consequences for forest regeneration and succession (as described in section 2.3), in other respects pioneers are just the extreme end of a continuum of species. However, for the purposes of making sense of the bewildering diversity of the rain forest, the recognition of pioneer and climax ecological species groups is useful.

Besides their light requirement for seedling success, pioneer and climax species have a whole group, or syndrome, of other associated

characters (Table 7.3), but these are just tendencies and exceptions are well-known. For example, *Aleurites* is a strong light-demander but has huge seeds (*c.* 8 g); *Milicia excelsa* and *M. regia* are African pioneers with dark, dense, durable wood, iroko, a teak substitute (much used for garden furniture); and *Casuarina* and *Securinega* are likewise heavy-timbered Eastern pioneers.

Pioneer tree species

Pioneer species germinate and establish in a gap after its creation (Figs. 7.6–7.9). They grow fast in height (Fig. 7.9) and laggards are suppressed so the canopy grows up with a strong tendency to be one-layered. Below the canopy seedlings of climax species establish and, as the pioneer canopy breaks up after the death of individual trees, these climax species are 'released' (to use foresters' parlance) and grow up as a second growth cycle. Succession has occurred as a group of climax species replaces the group of pioneer species. Secondary tropical rain forests have only a few tree species per hectare, and sometimes only one or two, by contrast to primary forests which are usually very rich (Fig. 2.27).

Pioneer species are also called light-demanders or (shade-) intolerants in reference to their seedling requirements for solar radiation. Sometimes they are called secondary species because they form secondary or regrowth forest on cleared surfaces. The terminology is confusing, and there is no general agreement.

Amongst pioneers, different species attain different heights at maturity, the larger ones are

Fig. 7.6. *Cecropia obtusa* (*left*) and *C. sciadophylla* (*right*). Brazilian Amazon, with J.N.M. Silva (1985). *Cecropia* is the biggest genus of pioneer trees in the neotropics, three or four species may occur together.

Fig. 7.7. *Macaranga gigantea*. East Kalimantan. This east Borneo variant has perhaps the largest palmate leaves of any plant.

Fig. 7.8. *Macaranga triloba*. Malaya.
This genus ranges from West Africa to Polynesia. Most of its *c*. 250 species are pioneers, which makes it easily the largest genus of pioneer trees in the world. Up to about ten species occur together.

longer-lived. Table 7.4 lists common pioneer species from the different parts of the tropics. Reference is often made to small or short-lived and to big or long-lived pioneers, but it must be recognized that these are only arbitrary subdivisions of continuous variation. A big gap is sometimes simultaneously colonized by pioneers of different mature heights. After the short-lived pioneers die a taller canopy develops of the larger, long-lived ones, sometimes together with light-demanding climax species (see below) that have become established.[169] This is particularly noticeable in tropical America where there is a big group of the latter (e.g. *Cavanillesia* spp., *Cedrela odorata*, *Simarouba amara*, *Swietenia macrophylla*, *Vochysia* spp.). All these species have similar timber (see below) and these tall secondary forests are of increasing commercial importance. Viewed from a roadside one sees two stages, early secondary (early seral) forests later turn into late secondary (late seral) ones.[170] Each has a different set of species dominant and succession seems to have occurred, but if one enters an early seral forest and makes an enumeration it is found that all the species colonized early on.

Climax tree species

Climax species usually germinate and establish below a canopy, therefore they can perpetuate

Fig. 7.9. *Ochroma lagopus*, balsa. A 3.5-year-old tree at La Selva, Costa Rica.
Most famous tropical pioneer tree species, widespread in the New World tropics, renowned for its very low density timber, where grown fast, which is much used for model-building and was formerly used for aircraft.

in situ. They are sometimes called primary species or, with reference to their seedlings, (shade-) tolerants or shade-bearers. This is the group of which climax (primary) forest is composed, and climax plant communities are defined as those that are self-perpetuating, in a state of dynamic equilibrium. Seedlings on the forest floor only grow slowly, and in most species seldom reach taller than about 1 m because they eventually die unless released.

Climax species may be subdivided in various ways, for example, by usual mature height, but one of the most useful subdivisions is on the degree of seedling shade tolerance, discussed above. The small trees that never reach the canopy probably only require enough growing space above and below ground to enable them slowly to attain maturity. Strongly shade-tolerant species typically grow slowly and have dark, dense, hard and often siliceous timber which is often naturally durable. Light-demanding climax species (Figs. 7.10, 7.11) grow faster and most have paler, lighter, softer timber which is rarely siliceous and is seldom durable without chemical preservation. The range of different sorts of climax species can often be found in a single forest (Table 7.5).

Character syndromes of pioneer and climax species

It can be seen from Table 7.3 that in essence pioneer species are aggressive. Typically they produce a large volume of low density wood by fast growth, with open-branched crowns whereby they rapidly pre-empt competition by filling a large space. They start to reproduce early in life and produce copious seeds frequently, which are small and easily dispersed. By these means pioneers efficiently exploit new big gaps as these develop scattered through the forest. Growth is opportunistic and indeterminate, and continues so long as mineral nutrients and water are available. Leaves are short-lived, often produced continually, and as their efficiency soon diminishes their nutrients are recycled to fresh flushes, so there is no great need to invest in mechanical (tough, fibrous leaves) or chemical protection against herbivores. Pioneers are unable to regenerate *in situ*, a mature stand has few small individuals, and the stand table is negative (Fig. 7.12a).

Climax species, by contrast, are slower growing and with denser timber and denser crowns. In order to establish successfully below a canopy where photosynthetically active radiation is low the seeds contain sufficient reserves for the seedling to build a root-system and the first photosynthetic organs. Seeds are fewer and larger, produced later and less often. There is

plants were put back into low light conditions took over 60 minutes. Furthermore, the rate of carbon gain varied with past history and intermittent sunflecks could be used more efficiently than continuous high light. The conclusion is that the temporal pattern of sunfleck distribution through the day (cf. Fig. 7.3) is of importance, not just the daily total PAR. Further studies like this are needed to analyse interspecific differences of shade tolerance.

Seedling release of climax species

The light climate

As he gets to know a forest the ecologist or forester soon develops a feeling for how light-demanding the different tree species are. Most obvious is the difference between pioneer and climax species. For the latter, there are also his perceptions on the size of gap in which seedlings are seen to have grown up from the seedling bank. General observations on the natural history of the forest such as these are an important part of more rigorous scientific analysis, initially as precursor to help the intelligent design of experiments, and later to assist interpretation of the results. Nowhere is this general background knowledge better developed than in the dipterocarp rain forests of Malaysia, where foresters have now nearly a century of accumulated experience. It has been found that amongst dipterocarps there are species whose seedlings grow up in small canopy gaps and others that are successful in big gaps. Correlated properties are that the first group have lengthy persistence in the seedling bank (Fig. 7.18) and, as trees, have slow diameter growth and high density, dark coloured, naturally durable timber. Seedlings of the second group die sooner (Fig. 7.18) and trees grow faster and have lighter, paler, non-durable timber. The two groups are known as shade-bearers and light-demanders or as medium to heavy hardwoods and light hardwoods, respectively. The silvicultural systems invented for dipterocarp rain forests, described later in section 7.6, depend on these responses to gaps. Medium/heavy hardwood species can be successfully regenerated by small canopy disturbances; light hardwoods by more substantial disturbances. Similar categories are recognized in other rain forests, and indeed the whole of rain forest silviculture everywhere depends upon it. It is the practical working out of the wide range of shade tolerance shown by different climax species that was discussed above in section 7.2.

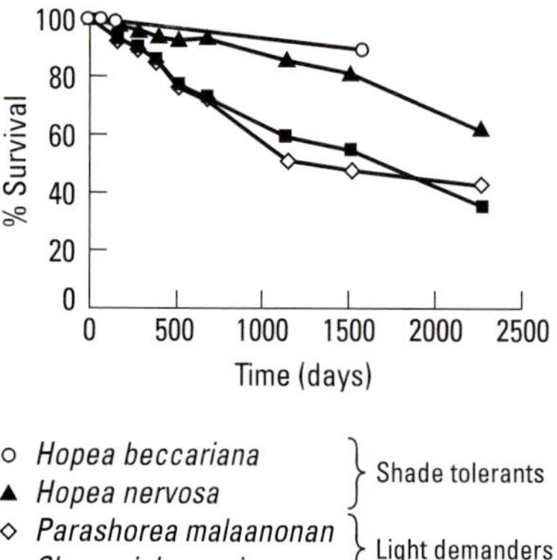

Fig. 7.18. Differential survival of dipterocarp seedlings on the forest floor over 6 years, Danum, Sabah.

Seedlings of shade-tolerant species, in this case the two *Hopea* spp., characteristically survive better in deep shade.

Why is there a gap switch size between these ecological groups? For the dipterocarp species that occur in the Danum forest, Sabah, this has now been elicited by observing over 6.4 years the growth, in artificial canopy gaps of different size, of seedlings from the forest floor seedling bank (Fig. 7.19). It can be seen in summary that the light-demander species *Shorea johorensis* was able to utilize for height growth the extra solar radiation received in larger gaps and over a longer period of time. This is the ecological consequence at population level of its ecophysiological flexibility at the leaf and single seedling level. By contrast, the shade-bearer *Hopea nervosa* lacked the flexibility to adjust to and make use of the extra PAR of the larger gaps. Success, however, is determined by many factors and *Parashorea* performed badly because it

Fig. 7.9. *Ochroma lagopus*, balsa. A 3.5-year-old tree at La Selva, Costa Rica.
Most famous tropical pioneer tree species, widespread in the New World tropics, renowned for its very low density timber, where grown fast, which is much used for model-building and was formerly used for aircraft.

in situ. They are sometimes called primary species or, with reference to their seedlings, (shade-) tolerants or shade-bearers. This is the group of which climax (primary) forest is composed, and climax plant communities are defined as those that are self-perpetuating, in a state of dynamic equilibrium. Seedlings on the forest floor only grow slowly, and in most species seldom reach taller than about 1 m because they eventually die unless released.

Climax species may be subdivided in various ways, for example, by usual mature height, but one of the most useful subdivisions is on the degree of seedling shade tolerance, discussed above. The small trees that never reach the canopy probably only require enough growing space above and below ground to enable them slowly to attain maturity. Strongly shade-tolerant species typically grow slowly and have dark, dense, hard and often siliceous timber which is often naturally durable. Light-demanding climax species (Figs. 7.10, 7.11) grow faster and most have paler, lighter, softer timber which is rarely siliceous and is seldom durable without chemical preservation. The range of different sorts of climax species can often be found in a single forest (Table 7.5).

Character syndromes of pioneer and climax species

It can be seen from Table 7.3 that in essence pioneer species are aggressive. Typically they produce a large volume of low density wood by fast growth, with open-branched crowns whereby they rapidly pre-empt competition by filling a large space. They start to reproduce early in life and produce copious seeds frequently, which are small and easily dispersed. By these means pioneers efficiently exploit new big gaps as these develop scattered through the forest. Growth is opportunistic and indeterminate, and continues so long as mineral nutrients and water are available. Leaves are short-lived, often produced continually, and as their efficiency soon diminishes their nutrients are recycled to fresh flushes, so there is no great need to invest in mechanical (tough, fibrous leaves) or chemical protection against herbivores. Pioneers are unable to regenerate *in situ*, a mature stand has few small individuals, and the stand table is negative (Fig. 7.12a).

Climax species, by contrast, are slower growing and with denser timber and denser crowns. In order to establish successfully below a canopy where photosynthetically active radiation is low the seeds contain sufficient reserves for the seedling to build a root-system and the first photosynthetic organs. Seeds are fewer and larger, produced later and less often. There is

Table 7.4 Some common rain forest pioneer tree species*

Stature	Neotropics	Africa†	Eastern tropics
Small, 2–7.9 m tall	*Cordia nitida* *Ocotea atirrensis* some *Piper* *Vernonia patens* *Vismia baccifera*	*Ficus capensis* *Leea guineensis* *Phyllanthus muellerianus* *Rauvolfia vomitoria*	*Commersonia bartramia* *Glochidion* spp. *Macaranga* >> 100 spp. some *Mallotus* spp. some *Melastoma* spp. *Phyllanthus* spp. *Pipturus* spp. *Trichospermum*, 8 spp.
Medium, 8–29 m tall	*Trema* *Alchornea triplinervia* *Cecropia*, *c.* 100 spp. *Cordia* spp. *Jacaranda copaia* *Muntingia calabura* *Ochroma lagopus* *Schefflera (Didymopanax) morototoni*	*Trema* *Anthocleista nobilis* *Psydrax arnoldiana* *Cleistopholis patens* *Macaranga* *Maesopsis eminii* *Musanga cecropioides* *Spathodea campanulata* *Vernonia conferta* *Vismia guineensis*	*Trema* *Acacia aulacocarpa* *Acacia mangium* *Adinandra dumosa* *Alphitonia petrei* *Anthocephalus*, 2 spp. few *Macaranga* spp. *Ploiarium alternifolium*
Large, > 30 m tall	*Ceiba pentandra* *Cedrelinga catenaeformis* *Goupia glabra* *Laetia procera*	*Aucoumea klaineana* *Ceiba pentandra* *Lophira alata* *Milicia excelsa* *Milicia regia* *Nauclea diderrichii* *Ricinodendron heudelotii* *Terminalia ivorensis* *Terminalia superba*	*Eucalyptus deglupta* *Octomeles Sumatrana* *Paraserianthes (Albizia) falcataria*

All the large and some of the medium sized species are important for timber
* African species from M.D. Swaine; American species from Finegan (1992) and G.S. Hartshorn
† For descriptions see Hawthorne (1995)

Fig. 7.10. Pole of *Entandrophragma angolense*, a climax species whose seedlings grow fast in high solar radiation but soon die in deep shade. Ghana.

Fig. 7.11. Pole of *Campnosperma auriculatum*, a climax species with similar high light requirements to *Entandrophragma* (Fig. 7.10). Malaya.

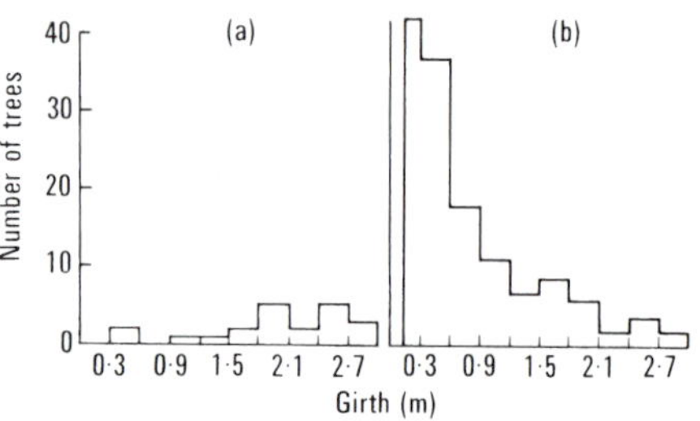

Fig. 7.12. Stand tables of (a) the pioneer *Endospermum medullosum* and (b) the climax species *Parinari papuana*. Lowland rain forest, Kolombangara, Solomon Islands. (Whitmore 1974, fig. 7.2.)

Pioneer species cannot reproduce within the forest canopy so the population consists mostly of big trees, whereas climax species reproduce *in situ* and have juveniles and adults present together and hence what foresters sometimes call a 'reverse J' or positive stand-table. See Figs. 2.22, 2.23.

not the imperative need for efficient dispersal over long distances because regeneration can occur below a canopy. Regeneration *in situ* means juveniles and adults usually occur together. In the jargon of foresters the stand-table is positive (Fig. 7.12b).

Succession

The shift from a secondary forest of pioneers to a primary forest of climax species is sometimes called succession by 'relay floristics' by analogy to a race.[171] However, in some big gaps pioneer and climax species grow up together, the former from seed the latter either from seedlings which survived gap formation or from stem or root sucker shoots. In this case, which is likely where the forest floor has not been completely disrupted, succession is by 'simultaneous colonization', with the pioneers growing fastest and initially dominant. Both modes of succession can be found in the same forest, dependent on the severity of disturbance.

Climax species arrive and establish seedlings under the canopy of a secondary forest. As the mature phase canopy of pioneers ages, individual trees or small groups die and create small gaps. In these the climax seedlings are released and grow up as a second growth cycle below whose canopy climax species establish again. As that mature phase canopy breaks up these seedlings are released as another growth cycle. Climax species as a group thus perpetuate themselves *in situ*, there is no directional change in species composition. This is called cyclic regeneration or replacement.

In a small gap, pre-existing climax seedlings are released. In a large gap pioneers, which appear after gap creation, form the next forest growth cycle. One of the puzzles which remains unsolved is what determines gap-switch size.

Table 7.5 Regeneration behaviour of the common big tree species, Kolombangara, Solomon Islands

Group	Species	Conditions to establish	Conditions to grow up	Timber
Climax	*Dillenia salomonensis*	High forest	High forest	Heavy hardwoods: high density, hard, dark, often with silica
	Maranthes corymbosa	High forest	High forest	
	Parinari papuana	High forest	High forest	
	Schizomeria serrata	High forest	High forest	
	Calophyllum peekelii	High forest/small gaps	High forest/small gaps	Medium hardwoods: medium density, hardness and colour
	C. neoebudicum	High forest	High forest/gaps	
	Pometia pinnata	High/disturbed forest	High forest/? small gap	
	Elaeocarpus angustifolius	High forest	Gaps	Light hardwoods: low density, soft, pale
	Campnosperma brevipetiolatum	High forest/gaps	Gaps	
	Terminalia calamansanai	High forest/gaps	Gaps	
Pioneer	*Endospermum medullosum*	Mostly gaps	Gaps	
	Gmelina moluccana	Mostly gaps	Gaps	

From Whitmore (1974)
The degree of canopy disturbance needed by seedlings increases down the Table. Four ecological species groups can be recognised.

Casual observations in western Malesia on dipterocarp seedlings show that in big gaps climax species become prone to attack by shoot borers or to partial defoliation or to leaf galls. This suggests that the plants become stressed and unable to resist insect attack, and that in these conditions pioneers replace them. A specimen tree of the extremely shade-tolerant dipterocarp *Neobalanocarpus heimii* (chengal), a huge timber species, was planted in 1955 to celebrate the creation of Templer Park near Kuala Lumpur, Malaya. In this open site repeated shoot attack by boring insects has kept it the size of an apple tree, a sort of bonsai dipterocarp. Chengal has also been shown to have its maximum rate of photosynthesis at intermediate levels of PAR and also to have a very low light compensation point (cf. section 7.5). In Kramer's experiment, discussed in section 2.3, which was conducted in lower montane rain forest, gap-switch size was at between 1000 and 2000 m^2. In the lowlands, where the climate is hotter and less humid, pioneers are commonly observed in gaps above about 200 m^2.

7.3 SEED AND SEEDLING ECOLOGY[172]

Pioneer species

Pioneer species germinate from seed in a big gap after it forms. The seeds may be borne in after gap creation, or may already be present in the soil. C.F. Symington in 1933[173] reported that at Kepong, Malaya, he had placed some forest soil in the open and observed germination of pioneer species from it. The experiment was repeated by R.W.J. Keay in Nigeria in 1960[174] and many times again since the early 1970s in all parts of the tropics[175] with the same results: wherever a soil seed bank has been sought under lowland rain forest one has been found. Moreover, it is always mostly of pioneers.

Seed physiologists have discovered that pioneer species have seeds which can withstand desiccation and then become dormant; longevity is extended at low temperatures. They have called this behaviour 'orthodox'[176] (cf. Table 7.3). It is this capacity for dormancy which enables seeds to accumulate in the soil.

However, beyond the demonstration that seed banks are (so far) ubiquitous, little is yet known about their variation in either space or time. These are still topics ripe for investigation. Pioneers have small, easily dispersed, copious seed and the seed bank may develop by seed rain, and be continually augmented. In this case a seed bank uniform over large areas, the more so with increasing seed longevity, should develop. Experiments such as Symington's tell us nothing about how long seeds have been in the soil. In north Thailand the soil bank contained more seed than one year's production of nearby pioneers. Several *Cecropia* species in tropical America have seeds that live more than one year in the soil.

The alternative possibility is that the soil seed bank is a record or archive of pioneer trees which formerly occupied the site. Despite good adaptations to dispersal many seeds are likely to fall below and near the parent tree. In this case the spatial pattern will be a series of patches or 'footprints'.

At Danum, Sabah, the seed bank below primary forest contained almost none of the usual expected pioneer trees (*Endospermum, Macaranga, Mallotus, Trema*), but the shrubs and climbers of adjacent unstable earth stream banks instead. This site had only recently become accessible by road and the nearest seed source of the Sabah pioneer tree flora was several kilometres distant, out of dispersal range.

Triggers to germinate

The dormancy of the seeds of tropical pioneer species has been shown to be broken by one of two features of gap microclimate. Many, so-called photoblastic, species are triggered to germinate by exposure to light in which the photon flux density of the red (660 nm) wavelength exceeds the far-red (730 nm). This is mediated by their possession of the pigment phytochrome. Photoblastic germination is easy to demonstrate by experiment (see Fig. 7.13). Red light is depleted below a forest canopy by absorption and reflection (Fig. 7.2) and far-red exceeds red.

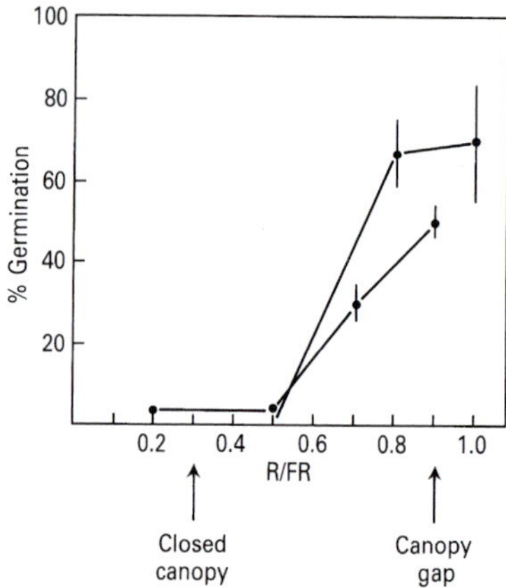

Fig. 7.13. The small pioneer tree *Urera caracasana* has photoblastic seeds, which germinate at high ratios of red to far-red light, i.e. in gaps, but not below the forest canopy. Los Tuxtlas, Mexico. (Orozco-Segovia *et al.* 1987, fig. 1; the bars indicate the standard errors.)

The seeds have evolved to detect the occurrence above them of a canopy gap with full white light, in which red exceeds far-red. Moreover, short bursts of white light, as in passing sunflecks, do not trigger germination.

Other species, which include *Ochroma lagopus* (balsa), have seeds triggered to germinate by either high temperature,[177] or by alternating high and low temperature. This so-called thermoplastic response enables them to detect the change from closed canopy to gap (Table 7.2). In parts of the neotropics, pure extensive stands of balsa develop where burning has followed forest clearance. It has been discovered that *Ochroma* seeds have an impermeable layer in the testa which is ruptured by heat to allow the seeds to imbibe water and to germinate.

Microsite for establishment

Soil seed banks characteristically contain many species. For example, small samples (2 m^2 × 50 mm deep) from six different forests in Ghana contained from 17 to 38 species each, nearly all of them pioneers. Equally characteristically, pioneer forests commonly comprise only one or a few tree species, a subset of the seed bank, and

Table 7.6 Different microsite preferences for establishment of seven pioneer species in a cleared Venezuelan forest site

	Soil features		Shrubs or trees				Forbs		Grass
	Maximum soil surface temperature (°C)	Mean water loss[a]	*Cecropia ficifolia*	*Vismia lauriformis*	*Solanum stramonifolium*	*Clidemia sericea*	*Eupatorium cerasifolium*	*Borreria latifolia*	*Panicum laxum*
Without dead litter of leaves, twigs and branches:									
Root mat	50	8.9	–	–	–	–	–	*	–
Charcoal	54	6.9	–	–	–	–	–	*	–
Bare soil[b]	44	14.8	–	–	–	–	–	*	*
With litter:									
Root mat	39	4.3	–	–	–	–	–	*	–
Charcoal	*c.* 35	2.9	*	*	*	*	*	*	–
Bare soil[b]	*c.* 35	6.7	*	*	+	+	*	*	–

From Uhl *et al.* (1981)
Established success: * very good; + good; –weak or nil
[a] Water loss from a can of soil at field capacity in grams during an 8-h day
[b] Soil was a cream coloured sand

we do not yet fully understand why. Part of the answer probably lies in differential success in seedling establishment, dependent on weather at the time of germination, or on microsite. Table 7.6 shows results of an experiment in Venezuela where six of the microsites commonly occurring in felled and burned forest were planted with seven of the local pioneer species. The microsites differed substantially in soil surface temperature and water loss. Only one species, a forb (viz. broad-leaved herb), established on all six. Only the grass established on bare soil. All species except the grass established on the two microsites that were coolest and most moist.

Climax species

Many rain forest climax species have seeds that germinate immediately or within a few days. They cannot withstand desiccation or low temperatures. These seeds have been called 'recalcitrant'[178] because they cannot be stored, and thus pose a problem to plant breeders and nurserymen. Examples are many fruit species, e.g. citrus, cocoa, mango, and durian (*Citrus, Theobroma cacao, Mangifera, Durio zibethinus*) and many timber species, including all rain forest dipterocarps. Not all climax species have recalcitrant seeds, for example, many

Leguminosae have hard testas and prolonged dormancy. No pioneer species has yet been found whose seeds are recalcitrant.[179]

Many climax species have bigger seeds than pioneers. By immediate germination these escape seed eaters and also degradation of their usually fatty storage tissues by micro-organisms. Seedlings develop, and climax species have a 'seedling bank' on the forest floor in contrast to the soil seed bank of pioneers.

Microsite for establishment

As with pioneers, different climax species may be more successful on some parts of the forest floor. A close study of the seedling ecology of some Malayan Dipterocarpaceae by P.F. Burgess explained several previously enigmatic aspects of the distribution of this important family.[180] Dipterocarps are confined to the lowlands and do not penetrate lower montane rain forest. The upper limit was shown to be set by the inability of the radicle of a germinating seed to penetrate peat, which develops on the surface above 1050 m elevation. Seedling establishment is most successful on flat microsites; seedling stocking diminishes with increasing slope, rapidly at microsites steeper than 45° and falling to nil at 65° slope. This explains the decrease in numbers of dipterocarps with elevation where the land becomes more rugged. The important timber species *Shorea curtisii* (seraya, Fig. 2.3) was studied in detail. After the 1968 seed year seedlings were recruited markedly better on granite-derived soils, which are coarse and sandy, rather than on the more clay-rich soils derived from shale, probably because the latter develop a hard surface skin under the impact of heavy rain. However, after several years both sorts of soil had about the same stocking of seedlings; the heavier mortality on granite-derived soils could be because their coarser structure makes them more prone to drought. This study serves to reveal the great complexity of intertwined factors which control seedling establishment and survival and that are important in determining the ecological range of different species. At Lambir, Sarawak, the dipterocarp *Dryobalanops aromatica* occurs mainly on ridges and upper slopes with sandy soil whilst *D. lanceolata* occurs on clay soils and in sandy-soil valleys. This pattern has been shown to result from differences in seedling establishment. For, although both species germinated everywhere, *D. lanceolata* suffered root predation on ridges and had higher establishment success in valleys.[181]

Seedling survival

Many seeds fall near the parent tree and dense carpets of seedlings form. In several forests mortality has been shown to be density dependent and, therefore, is greatest near the parent, due to pathogens or herbivores.[182] The fewer seeds which disperse to a greater distance are most likely to grow into seedlings that survive. This so-called 'escape hypothesis'[183] has been invoked as a mechanism which prevents rain forest trees forming single-species stands, although there are exceptions. For the dipterocarps *Shorea leprosula* and *S. macroptera* it was found[184] that mortality depended more on microsite, and had no relationship to density or to distance from the parent. In the case of *S. curtisii* ants destroyed isolated seedlings, thus accentuating the clumping created by seed falling mainly near the parent.[185]

By whatever means they arise and are maintained, species patterns amongst seedlings on the forest floor are a powerful, though little studied, determinant of floristic pattern in the following mature phase forest.

Dipterocarps bear fruit heavily only once every several years (p. 63). Seedling populations are then augmented and many attain a density of over one million per hectare. There is a subsequent decline, which may reach zero, before the next gregarious fruiting event (see Fig. 7.14).

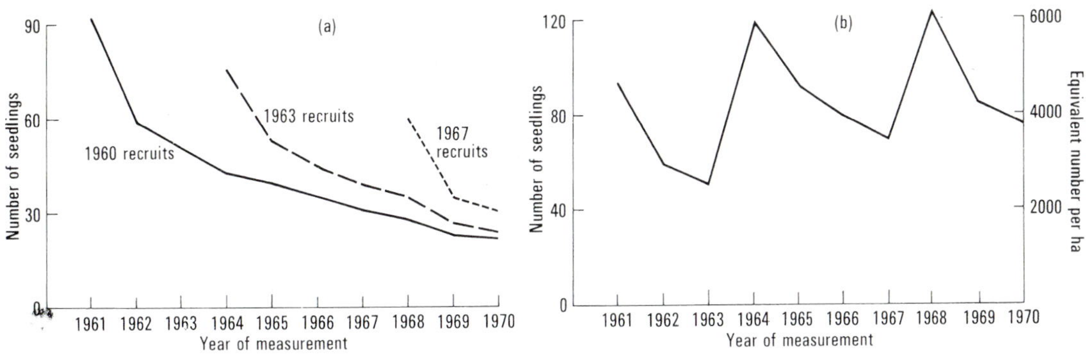

Fig. 7.14. Fluctuation of seedling populations of the light hardwood dipterocarp *Parashorea tomentella* at Sepilok, Sabah. (a) Fate of individual flushes; (b) total population. Numbers on eight plots of 4 m^2. (Fox 1972 in Whitmore 1984*a*, fig. 7.7.)

In Malaya, by contrast to Sabah, the populations of dipterocarp seedlings commonly drop to zero between successive flushes. This has important consequences for silviculture.

7.4 COMPONENTS OF SPECIES RICHNESS

The dynamic nature of forest canopies provides many different regeneration niches to which different species have become specialized. A major niche is the light climate and there are species with all degrees of shade tolerance, as discussed above, divisible into pioneer and climax ecological groups, which are themselves heterogeneous but less clearly divisible.

In all tropical rain forest floras there are fewer pioneer than climax species, and they mostly belong to a few families; for trees these are Euphorbiaceae, Malvaceae, Moraceae, Sterculiaceae, Tiliaceae, Ulmaceae and Urticaceae. The Eastern tropics are richest in pioneers, the biggest pioneer genus, *Macaranga* (Euphorbiaceae, Figs. 7.7, 7.8), is concentrated there though with a few outliers in Africa. Most of its *c.* 250 species are pioneers and many are confined to only one or a few islands of the Malay archipelago. *Cecropia* (Moraceae, Fig. 7.6) is the second largest pioneer tree genus. It has *c.* 100 species, most very widespread, and is entirely neotropical. *Musanga cecropioides* (Moraceae, Fig. 7.15) of Africa has similar crown form to these two. It is monotypic. *Trema* (Ulmaceae), with 10–15 species, occurs

Fig. 7.15. *Musanga cecropioides* (palmate leaves) and *Anthocleista nobilis* (oblong leaves), pioneers of West African rain forests. Ghana.

in all three rain forest regions. Pioneer genera do not have narrowly endemic species of very restricted geographic distribution, and most of them have only a few species. These features are probably a reflection of efficient dispersal which militates against species evolution via the development of localized genetically distinct populations. Moreover, some pioneers, including the largest two genera *Cecropia* and *Macaranga*, are dioecious, which has the same result. It is noteworthy that the region with the greatest number of pioneer species is the Eastern tropics. This is probably because partial isolation on the islands that comprise this region has allowed more species evolution than has occurred on the continuous landmasses of the African and American tropics.

There has been much speculation about the extent to which the co-existence of numerous species in a given forest is due to specialization to different facets of the regeneration niche, thus avoiding direct competition. A little evidence of specialization has been gathered. When a tree blows over its roots are upended and a hollow is formed plus a mound of mineral soil, sometimes called the root plate (Fig. 2.21). *Cecropia obtusa* in Guyana and *Trema tomentosa* in Penang have been shown[186] to establish preferentially on root plates. It has also been shown at La Selva and at Los Tuxtlas, Mexico, that different species successfully establish seedlings in different parts of treefall gaps, the crown, trunk or butt regions.[187] Differential seedling performance in the changing microclimate from gap centre outwards was shown by rooted cuttings of seven *Miconia* and *Piper* species at La Selva[188] and by wild seedlings of three dipterocarps at Danum (Fig. 7.16).

Degrees of disturbance

The most species-rich forested landscape will be one that includes both patches of secondary forest recovering from a big disturbance and consisting of pioneers, and also patches of primary forest composed of climax species. When succession is completed and the whole landscape is primary forest again there will be

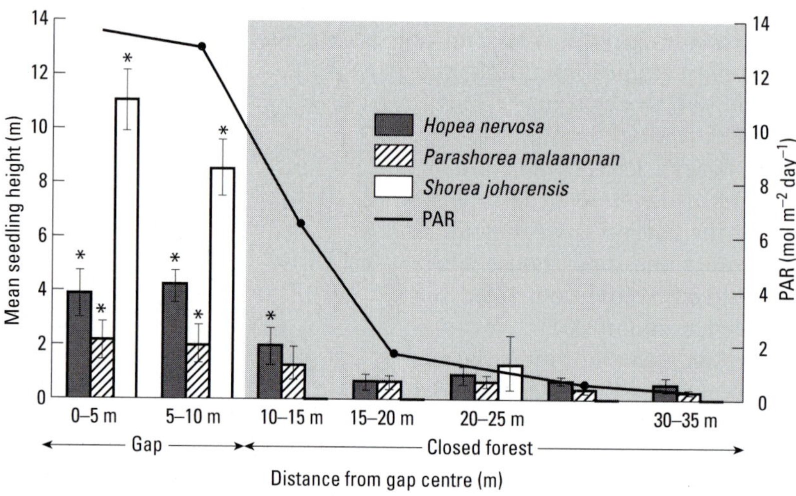

Fig. 7.16. Differential seedling height growth of three dipterocarps along a transect from gap centre to closed forest. Danum, Sabah. (Brown 1996, fig. 1; error bars are 95 per cent confidence level.)

*Means height growth over 6 years was significantly larger than of same species in closed forest control plots. *Hopea* grew faster in both the gap and its penumbra, *Parashorea* and *Shorea* in the gap (though the latter was badly damaged by an insect, see text). Mean daily PAR up to 10 m from the gap edge was significantly greater than under closed forest.

no pioneers and landscape-scale floristic richness will be reduced. This is the *intermediate disturbance hypothesis* of J.H. Connell.[189] The concept is bound to be true at certain scales of dimension. But if we were to look at tiny patches of the landscape we would probably find species-poor stands of pioneers and also stands of high species richness where the more numerous climax species occur in greater mixture and which have, moreover, accumulated dependent synusiae of epiphytes and climbers, so at this smaller scale the climax patches are richest. The catch-all hypothesis of Connell includes intermediate intensity and frequency of disturbance, as well as intermediate stage of recovery. It caused scientists to look at species composition in a particular way and was a salutary correction to prevalent ideas tying diversity to maturity and stability. But it is not so much an hypothesis as an inevitable property of vegetation at certain scales of space and time.

7.5 GROWTH AND SHADE TOLERANCE ANALYSED

The physiological basis of growth rates and of shade tolerance amongst tropical rain forest tree species is now fairly well understood.[190] This is the interface between ecology and physiology, so-called ecophysiology. It is important to conduct experiments in the forest, despite the great complexity of the natural environment, as well as in the laboratory and shade-house where conditions can be specified and controlled but are inevitably somewhat unnatural.[191]

The basis for the fast growth of pioneer trees is their high unit leaf rate, *E*, the rate of dry weight created by photosynthesis per unit area of leaf. Table 7.7 shows that of six species studied at La Selva, Costa Rica, *Ochroma lagopus* (balsa) had unit leaf rate not significantly lower than of the herb *Helianthus annuus* (sunflower) grown as a control. Other species, subjectively ranked as progressively more shade-tolerant, had progressively lower unit leaf rate. Relative growth rate (dry weight increase per unit of dry weight) also decreased with increasing shade tolerance but is a less basic measure because it changes with age. Other research has shown that the rate of carbon dioxide uptake is also higher in light-demanding than in shade-tolerant species, despite high variability due to leaf age and past history (Fig. 7.17). This is due to higher apparent quantum efficiency and mesophyll conductance (Table 7.8 a,b). Pioneers also have higher dark respiration and light compensation point than climax species (Table 7.8 c,d). Numerous experiments have shown that, overall, pioneers are more flexible in adapting to changing conditions: they are better able to adjust architecture, morphology, leaf anatomy or various ecophysiological attributes.

Within the forest, most of the photosynthetically active radiation (PAR) is in the sunflecks (section 7.1). Research[192] on the herb *Alocasia macrorrhizos* and on seedlings of the tree *Toona australis*, both shade-tolerant species from the Australian rain forests, showed that photosynthesis took 25–40 minutes to reach steady state in bright light after a prolonged, low-light period. That is to say, there is an induction period. For *Alocasia*, loss of induction when the

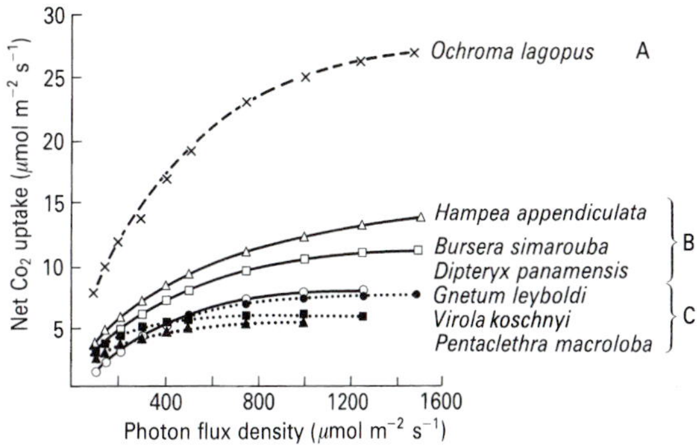

Fig. 7.17. Photosynthesis rate of seven tree species at La Selva, Costa Rica. A, pioneer; B, gap-dependent; C, shade-tolerant canopy species. (After Oberbauer and Strain 1984, fig. 1.)

Rate of photosynthesis (measured as net CO_2 uptake) increases to an asymptote with increasing light. The maximum decreases with increasing species shade tolerance.

Table 7.7 Growth in an open site expressed as unit leaf rate (E) and relative growth rate (RGR) of seedlings of six tree species of increasing shade tolerance compared to sunflower *Helianthus annuus*. (La Selva, Costa Rica)

	E (gm^{-2} $week^{-1}$)	*E* as % of *Helianthus*	*RGR* (gg^{-1} $week^{-1}$)
Helianthus annuus (i)	57[a]		1.00
(ii)	65[a]		0.75
Pioneer species			
Ochroma lagopus (i)	57[a]	87	0.85
(ii)	52[ab]	79	0.81
Heliocarpus appendiculatus	42[b]	65	0.54
Climax species			
'Gap-dependent spp.'			
Cordia alliodora	41[b]	64	0.48
Terminalia oblonga	10[c]	15	0.13
'Shade-tolerant canopy spp.'			
Brosimum alicastrum	26[bc]	40	0.15
Pentaclethra macroloba	22[c]	34	0.15

From Oberbauer and Donnelly (1986) table 2
a,b,c Different letters following *E* show significantly different values

Table 7.8 Respiration and photosynthesis of seedling leaves of nine species, Gambari forest, Nigeria

Species group	Photosynthesis		Respiration	
	a Apparent quantum efficiency (mol CO_2 $photon^{-1}$)	b Mesophyll conductance (mol CO_2 m^{-2} s^{-1})	c Dark respiration (μmol m^{-2} s^{-1})	d Light compensation point CO_2 (c/a)
Weedy shrubs of gaps (2 spp.)	0.037	0.028	1.75	47
Pioneer trees (3 spp.)*	0.034	0.028	1.10	32
Light-demanding climax trees (2 spp.)†	0.018	0.010	0.35	19
Shade-tolerant climax trees (2 spp.)	0.028	0.008	0.5	18

Whitmore (1996) table 1, based on data of Riddoch *et al.* (1991)
* *Ceiba, Milicia, Ricinodendron*
† *Pterygota, Sterculia*
Weeds and pioneers have more efficient photosynthesis (a, b), higher dark respiration (c) and light compensation point (d). They contrast in these respects with climax species.

plants were put back into low light conditions took over 60 minutes. Furthermore, the rate of carbon gain varied with past history and intermittent sunflecks could be used more efficiently than continuous high light. The conclusion is that the temporal pattern of sunfleck distribution through the day (cf. Fig. 7.3) is of importance, not just the daily total PAR. Further studies like this are needed to analyse interspecific differences of shade tolerance.

Seedling release of climax species

The light climate

As he gets to know a forest the ecologist or forester soon develops a feeling for how light-demanding the different tree species are. Most obvious is the difference between pioneer and climax species. For the latter, there are also his perceptions on the size of gap in which seedlings are seen to have grown up from the seedling bank. General observations on the natural history of the forest such as these are an important part of more rigorous scientific analysis, initially as precursor to help the intelligent design of experiments, and later to assist interpretation of the results. Nowhere is this general background knowledge better developed than in the dipterocarp rain forests of Malaysia, where foresters have now nearly a century of accumulated experience. It has been found that amongst dipterocarps there are species whose seedlings grow up in small canopy gaps and others that are successful in big gaps. Correlated properties are that the first group have lengthy persistence in the seedling bank (Fig. 7.18) and, as trees, have slow diameter growth and high density, dark coloured, naturally durable timber. Seedlings of the second group die sooner (Fig. 7.18) and trees grow faster and have lighter, paler, non-durable timber. The two groups are known as shade-bearers and light-demanders or as medium to heavy hardwoods and light hardwoods, respectively. The silvicultural systems invented for dipterocarp rain forests, described later in section 7.6, depend on these responses to gaps. Medium/heavy hardwood species can be successfully regenerated by small canopy disturbances; light hardwoods by more substantial disturbances. Similar categories are recognized in other rain forests, and indeed the whole of rain forest silviculture everywhere depends upon it. It is the practical working out of the wide range of shade tolerance shown by different climax species that was discussed above in section 7.2.

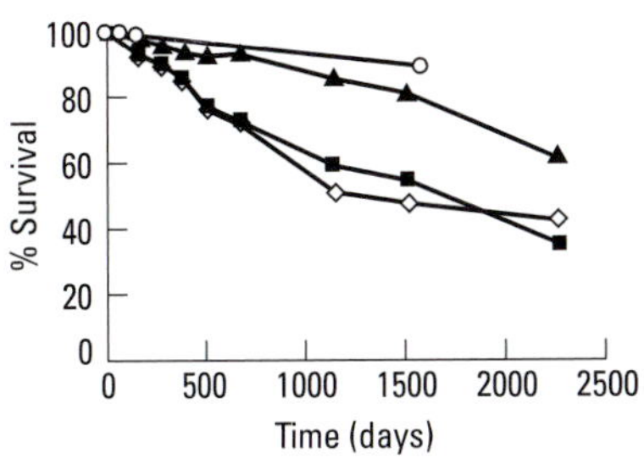

Fig. 7.18. Differential survival of dipterocarp seedlings on the forest floor over 6 years, Danum, Sabah.
Seedlings of shade-tolerant species, in this case the two *Hopea* spp., characteristically survive better in deep shade.

Why is there a gap switch size between these ecological groups? For the dipterocarp species that occur in the Danum forest, Sabah, this has now been elicited by observing over 6.4 years the growth, in artificial canopy gaps of different size, of seedlings from the forest floor seedling bank (Fig. 7.19). It can be seen in summary that the light-demander species *Shorea johorensis* was able to utilize for height growth the extra solar radiation received in larger gaps and over a longer period of time. This is the ecological consequence at population level of its ecophysiological flexibility at the leaf and single seedling level. By contrast, the shade-bearer *Hopea nervosa* lacked the flexibility to adjust to and make use of the extra PAR of the larger gaps. Success, however, is determined by many factors and *Parashorea* performed badly because it

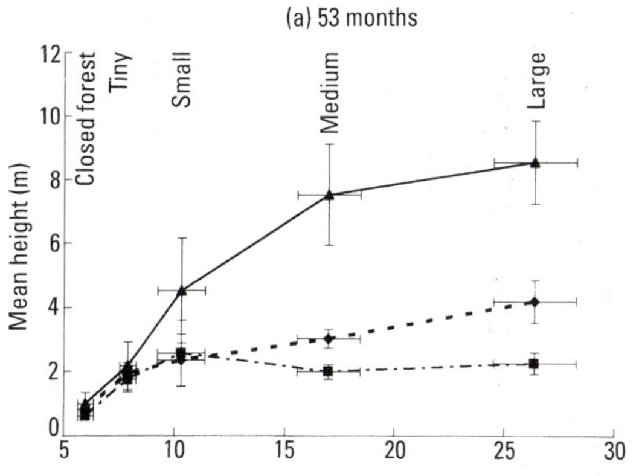

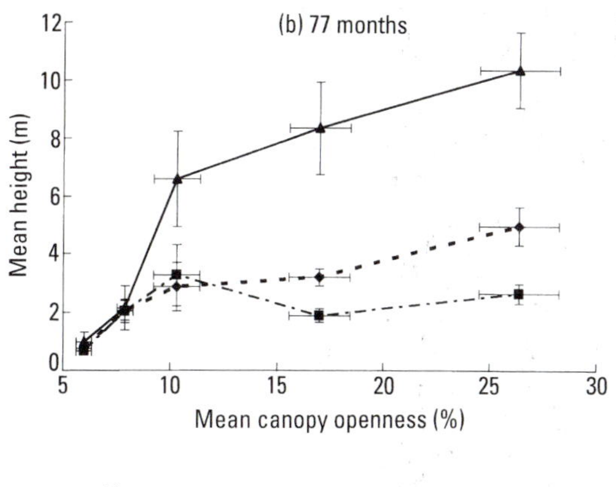

Fig. 7.19. Growth of dipterocarp seedlings in canopy gaps of different size over (a) 53 months, (b) 77 months. Danum, Sabah. (Whitmore and Brown 1996, fig. 2.)

In closed forest and tiny gaps, height growth was similar but then there was a switch and in larger gaps *Shorea johorensis* grew progressively faster with increasing gap size. The effect was stronger at 77 than 53 months. *Parashorea*, which natural history, timber density and leaf ecophysiology suggest has similar ecology to *Shorea*, grew slowly because of serious repeated damage to its apex by the cricket *Nisitrus vittatus*.

occurred in very dense populations and these succumbed to insect damage.

Nutrients

The role of irradiance in seedling growth and release is easy to observe and has been much investigated. By contrast, little attention has been given to the potential role of plant mineral nutrients.[193] The extent to which these increase in a gap will vary with the kind of gap independent of its size (e.g. root plates expose subsoil and eliminate living roots). So far, nutrients seem unimportant compared to radiation. For example, four independent experiments on dipterocarp seedlings have shown no effect of added nutrients on growth; only one of four climax species fertilized at Bukit Timah, Singapore responded; upper montane species in Jamaica failed to respond; and in Ghana only three of fifteen species tested showed a clear response to soil fertility.

Overall the shade/nutrient interaction story remains unresolved. One part of the picture is likely to be that there is no response to nutrients in dark conditions where irradiance is limiting, but a response at higher irradiances.

Up to 50 species of *Piper* can co-exist in a single central American rain forest. These are sorted mainly on radiation gradients with soil moisture and nutrients sometimes playing a role.

The concept of seedling shade tolerance dissected

The preceding discussion has shown that shade tolerance is a useful and easily understood concept that helps frame analysis of how the forest works. It does, however, have three distinct, though overlapping, meanings. Seedlings may differ in the levels of PAR required for seedling establishment (as in the climax : pioneer dichotomy), or they may have different longevity at a given level of PAR (Fig. 7.18), or they may differ in the amount of PAR required for release. This third meaning has now been analysed (Fig. 7.19) in an experiment which revealed that shade-bearers do not increase their

growth much as gap size increases, wherease light-demanders do, so are likely to win the growth race in all but the tiniest gaps.

Growth of saplings, poles and trees

The relative performance of the seedlings of different species at the gap phase is the major determinant of the composition of the growth cycle. For example, at Danum by 6.4 years from gap creation seedlings had become substantial poles[194] in the larger gaps, and the forest was into building phase, its species composition probably determined for the next century or more. Thus it is natural that there has been such strong attention paid to seedling behaviour.

It has now been shown several times that species may change their ranking as they grow up. For example, in Queensland six species of the pioneer, 'intermediate' and 'shade-tolerant' guilds planted as seeds in gaps of various sizes changed their rank order of height over seven years.[195] In Ghana there is a group of species that need high light at first but later grow in deep shade.[196] Observations on Kolombangara, Solomon Islands, of 1700 trees of 12 species over 30 years has shown they change their ecological characteristics in a complex way through life (Fig. 7.20).

Seedlings may show a burst of growth then stagnate as the canopy closes over them and by the time they reach tree size will have had several such spurts. This phenomenon is called multiple release. In temperate forests it is easy to detect from changes in the width of annual growth rings in the wood. It is more difficult to demonstrate in tropical rain forests but undoubtedly occurs. Polycylic silviculture (section 7.6), for example, makes use of it. Side light as well as light from gaps overhead has been shown to stimulate growth.[197]

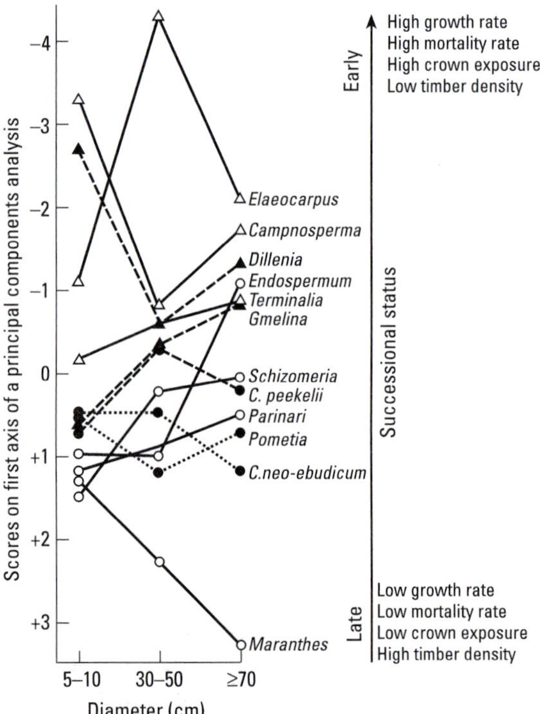

Fig. 7.20. Changes in successional status through life, Kolombangara, Solomon Islands, twelve species monitored 1964–93. (Burslem and Whitmore unpublished.)

This complex diagram shows how the 12 common big tree species change successional rank through life. The four ecological groups they form as seedlings (Table 7.5) have already lost their integrity by 5–10 cm diameter, but by 70 cm there is still a roughly similar rank order.

Ecophysiology of dipterocarps[198]

Those dipterocarp species investigated have been found all to have similar maximum photosynthesis rates at low levels of irradiance, with the light-demanding, light hardwood species responding more strongly to high irradiance. Ten species tested were, as a group, less responsive to irradiance than the pioneer *Macaranga depressa*, but much more responsive than a shade-dwelling *Begonia*.

Dryobalanops lanceolata, a shade-tolerant, has been shown to have a higher carbon fixation capacity at low irradiance than *Shorea leprosula*, a well-known light-demander. This was due to lower compensation point and dark respiration (cf. Table 7.8), and higher mean assimilation rate. *Dryobalanops* has different

Fig. 7.21. Seedling architecture of a light-demander and a shade-bearer, the dipterocarps *Shorea leprosula* and *Dryobalanops lanceolata*. (Zipperlen and Press 1996, fig. 2.)

These were grown at *c.* 10 per cent canopy openness and have the same total axis length. These two growth forms are common amongst rain forest seedlings. They have been dubbed optimists and pessimists, respectively, by Kohyama (1987) and represent a trade-off between success after a gap opens in the future and maximal interception of scattered sunflecks in the present.

architecture (Fig. 7.21) putting much more of its growth into lateral branches, an efficient seedling shape for exploiting patchy sunflecks. Similar differences in architecture existed between the species shown in the growth experiment in Fig. 7.19.

Different dipterocarps have also been found to differ in their ability to tolerate drought periods. At the physiological level, *Dryobalanops lanceolata* seedlings were shown to have lower stomatal conductance than other species yet about the same level of carbon assimilation. Over a day the stomata were open for a shorter period and so the plants lost less water by transpiration. Thus, the species had higher water-use efficiency. Foresters know that nursery-raised *Dryobalanops* seedlings survive better on planting out. Ability to survive drought periods has been shown to be the major factor determining the success of *Shorea curtisii* on coastal hills and inland ridges in Malaya (Fig. 2.3) and differential mortality of three *Shorea* species in the regular early year drought has been demonstrated in Penang. In the fairly seasonal rain forest climate of Sri Lanka the interplay of irradiance and drought has been shown to account for the partitioning of four *Shorea* species between ridge, slope and valley.[199]

7.6 SILVICULTURE

Rain forest silviculture is the manipulation of the forest to favour certain species and thereby to enhance its value to humans. For example, around the turn of the twentieth century the Malayan rain forest was treated to increase the abundance of *Palaquium gutta* (Sapotaceae) because there was a strong market for its latex, gutta percha, which was used as the insulator in submarine telegraph cables.[200] More commonly, manipulation favours particular timber species. Timber properties, whether heavy or light, dark or pale, durable or not, are strongly correlated with growth rate and thus to the extent to which the species is light-demanding (Table 7.5). Thus, the ecological basis of natural forest silviculture is the manipulation of the forest

canopy. The biological principle of silviculture is that by controlling canopy gap size it is possible to influence species composition of the next growth cycle. The bigger the gaps the more fast-growing light-demanders will be favoured. This concept has been known in continental Europe since at least the twelth century. Indeed, our knowledge of forest dynamics, the forest growth cycle and the significance of canopy gap size, was discovered by foresters empirically before the formal science of ecology came into existence.[201] The practice of silviculture reached tropical rain forests via India and Burma then Malaya and was strongly influenced by Germanic tradition.[202]

Silvicultural systems

The silvicultural systems that have been applied to tropical rain forests belong to one of two kinds: the polycyclic and monocyclic systems, respectively (Figs. 7.22–7.24). As the name implies, polycyclic systems are based on the repeated removal of selected trees in a continuing series of felling cycles, whose length is less than the time it takes the trees to mature (which foresters call the rotation age). The aim is to remove trees before they begin to stagnate and deteriorate from old age, leaving all appreciating stems to swell the future yield. Because of the very species-rich nature of most tropical rain forests, and the relatively small number of species with timber that is commercial by current standards, extraction on a polycyclic system tends to result in the formation of scattered small gaps in the forest canopy.

By contrast, monocyclic systems remove all saleable trees at a single operation, and the length of the cycle more or less equals the rotation age of the trees. Except in those cases where there are few saleable trees, damage to the forest is more drastic than under a polycyclic system, the canopy is more extensively destroyed, and bigger gaps are formed. Furthermore, new crop trees all grow up together and this can lead to overcrowding, competition between their crowns, and reduced growth rate.

It can be seen at once that the two kinds of system will tend to favour shade-bearing and light-demanding species, respectively, but the extent of the difference will depend on how many trees are felled at each cycle in a polycyclic system. The nub of the difference is that, on the one hand, polycyclic systems retain advanced growth of half-grown trees to produce marketable trees at subsequent intermediate

Fig. 7.22. Lowland rain forest, Windsor Tableland, Queensland, twelve years after logging on a polycyclic system.

Only selected big trees were felled. The canopy is more broken than in virgin forest, and the logging road (which will be used again next felling cycle) is clearly visible.

Fig. 7.23. Lowland dipterocarp rain forest, in Pahang, Peninsular Malaysia, shortly after logging on the monocyclic Malayan Uniform System.
All mature commercial stems have been removed, mostly dipterocarps, and the yield would have been 20–30 m^3 ha^{-1}. The forest matrix remains intact. Note lush regrowth along the logging track.

felling cycles within the rotation. This is done by taking good care of these 'adolescent' trees at each felling. On the other hand, monocyclic systems forego the increment already accumulated in these adolescents and rely almost entirely on seedlings to produce the next crop, which will be ready to harvest only at the full rotation age. If there does happen to be any advanced growth left from the previous forest it is a bonus, over and above the seedling growth on which monocyclic systems are based.

Under the best conditions, polycyclic systems can greatly increase the timber yield over a full rotation. Their success is utterly dependent on advanced growth and this means that the forest must be adequately stocked with adolescents before felling begins, and damage to them by logging kept acceptably low. If logging damage at each felling exceeds a low minimum, regeneration will be lost and the yield will progressively fall off through a series of cycles. Control can be achieved either by marking trees for retention before felling begins, which is labour-intensive, or by felling down to a rather high diameter limit.

The often-claimed disadvantage of polycyclic systems is that the faster-growing genotypes are progressively removed, because at each felling

Fig. 7.24. Lowland dipterocarp rain forest in Sabah, East Malaysia, shortly after logging on the variant of the monocyclic Malayan Uniform silvicultural system used there, M.D. Swaine observes (1987).

The forests of this region have a very high proportion of commercial stems. Removal of these, as here, commonly yields 100 m^3 or more of timber ha^{-1}, and only scattered big trees remain.

many or all of the stems that have reached a minimum prescribed girth are removed. The small stems retained may be either young and potentially fast-growing or old, stunted, and unable to respond to release. In fact, both fast and slow growth are also determined partly by phenotype (i.e. microsite) as well as by genotype. The magnitude of the disadvantage[203] is therefore anybody's guess. It can be mitigated, as was done in Queensland, by flexibility in the rules used to determine whether a tree is felled or not. In Queensland every species had a specified diameter at which it was normally felled, but vigorous individuals were retained for a later cycle and poorly formed feeble ones were felled at smaller size.

Silvicultural treatment before felling has sometimes been practised in the past in order to reduce numbers of climbers or to boost the growth of juveniles. Post-felling silviculture to remove deformed stems or to free regrowth from competition was also widely practised until recently. At its fullest development several treatments were given over the first decade or so after felling. Post-felling treatments have been largely abandoned, because of the high cost. It was demonstrated in Sarawak that a given sum of money gave a better return in terms of growth increment in the regrowing forest if invested in reducing damage by the logging operation itself rather than in post-felling silviculture. Nowadays it is generally the case that the felling operation provides the only canopy manipulation the foresters can afford.

Criteria for success

Silviculture (Figs. 7.22–7.24) will be successful so long as it is practised within the biological limits of the forest. The desired species must be capable of growing in the size of gap that is created.

Forests are a potentially renewable source of timber, and their use for timber production is not simply a quarrying operation. The success of silviculture may be defined as the bringing to maturity of a stand of timber trees with regeneration below them to form the next silvicultural cycle, and without site deterioration, such as soil erosion or degradation.

The failure of silviculture follows from working beyond the limits of the inherent dynamic capabilities of the forest ecosystem. This is commonly because rules drawn up by silviculturists are not enforced, often because of political intervention. It may also be because economists, eager to enrich a nation, enforce their dismal pseudoscience to override basic bio-

Fig. 7.25. Climber towers, Enoghae, Solomon Islands. This very heavily logged forest has become carpeted with the wiry big-leaved climber *Merremia*.

logical principles and dictate the removal of a larger harvest than the forest can sustain without degradation.

Woody climbers

Unwanted 'weed' tree species can be a problem. So too can woody climbers. The natural behaviour of many of the latter is to grow vigorously in strong illumination. Thus they are common at forest fringes, such as river banks (and give rise to the myth of the impenetrable jungle), and in big canopy gaps. In many tropical rain forests the removal by logging of many trees per hectare creates large openings ideal for vigorous climber growth. In many Eastern rain forests dense carpets of the big-leaved wiry convolvulaceous climbers *Merremia* spp. smother the residual forest and form climber towers up remaining isolated trees (Fig. 7.25). Trees eventually grow up and carry the carpet skywards but forest recovery is delayed and the trees may develop kinked stems (Fig. 7.42). Very little is known about the ecology of woody climbers. For example, in the Solomon Islands climbers do not proliferate excessively after cyclones but become extremely abundant after logging which causes similar canopy damage. In Sabah, *Merremia*, *Mezonevron* and *Uncaria*, which are very serious impediments to forest regeneration, are uncommon in virgin forest, perhaps because in this country massive canopy opening does not occur naturally.

Damage

The amount of damage to the forest depends more on how many trees are felled than on timber volume extracted. It is commonly the case that for every tree removed for timber (logged) a second tree is totally smashed and a third tree receives damage from which it will recover (Fig. 7.26). In Queensland (Fig. 7.27) the number of stems smashed was a similar proportion but before strict rules were introduced in 1982 the proportion damaged was much higher.

The pattern of damage differs between extraction by tractor and by overhead cable, so-called high-lead logging. The latter causes extreme devastation along the paths of the cables and little between. Seedlings survive better than bigger plants. In tractor logging the damage is spread more uniformly and seedling numbers can be very seriously depleted.

For successful recovery from logging the single most important criterion is to minimize damage to the forest floor, for several reasons. On the floor occur the seedlings of climax species. Within the surface soil lies the seed

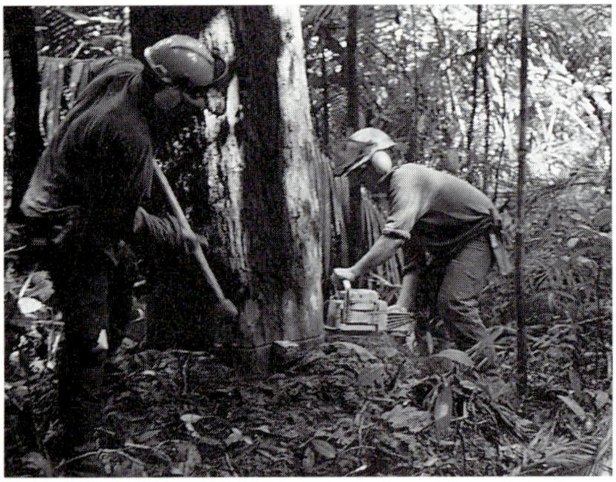

(a)

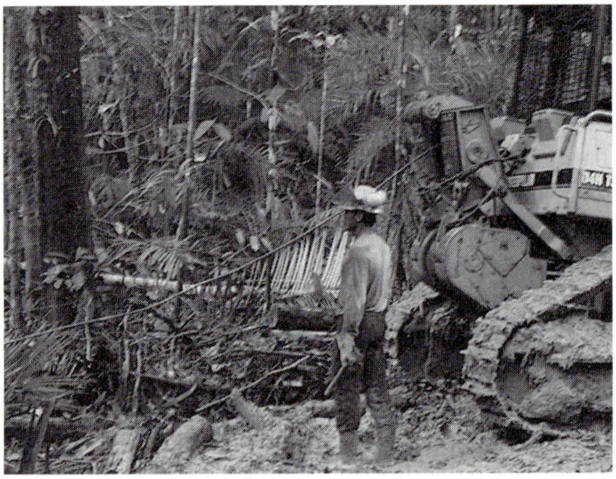

(b)

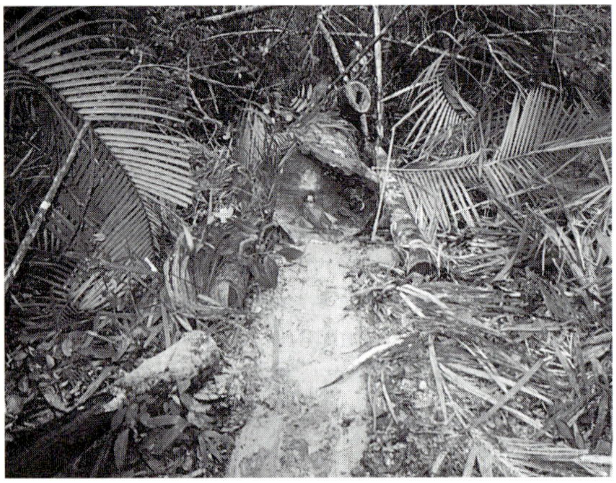

(c)

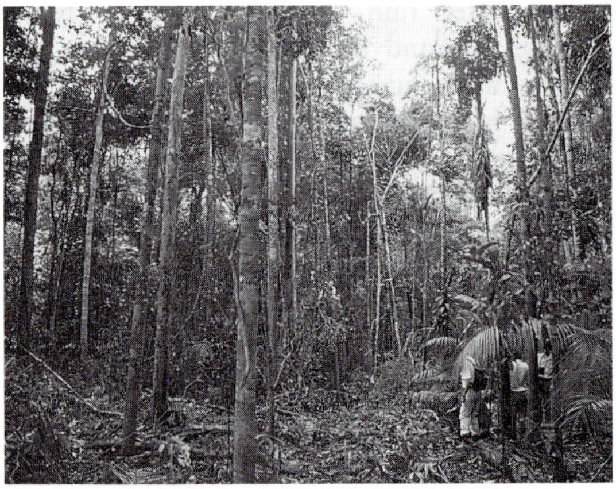

(d)

(e)

Fig. 7.29. Low impact logging, Precious Woods Ltd. Itacoatiara, central Brazilian Amazon.

(a) Trees are felled close to the ground to avoid waste, and in a predetermined direction (aided by wedges) so as to minimize damage to the forest.

(b) Skidder winches log into track, causing

(c) minimal damage to the forest floor.

(d) Forest after felling and log removal looking down a skid track; note abundance of juvenile trees whose growth will accelerate due to canopy opening and reduction in competition.

(e) The main extraction road; note minimal width.

Fig. 7.25. Climber towers, Enoghae, Solomon Islands. This very heavily logged forest has become carpeted with the wiry big-leaved climber *Merremia*.

logical principles and dictate the removal of a larger harvest than the forest can sustain without degradation.

Woody climbers

Unwanted 'weed' tree species can be a problem. So too can woody climbers. The natural behaviour of many of the latter is to grow vigorously in strong illumination. Thus they are common at forest fringes, such as river banks (and give rise to the myth of the impenetrable jungle), and in big canopy gaps. In many tropical rain forests the removal by logging of many trees per hectare creates large openings ideal for vigorous climber growth. In many Eastern rain forests dense carpets of the big-leaved wiry convolvulaceous climbers *Merremia* spp. smother the residual forest and form climber towers up remaining isolated trees (Fig. 7.25). Trees eventually grow up and carry the carpet skywards but forest recovery is delayed and the trees may develop kinked stems (Fig. 7.42). Very little is known about the ecology of woody climbers. For example, in the Solomon Islands climbers do not proliferate excessively after cyclones but become extremely abundant after logging which causes similar canopy damage. In Sabah, *Merremia*, *Mezonevron* and *Uncaria*, which are very serious impediments to forest regeneration, are uncommon in virgin forest, perhaps because in this country massive canopy opening does not occur naturally.

Damage

The amount of damage to the forest depends more on how many trees are felled than on timber volume extracted. It is commonly the case that for every tree removed for timber (logged) a second tree is totally smashed and a third tree receives damage from which it will recover (Fig. 7.26). In Queensland (Fig. 7.27) the number of stems smashed was a similar proportion but before strict rules were introduced in 1982 the proportion damaged was much higher.

The pattern of damage differs between extraction by tractor and by overhead cable, so-called high-lead logging. The latter causes extreme devastation along the paths of the cables and little between. Seedlings survive better than bigger plants. In tractor logging the damage is spread more uniformly and seedling numbers can be very seriously depleted.

For successful recovery from logging the single most important criterion is to minimize damage to the forest floor, for several reasons. On the floor occur the seedlings of climax species. Within the surface soil lies the seed

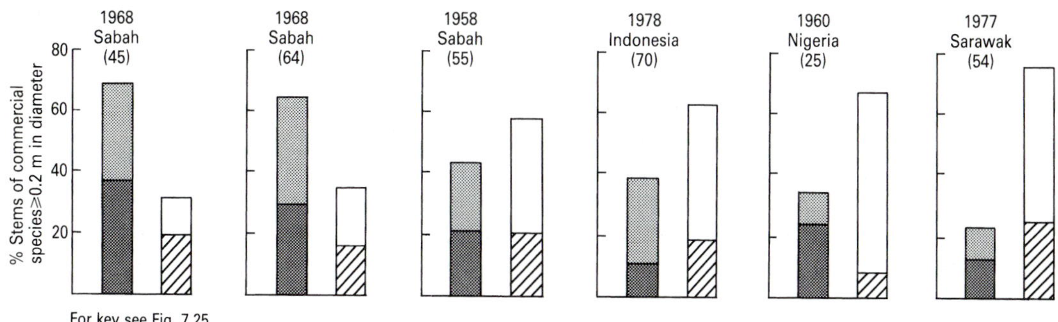

Fig. 7.26. Effects of logging (total commercial stems ha^{-1} in parentheses) in various forests. (After Nicholson and Keys unpublished.)

As the number of stems lost per hectare (by logging or smashed during logging) decreases the number that survive (intact or damaged) increases. Note that in the first three forests as many trees were smashed as were logged and that the Nigerian site, which had fewest commercial stems, had fewest smashed or damaged trees.

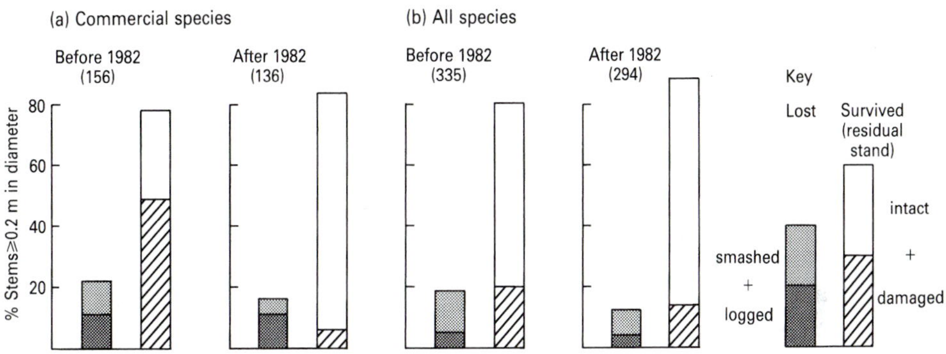

Fig. 7.27. Effects of selective logging, Queensland, on (a) commercial species and (b) all species, before and after introduction of strict silvicultural rules in 1982. Total stems ha^{-1} in parentheses. (From Nicholson and Keys unpublished.)

The new rules reduced the numbers of stems smashed or damaged. Their effect was about the same on the whole forest as on just the commercial species.

bank. Most of the roots and mineral nutrients are usually in the top 0.1–0.3 m of the soil. The surface root and humus layers allow good infiltration of rainfall. It has been clearly demonstrated several times that damage to the canopy and forest floor can be minimized by the careful planning of extraction roads and tractor paths (so-called snig tracks). In Sarawak such planning enabled 36 per cent more timber to be moved per hour; overall costs were 19 per cent down and open spaces in the forest were reduced from 40 to 17 per cent, with 60 commercial stems per hectare surviving rather than 40. In Queensland the new strict rules of 1982 (above) not only reduced the numbers of trees smashed or damaged (hence increasing the number that survived intact; Fig. 7.27) but also reduced snig tracks to 9 per cent of the area, only 2 per cent with subsoil exposed.

Soil compaction along tracks is an ecological problem. Infiltration of rainfall is strongly reduced. Plants are unable to establish on these surfaces which they do not encounter naturally. Tracks can be raked to break up the compacted soil, or specially designed tyres used on wheeled vehicles which have a low load per unit area. Wheeled vehicles cause more compaction than tracked ones.[204]

Erosion and hydrology

Another problem created by logging is erosion, which can be disastrous (Fig. 7.28) and whose extent has now been well studied.[205] Streamflow increases for 5–10 years after logging and is greater after rain than in unlogged forest because overland flow rises. Peak water discharges also increase after logging, but the increase is rarely statistically significant. There is a pulse of increased sediment of two- to tenfold by road construction and up to 20-fold by logging, but by 5 years sediment loads are only twice those from the original forest. Simple precautions can reduce extra sediment down to a quarter of these figures, for example, by appropriate alignment and design of extraction tracks, including the construction of cross drains.[206] It is also important not to block streams or the forest is killed in the lakes that then develop.

Low impact logging

The management of the rain forests of Queensland in the early 1980s most nearly approached these ideals until, after intense pressure from environmentalists, all logging was stopped at the time they were declared a World Heritage area (Table 10.15).

In the mid-1990s a change in sentiment occurred and throughout the tropics the will began to emerge to harvest timber with minimal damage to the forest. The rising international concern about the heavy human impact on tropical forests and the mechanisms evolving to control it are analysed in Chapters 10 and 11. Good logging practice, procedures well known to foresters but previously usually unenforceable, are ideas whose time had now come (Figs. 7.29, 7.30). Pre-planning of snig tracks, directional felling to minimize damage to the canopy and facilitate log removal, and winching in logs by 20–30 m of cable to reduce damage to the forest floor, are essential. Climber cutting some months before felling can reduce damage to forests with abundant big woody climbers but is

Fig. 7.28. Erosion along tracks built in rain forest climates can be spectacular as here in Malaya.

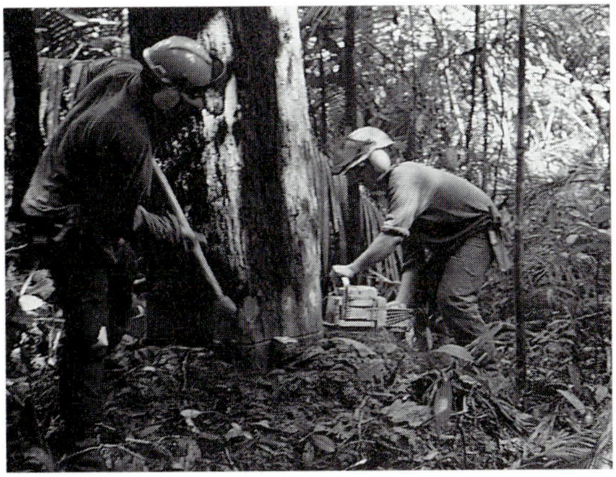

(a)

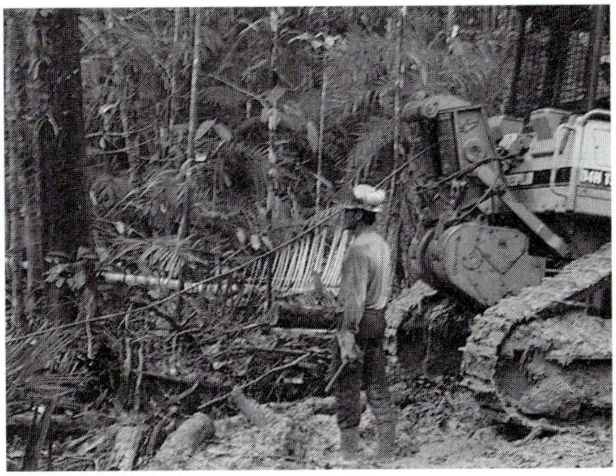

(b)

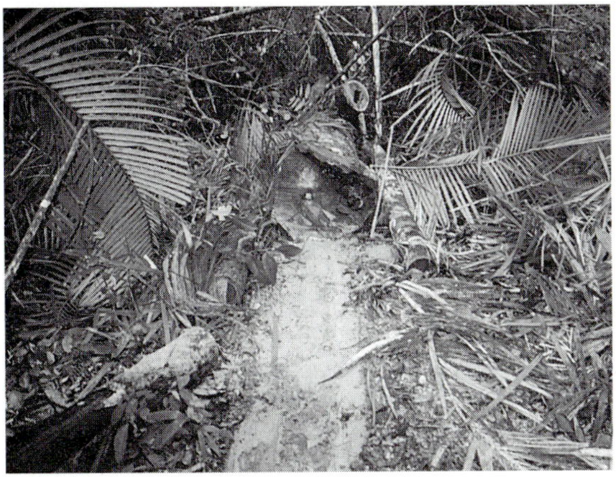

(c)

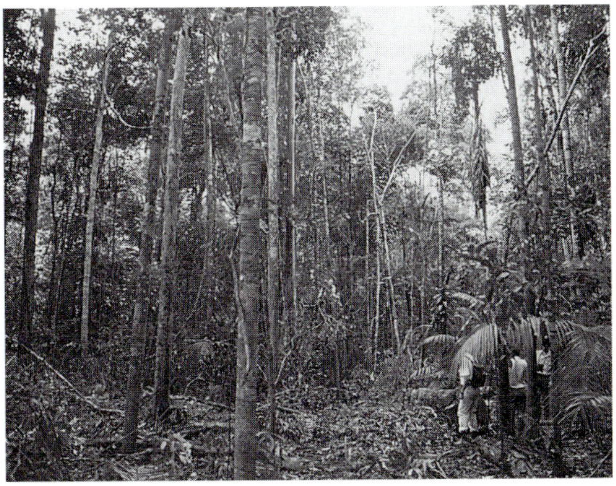

(d)

(e)

Fig. 7.29. Low impact logging, Precious Woods Ltd. Itacoatiara, central Brazilian Amazon.

(a) Trees are felled close to the ground to avoid waste, and in a predetermined direction (aided by wedges) so as to minimize damage to the forest.

(b) Skidder winches log into track, causing

(c) minimal damage to the forest floor.

(d) Forest after felling and log removal looking down a skid track; note abundance of juvenile trees whose growth will accelerate due to canopy opening and reduction in competition.

(e) The main extraction road; note minimal width.

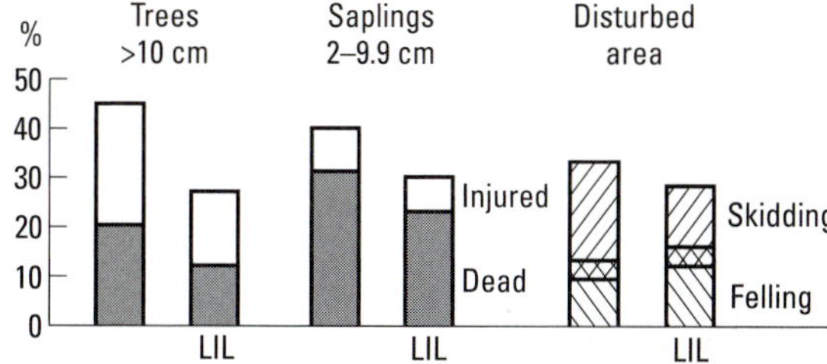

Fig. 7.30. Damage reduction by low-impact logging, East Kalimantan (Bertault and Sist 1995.)
8–10 trees ha^{-1} > 50 cm diameter were felled. Low impact logging (LIL) (involving climber cutting, pre-planning of skid-tracks and directional felling) reduces death and injury to both remaining trees and saplings; the area disturbed by felling is similar but LIL reduces the area disturbed by skidding.

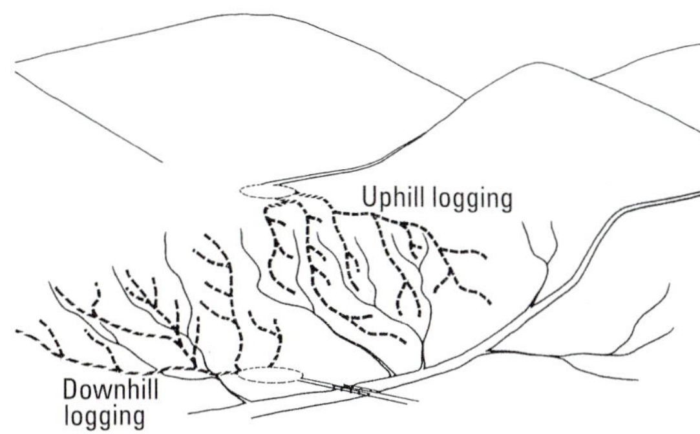

Fig. 7.31. Erosion is minimized by logging uphill (Bruijnzeel and Critchley 1994, fig. 8.)

not always cost-effective. Erosion control (above), including extraction uphill rather than downhill (Fig. 7.31), is also important. There have recently been experiments and demonstrations in Sabah, Kalimantan and Brazil[207] of good practice.

Silviculture in practice[208]

Eastern tropics

The dipterocarp forests of western Malesia are a silviculturist's dream.[209] This is because Dipterocarpaceae of that region have the unique property amongst rain forest trees of periodic mass fruiting (p. 63) which leads to dense seedling carpets. In addition there is an unusually high proportion of the species in the forest with valuable, readily marketed timber. Amongst the dipterocarps there are both heavy hardwoods, produced by slow-growing species with very shade-tolerant seedlings, and numerous light hardwoods, produced by faster-growing species with more light-demanding seedlings. Although the forests are extremely species-rich (Fig. 2.27) with up to twenty different dipterocarps growing together in any one place, their timbers are sufficiently similar to be grouped for sale into only a handful of classes (Table 10.8). This gives a continuous supply to the consumer and ensures a steady demand.

Silviculture of dipterocarp forests developed earliest in Malaya and has shown an interesting historical development, driven by market forces. The early demand was for firewood, poles and for naturally durable constructional timber, the heavy hardwoods. This favoured management on a polycyclic system with very few stems removed at a felling; the forests were manipulated to yield these products and at the same time to release from competition small specimens of the desired species. All the more accessible forests were systematically worked for heavy hardwoods, especially *Neobalanocarpus heimii*, *Shorea laevis*, and *S. maxwelliana*, vast quantities of these being required for building construction and railway sleepers, before pressure impregnation made timber of other species durable when in contact with the ground. These heavy hardwoods were felled by axe and converted by hand-sawing into squares, baulks, and sleepers, which were then manhandled or extracted by buffalo (Figs. 7.32, 7.33).

Reconstruction after the Second World War, followed by rapid national development and the increased international trade, gave a big boost to demand. To start with, for the supply of sawmills, light hardwoods (*Shorea*, *Dipterocarpus* and *Dryobalanops aromatica*) were also felled by axe and either hand-sawn into halves and extracted by buffalo or extracted as whole logs by hand-hauling on

Fig. 7.32. Logs are still manhandled out of the peat swamp forests of Borneo.

Fig. 7.33. Log extraction by elephant.

Elephants cause minimal environmental damage compared to machines and are still used extensively in Burma. In Thailand, because of the late 1980s logging ban, an elephant unemployment problem has arisen. Here, an elephant is at work in the Malaysia–Thailand border region. Most rain forest logs are too big for them.

wooden sledgeways. These methods were still practised until the late 1950s.

Since then the situation has completely altered. Trees are now felled by chain-saw and extracted by winch lorry or tractor. Hard, heavy, naturally durable timbers became less in demand than medium and light hardwoods, many of which can be impregnated with chemical preservatives if durability is required. An increasingly long list of species could be utilized as world-wide demand increased and prices rose. The necessity to get an adequate return on the substantial capital invested in vehicles, roads, sawmills, and plywood mills, made it

desirable to extract all marketable species in a single felling cycle. The consequence was to change to a monocyclic system, the Malayan Uniform.[210] The extensive canopy damage, which a uniform system causes, favours regrowth of light-demanding dipterocarp species, whose rapid growth and pale, light timber makes them the desirable species to cultivate. They take *c.* 70 years to attain timber size. The first managed rotation under such a system will almost certainly contain a higher fraction of these economically desirable species than the original primary rain forest. Moreover, in a well-managed forest under a uniform system, the basal area per hectare will be lower than in only mildly disturbed virgin forest; crowns of the crop trees will have greater exposure to light and space to develop better form. All these factors favour high growth rates.

One reason for the success of the Malayan Uniform silvicultural system, where it was applied correctly, is a consequence of the unique fruiting pattern of western Malesian Dipterocarpaceae, discussed above, and which is not shown by any other major group of big rain forest trees. It will be recalled that dipterocarps fruit gregariously once every several years. Dense carpets of seedlings then develop on the forest floor which subsequently die away (Fig. 7.14), sometimes to zero before the next gregarious fruitfall. Good regeneration is dependent on felling the forest when there are abundant seedlings. If an area is logged at a time in the cycle when seedlings are sparse, disaster may result because no seed source is retained, as all the trees big enough to bear fruit are felled. The foresters' rule of thumb in Malaya was 'felling must follow seeding'. Polycyclic systems do not suffer from this problem.

By the mid-1970s the Malayan forests, which had seemed limitless, came within sight of exhaustion. Economic pressures led to the replacement of the Malayan Uniform System in forests well stocked with adolescent trees by a polycyclic system.[211] Because only large trees are removed at each cutting cycle it is possible to return for another cut every 20 or 30 years. The political attraction was that the sustainable annual coupe becomes 1/20 or 1/30 instead of 1/70 of the total forest estate.

Modified Malayan Uniform Systems were introduced into the dipterocarp forests of Sabah and Sarawak.[212] The Philippines and Indonesia both employ polycyclic systems[213] with the amount removed controlled by allocation of an annual coupe and by tree diameter, but this cutting limit is difficult to enforce as are the rules on road construction, so unnecessarily heavy damage commonly occurs.

The Eastern rain forests beyond the dipterocarp forest zone, east of Wallace's Line, are almost as ecologically robust because they contain numerous light-demanding species which regenerate well after heavy logging damage. Papua New Guinea suffers many natural cataclysms (see p. 145) and the islands of Melanesia are subject to cyclones. Logging merely adds a new sort of catastrophe.

The rain forest finger that extends southwards into Queensland is also subject to cyclones and most, if not all, tree species in these subtropical rain forests regenerate vigorously after logging. A polycyclic system with selective felling on a roughly 40-year cycle was perfected in 1982 but all logging was stopped in the late 1980s (see above).

Neotropics

Silviculture of the New World rain forests is still in its infancy.[214] Investigation of the forests of Suriname[215] and of 63 000 km^2 along the lower Amazon river in Brazil[216] (Table 10.9) found them to be dominated by species with very heavy timber, which are therefore slow growing and shade-tolerant. In the natural forest there is a paucity of more light-demanding species with low density timber. The heavy hardwoods respond too slowly to canopy opening to be attractive for silviculture. When the canopy is opened more than very slightly there are few seedlings able to respond by vigorous growth. In Suriname this degree of canopy disturbance leads to the forest 'tumbling down' to a mass of woody climbers and the commercially useless

pioneers *Cecropia* spp. Great silvicultural skill is required to open the canopy just enough to favour *Ocotea*, the timber species with fastest growth, but not to cause this degradation. In Brazil the main study has been at Tapajós National Forest.[217] Here, 13 years after selective logging with no silvicultural treatment, so-called 'log and leave', there were fewer shade-tolerants and light-demanders (including large pioneers such as *Goupia glabra* and *Jacaranda copaia*) had increased. As is usual, growth had been stimulated for a few years but had then fallen back, to 0.8 m^3 ha^{-1} year for commercial species. A cutting cycle of 30–35 years was tentatively recommended. Some 7000 km^2 of the Brazilian Amazon are now being exploited under approved management plans that involve selective felling.

The CELOS polycyclic system developed in Suriname has been applied since 1995 on a 800 km^2 concession near Itacoatiara in Amazonian Brazil by a private company committed to sustainability (Fig. 7.29). The plan is to export processed timber and manufactured goods of a wide range of species, the value added by processing being essential to pay for the intensive management and silviculture.

The mahogany, *Swietenia macrophylla*, occurs from Middle America southwards to western South America and in a broad belt across the southern part of Amazonia to southeast Pará where it occurs in forests that experience a strong dry season. It is one of the best known rain forest timbers, unusual in its very high value, up to three times that of most others. Mahogany typically occurs as very scattered giant trees with little or no regeneration. Very little is known about its regeneration.[218] Recent studies in Mexico, Bolivia and Pará, Brazil, show that it is able to germinate and establish under closed canopy but the seedlings do not persist for long unless released by creation of a canopy gap. In Mexico it also regenerates along tracks and on abandoned fields. Because the seedlings are so short-lived it is essential to preserve seed trees. Silviculture of this valuable species has still to be perfected. Clearly, it is favoured by massive canopy opening, so a low intensity polycyclic system is unlikely to be successful.

West Africa[219]

As in America there are some forests which are predominantly of heavy hardwood species. The ecological equivalents of the light hardwood dipterocarps are certain Meliaceae (the West African mahoganies), *Entandrophragma* and *Khaya*, but they are less abundant than dipterocarps. Shade-tolerant, slow-growing, heavy hardwoods (Fig. 7.34) include other Meliaceae, *Guarea* and *Lovoa*, and also (as in America) numerous caesalpinoid Leguminosae (e.g. *Anthonotha*, *Berlinia*, *Cynometra*, *Gilbertiodendron*).

Complex silvicultural systems were developed in Ivory Coast, Ghana and Nigeria, which involved considerable treatment before and after felling. Over 30 000 ha in Ivory Coast and over 200 000 ha each in Ghana and Nigeria were treated in the two decades after the mid-1940s. The systems were then abandoned partly because of the cost and the amount of skilled labour needed, partly from the realization that trees which later proved to be valuable, for example *Pericopsis elata* and *Pycnanthus angolensis*, were being poisoned. Felling was controlled by prescription of area and girth limits. Today in West Africa polycyclic systems are the declared choice.

7.7 GROWTH RATES AND LONGEVITY

The wood of most rain forest species does not have growth rings and where they do occur they may not be annual so tree age cannot be measured directly. It has been found that the fastest growing juvenile trees in a forest are the ones most likely to succeed, so growth rates averaged from a number of stems are misleading. Longevity was calculated for a number of species at La Selva, Costa Rica, making due allowance for this (Fig. 7.35). However, we

Fig. 7.34. Logs of *Tieghemella heckelii* for sliced veneer, felled under a highly selective polycyclic silvicultural system. Kumasi, Ghana. The author stands by (1987).

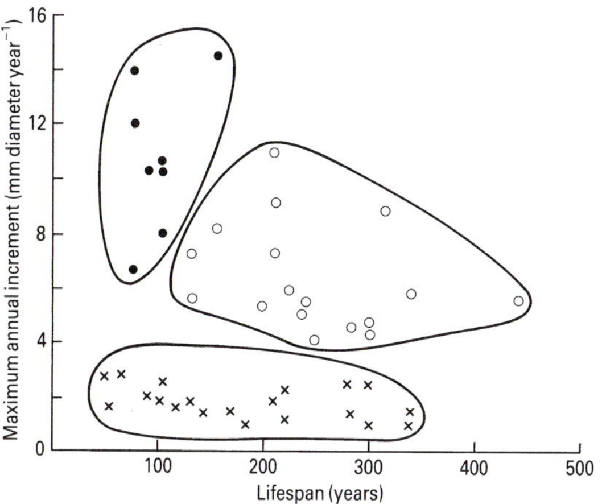

Fig. 7.35. Estimated life span and growth rate of the 46 common tree species at La Selva, Costa Rica. (Data of Lieberman *et al.* 1985.) X, Climax species with slow maximum growth rate, none reach the canopy top. ○, Climax species with medium to fast growth rate. Growth is stimulated by a canopy gap. Most reach the canopy top.
• Pioneer species, growth rate fast to very fast. All except the slowest-growing reach the canopy top.

have very little reliable information on how long trees can live.

Height growth rate can be spectacular. *Ochroma lagopus* (balsa) often grows 5–6 m year^{-1} (Fig. 7.9). *Paraserianthes falcataria* holds the world record, 9.91 m year^{-1} for the fastest individual in a Sabah plantation and 9.14 m year^{-1} mean value for all 199 trees, measured at age 1.8 years.[220] Bamboo culms may grow at 1 m day^{-1} for a few days by expansion of the internodes of previously formed shoots.

The forester is more interested in biomass increment. Natural lowland rain forest commonly adds 2–3 tonne year^{-1} of dry weight of bole timber, which may increase to 3.6–12 tonne year^{-1} in forest under good silvicultural management. Growth rate varies with the phase of the forest growth cycle (Fig. 7.36), from which it follows that care must be taken in obtaining or interpreting biomass increment figures from natural forests.[221] The forester aims to harvest trees in both natural forest and plantations at about the time they mature and growth slows down. In plantations, tropical conifers grow faster than most broadleaf species (Fig. 7.37).

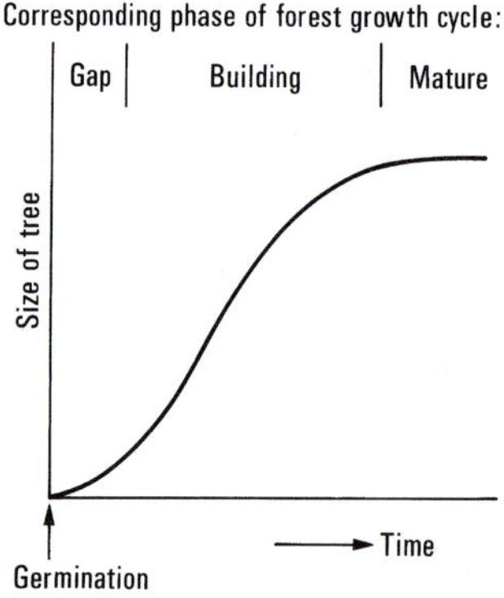

Fig. 7.36. Growth of a freely growing tree and its relation to the forest growth cycle. (Whitmore 1984*a*, fig. 9.3.)

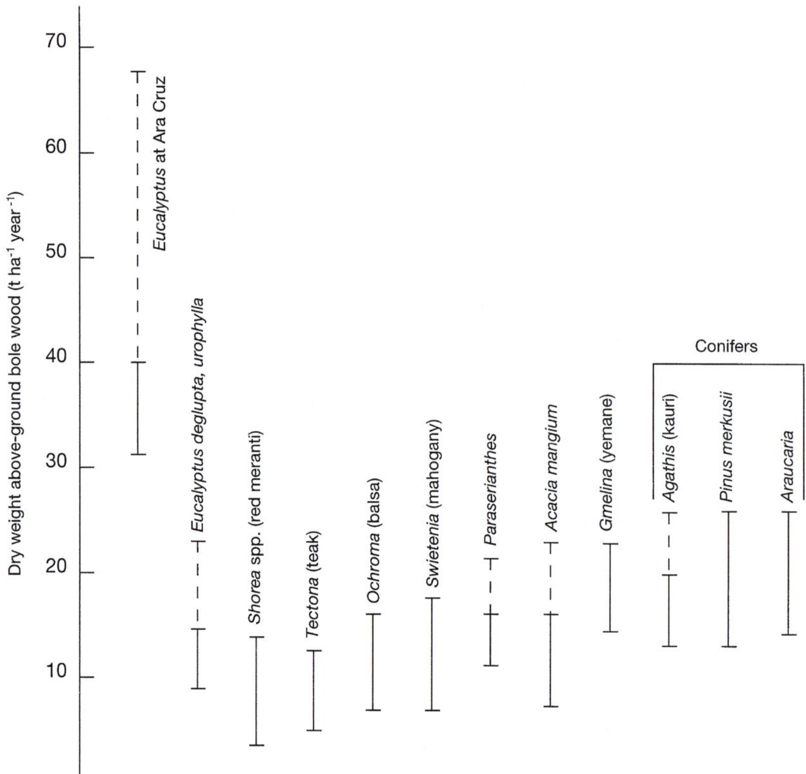

Fig. 7.37. Timber yield of various species grown in plantations in the tropics. (Whitmore 1984*a*, fig. 9.1 updated.)

Trees grow for all or most of the year and in consequence yields far exceed those attained in other parts of the world. The highest yields of *Gmelina* (45 m^3 ha^{-1} yr^{-1}) are from strains selected by the Solomon Islands Forestry Division and grown on Kolombangara. The phenomenal growth at Ara Cruz (54–144 m^3 ha^{-1} yr^{-1}), from 1320 km^2 supplies *c.* 4 per cent of the world's paper pulp. Here, at 20 °S in Brazil in a perhumid climate with 1400 mm annual rain, eucalypt trees reach 20 m tall in 1000 days.

7.8 COMMUNITY-WIDE DISTURBANCES[222]

Some parts of the humid tropics are much more prone than others to extensive, catastrophic disturbance, betokened by a coarse-scale structural mosaic of the forest and the abundance of big-gap species. The belts of latitude where cyclones are common were mentioned in section 2.3. Two further examples of places with extensive large-scale natural disturbances may be given.

Papua New Guinea

This is a land wracked by continual catastrophe.[223] The mountains are young and continuing to uplift as the Australian plate subducts below the Pacific plate (see Chapter 6), so earthquakes with associated landslides are frequent on the young steep slopes. There are numerous active volcanoes which create lava flows and mudflows (lahars) and thick ash deposits. Strong destructive winds occasionally occur. In exceptionally dry years those forests that are always slightly seasonal become unusually dry and may catch fire. The big rivers which run on to the coastal plains have unstable courses. Shifting cultivation and associated regrowth forest (cf. section 8.1) is also extensive. It is no surprise that lists of timber tree species for a tract of lowland rain forest in Papua New Guinea usually include a considerable proportion of pioneers, such as species of *Albizia*, *Paraserianthes* and *Serianthes*, or *Eucalyptus deglupta*, besides strong light-demanders such as *Campnosperma* spp., *Pometia pinnata* and *Terminalia* spp.

Amazonia

Peru

Aerial photographs or satellite images of Amazonian Peru reveal curious and marked stripes in the forest canopy. Ground inspection shows these stripes represent forests of different species composition.[224] The upper tributaries of the Amazon which traverse the region run rapidly off the slopes of the Andes and are subject to violent annual flooding. They have a high silt load. Every year they change course, easily eroding the soft alluvial river banks, and move laterally many metres, in one case 180 m. Fresh alluvial banks are created on which primary succession occurs (Fig. 7.38). The stripes seen from the air are different stages of this succession. Climax is never reached, a new flood always intervenes to destroy the forest. About one-eighth of the Peruvian Amazon forests show these characteristics. The disturbance creates a mosaic of forest facies with strong species differences between them; this is diversity from place to place within the same

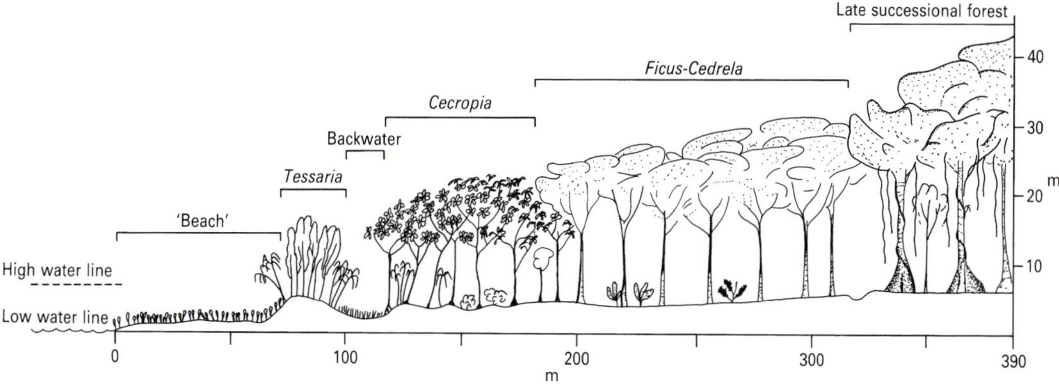

Fig. 7.38. Primary succession on newly deposited riverine alluvium, at Cocha Cashu on the Rio Manu in Peruvian Amazonia. (Salo *et al.* 1986, fig. 3.)

Fig. 7.39. Primary succession on an accreting river bank, south Papuan coast. Swamp grassland is replaced by a narrow belt of the pioneer tree *Octomeles sumatrana* as the levée builds up. On the right climax species have grown up under the *Octomeles* which persists as big scattered individuals.

community and is sometimes called beta diversity (cf. section 2.4). The only other place where this pattern has been discovered is in much smaller areas of Papua New Guinea, already mentioned (Fig. 7.39).

Brazil

The Amazonian rain forest in Brazil covers 3.3 million km^2. About 12 per cent of this huge forest, 390 000 km^2, has now been found to show signs of past human impact,[225] to be discussed below, p. 000. The most extensive signs are 196 000 km^2 where babassu palm, *Attalea speciosa*, is dominant or common (Fig. 7.40), 100 000 km^2 of liane-dominated forest, and 85 000 km^2 of bamboo-dominated forest. Babassu owes its abundance and extensive range to man.[226] Dense seedling carpets form (over 6000 ha^{-1}) and persist in forest shade. The young plant has a subterranean growing point and can survive fire. The palm can live for nearly 200 years. It has many useful products for sale and subsistence and is a major component of the rural economy in Maranhão state in the lower Amazon. The bamboo forest is in the south west state of Acre (Fig. 7.40) and extends across the frontier into Peru and Bolivia. Its total area is *c.* 120 000 km^2. It was unknown until the RADAM remotely sensed mapping survey of the 1970s and still remains little known. Principal bamboos are *Guadua sarcocarpa* and *G. weberbauri*. They fruit gregariously and then die at 26–29 year intervals (cf. section 3.3) and are believed to burn after death. This remarkable forest appears to be spreading up valleys.[227] Another widespread sign of former denser populations of Amazonia are the areas of soil known as *terra preta*, black from an abundance of fine carbon particles. These are mostly *c.* 4 ha in extent, contain potsherds, and are of anthropogenic origin. *Terra preta* is fertile and is preferentially used for agriculture by con-

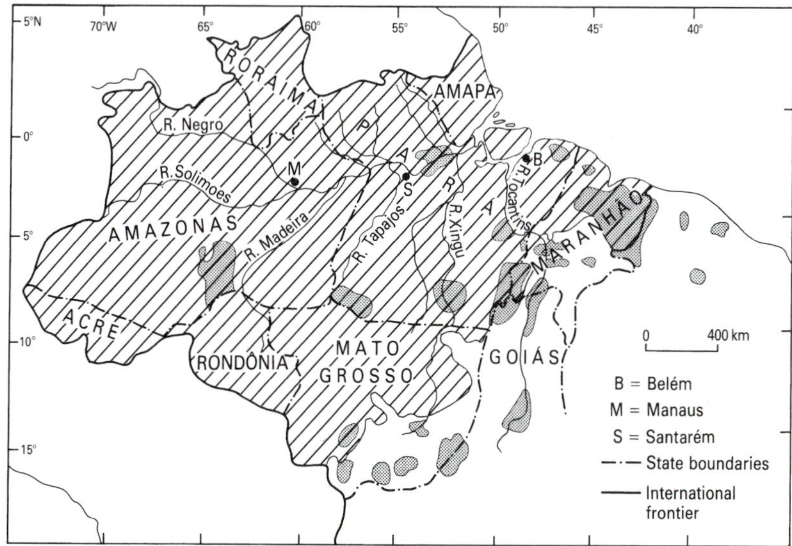

Fig. 7.40. Potential extent of the Amazonian rain forests in Brazil (Brown in Whitmore and Prance 1987, fig. 2.4) with the main occurrences of the babassu palm (*Attalea speciosa*) shown (Anderson *et al.* in Lugo *et al.* 1987). States, principal rivers, and cities shown.

temporary inhabitants. Damage by wind also occurs. Natural blowdowns of total area *c.* 900 km^2 were detected in the early 1990s from satellite images, as fan-shaped areas of secondary forest, later confirmed by inspection on the ground. These are mostly 30–100 ha in extent. They are created by down-gusts from moving convectional storms (well-known to the pilots of small aeroplanes) and are densest in a north–south belt passing through Manaus where thunderstorms are very frequent.[228]

Big disturbances codified

The various kinds of community-wide disturbances can be codified into three groups. A broad distinction can be made between disturbances followed by primary succession, by secondary succession and, as a minor category, by regeneration from resprouts and pre-established seedlings.

The first group includes disturbances where the forest is destroyed and the soil profile disrupted, resulting in loss or deep-burial of the soil seed bank. Landslides, volcanoes and mobile rivers are in this group, so are some human activities such as clearance for continuous cultivation or construction of logging roads and log landings.

The second group includes most other large-scale disturbance factors. The canopy is opened and the soil profile may be disrupted (severe windstorms, shifting cultivation, careless logging), or not (drought, fire, careful logging).

The rate, pattern and species composition of regeneration after large-scale disturbance depends on the causal factor, but there are very few long-term observations. Three of the most important factors warrant further discussion.

Drought and fire

Between July 1982 and April 1983 east Kalimantan received only 32 per cent of its usual 1820 mm of rain (Fig. 7.41). This was part of a pantropical climatic perturbation, called the E1 Niño Southern Oscillation, which results from anomalous warming of the usually

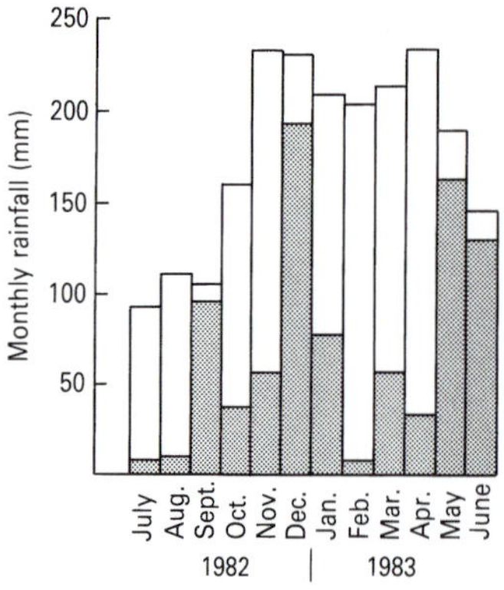

Fig. 7.41. The 1982–83 drought in northeast Borneo. Long-term mean monthly rainfall, and 1982–83 rainfall (*shaded*) at Kota Bangun, Kalimantan. (Lennertz and Panzer 1983, fig. 1.)

cool surface of the eastern equatorial Pacific Ocean.[229] In early 1983 fires started and burned for many months. In total over three million hectares of lowland rain forest were destroyed in Kalimantan and over one million more further north in Sabah.[230] Subsequent investigations revealed that in Sabah 85 per cent and in Kalimantan about one-third of the forest destroyed had been recently logged, and would have had a lot of readily flammable dry dead wood. Only about one-third (1.35×10^6 ha) of the destroyed Kalimantan forest was primary and part of that (1×10^5 ha) was killed by drought not by fire: the big trees, especially dipterocarps, had been singularly susceptible.[231] Research into the climatic record revealed that similar droughts have occurred in eastern Borneo two or three times in the last century. The still grass-covered 800 km^2 Sook Plain of Sabah originated from a fire in 1915 after an exceptionally dry period. The affected area of Kalimantan had previously been almost uninhabited. Immigrant Buginese settlers from Sulawesi had recently moved in illegally along newly built timber extraction roads. When the drought occurred fire escaped from their fields and set light to the dead trash left behind after lumber extraction. It appears eastern Borneo has always been susceptible to rare droughts some, but not all, associated with E1 Niño events.[232] Indeed, charcoal provides evidence of fires at 17 510 and 350 years BP probably persisting for years in superficial coal seams and spreading out at times of drought.[233] The difference was that in 1982 human activity dramatically exacerbated the consequences.

Several droughts occurred in the later 1980s and the 1990s. 1997 was the next strong El Niño event. These dry spells were used by agribusinesses in Sumatra and Kalimantan to burn off logged forest, clearing the land for plantation crops (see Figure p. 238), although after protests from neighbouring countries Indonesia declared this practice illegal in 1995. The fire often escaped. Where peat swamp forests had dried out, exacerbated by logging, the peat also caught fire. Massive palls of smoke (clearly visible from satellites) spread across the region. Airports were closed, sometimes for days on end, in Indonesia, Malaysia and Singapore. In September 1997 the situation became particularly severe. The smoke reached as far north as Phuket, 7° 52′ North, in Thailand, and east to the Philippines, and was a serious health hazard. Schools and airports were closed and there was talk of evacuating particularly vulnerable people or settlements. These massive man-made fires in logged forest dramatically reminded the world about current human impact on the Eastern rain forests. They were a once-and-for-all destruction of these species-rich ecosystems plus the release of *c.* 150 tonne sequestered carbon ha^{-1} into the atmosphere, and up to *c.* 400 tonne where peat was also burned off.

The 1982–83 El Niño event caused an exceptionally strong dry season in Panama and on a permanent sample plot of 50 ha on Barro Colorado Island unusually high tree mortality occurred.[234] At the time of writing the 1997 event is still in process. Its impact cannot yet be evaluated.

The 1982–83 Great Fire of Borneo was unexpected (although in fact it was mainly freshly logged forest that burned) because until recently it had been believed that primary rain forest, in contrast to secondary regrowth, is not

flammable. This view was also challenged by the discovery in the early 1980s of charcoal in the soil under rain forest in the north central Amazon basin.[235] ^{14}Carbon dates show numerous fires have occurred since 6260 BP, the mid-Holocene, and up to a few hundred years ago. Some of them coincide with known dry phases. More recently, soil charcoal has been found in several places near Manaus[236] and the 1983 El Niño event is believed to have resulted in the burning of over 50 000 ha forest and loss of foliage from a further 100 000 ha.[237] At the seasonal-climatic margin of rain forest fire may determine its limit. In Queensland rain forest is today invading the adjacent dry sclerophyll forest dominated by *Eucalyptus grandis* (Fig. 2.6) because there are no longer aborigines to set fires. And in Ghana the northern edge of the rain forest burns in unusually dry years, such as 1982–83.[238]

Wind

The important impact of cyclones on rain forests between 10°–20° north and south of the equator has been well known for over 20 years (section 2.3). There is evidence of other big winds influencing species composition outside the cyclone belts. Examples from the Brazilian Amazon were described above. The dipterocarp rain forests of Malaya have various indicators of rare wind storms (Fig. 7.42). Corridors of storm-felled trees are fairly common in the Ituri forest, Congo.

Human disturbance

Signs of past human disturbance have now been found in numerous rain forests, many in places uninhabited and, until recently, believed untouched by human hand (above, and section 10.1). The most dramatic case is probably the Brazilian Amazon. Middle America has numerous signs. Borneo has forests with high concentrations of fruit trees. The upper canopy of the forests of north Kolombangara, Solomon Islands, consists of species present solely as big trees, viz. not replacing themselves which suggests these forests have regrown after a catastrophe about a century ago (Fig. 2.22). Studies of tree demography over 30 years 1964–94 include the effects of a cyclone in 1967 and show that, although it caused massive canopy damage, it has not resulted in any increase by recruitment of these disturbance-favoured species.[239] It seems probable that the catastrophe that preceded forest establishment was cultivation, not an earlier cyclone, and there is now evidence that the currently empty landscapes were indeed once populated.

In Asia the very extensive *Adinandra-* or *Ploiarium*-dominated forest of Singapore, nearby islands and south Malaya is regrowth on areas exhaustively farmed for gambir (*Uncaria gambir*, a source of tannin), with soil degradation and loss of organic matter, a century and more ago. These vast secondary forests, and the even more extensive ones spread through Malesia today, mostly dominated by small pioneer trees, mainly *Commersonia*, *Macaranga*, *Mallotus*, *Trema*, (and *Alphitonia* and *Trichospermum* in the east), have no natural analogue (Fig. 7.43). They have arisen largely as a result of shifting agriculture and are becoming ever more extensive as logging is opening fresh forests and shifting agriculture follows. Similar anthropogenic secondary forests occur in Africa (*Musanga*, *Trema*) and South America (*Cecropia*, *Ochroma*, *Trema*).

Starting in the early 1970s huge pastures covering thousands of square kilometres have been created in Pará state on the southeastern fringe of the Amazonian rain forest. Cattle raising proves unprofitable, many holdings are abandoned after a few years, and secondary forest develops.

Conclusions

Evidence of past community-wide disturbance has now been found right across the humid tropics, either directly in species composition and community structure or, where older than the trees' life span, indirectly as potsherds, charcoal

Fig. 7.42. Storm forest at Kemahang, northeast Malaya, in 1968, 88 years after devastation by a freak storm in November 1880. Many trees have a kink believed to result from regeneration through a dense tangle of climbers.

The two central trees are *Shorea parvifolia*. This and other similar light-demanding dipterocarp species dominated. Forests of similar composition elsewhere in Malaya develop after windstorms that destroy a few hectares and line squalls that cause corridors of destruction. These need only recur once a century or less. The Sungai Menyala forest experienced two such storms in 1948 and 1958 and continues to be dominated by this species guild. (Whitmore 1975; Manokaran and Swaine 1994.)

Fig. 7.43. The tree fern *Cyathea contaminans* colonizing bare ground at the margin of lower montane rain forest, Cibodas, west Java.

or earthworks. There are by contrast some forests with no such signs. These include the 63 000 km^2 of the lower Amazon dominated by shade-tolerant species with very dense timber (section 7.6, Table 10.9), the dipterocarp forests dominated by strongly shade-tolerant *Neobalanocarpus heimii* and balau group *Shorea*, and the consociations scattered across Africa of shade-tolerant caesalpinoid legumes, for example, *Gilbertodendron dewevrei* around the Congo basin and others in Cameroon.[240]

Occasional or rare catastrophic events leave their mark on the forest perhaps for a century or longer. They can be more critical to the forest than average conditions: an instance when 'extremes are more important than means'. Another nail is driven into the coffin of the supposition that tropical rain forests are ancient and immutable. Moreover, these rare catastrophes provide an important context for the perpetual debate amongst some academic ecologists as to whether a given rain forest is or is not in equilibrium.

7.9 FOREST RECOVERY AFTER HUMAN DISTURBANCE

The time it takes climax rain forest to return on a site depends on the severity of the disturbance.

Low intensity selective logging on a polycyclic system closely mimics the natural processes of forest dynamics and scarcely alters the composition. Monocyclic silvicultural systems, and polycyclic systems with many stems felled per hectare, shift species composition to increase the

proportion of the more light-demanding, faster-growing tree species. Areas which are laid bare by timber extraction develop a secondary forest of pioneers. Compacted soil will be only slowly colonized by trees but will usually soon be covered by creeping plants or thicket-forming resam ferns (*Gleichenia* s.l.). Where streams have been dammed by tracks without culverts, swamps develop and the forest is killed. Thus a logged forest becomes a very complex mosaic of vegetation types, including true secondary forest, relict patches of primary forest, and depleted forest from which some species have been removed for timber. Repeated logging increases depletion and eventually all or most of the big trees may be lost. A dense, low, climber-tangled forest develops. Such derelict forests are common near to towns, where timber theft is a serious problem. In some ways they can be viewed as green deserts.

If land is clear-felled and then abandoned it is quickly re-occupied by forest. Regrowth is partly of pioneers germinated from seed, partly of surviving seedlings, and partly from root and stem sucker shoots. If the cleared site is burned before being abandoned, vegetative sprouts and surviving seedlings are reduced, thus altering the composition of the regrowth forest.

If land cleared and then used for pasture or for growing crops is abandoned forest will also soon develop. However, the pioneer woody species may differ from those that colonize land that has not been farmed. For example, *Melastoma* becomes common in the Eastern tropics. This is partly because the soil has lost nutrients and becomes more acid and compacted (section 8.1). In addition, during the cultivation period the soil seed bank becomes progressively depleted as does the capacity of roots and trunks to develop coppice shoots. Regrowth in these circumstances is from seed rain. Pioneers soon arrive, but the rapidity of later succession is strongly dependent on there being climax species nearby. In Mexico it was shown that climax species re-invaded around relict trees left on the farmers' fields because these provided bird perches.[241] A few years after the fields were abandoned the area became covered by secondary forest of pioneers within which were nuclei of climax species. At Kepong, Malaya, the first dipterocarp colonized in 1976 a field abandoned in 1944. The nearest parent tree was only 180 m distant.[242]

The nature of succession in the extensive man-made forests of pioneers described in section 7.8 remains enigmatic. Woody vegetation soon returns, but the prognosis for the restoration of climax forest without human assistance is gloomy, because of the absence of seed trees. Such forests did not exist in the past. Before humanity reached its present numbers and potency pioneers were confined to patches of a few hectares on landslips and river banks. It seems likely that in these situations pioneers and strongly light-demanding climax species perpetuate themselves in the absence of more shade-tolerant species. This is seen at Darién, Panama, where *Cavanillesia platanifolia* forms 60 per cent of forests known to have been pasture land at the time of the Spanish conquest four centuries ago.[243]

7.10 THE DYNAMICS OF CLIMAX RAIN FOREST

The forest canopy is in a continual state of flux as this chapter has sought to demonstrate. Scientists have wondered whether composition changes with time or fluctuates about a mean. Is the forest in equilibrium? Trees certainly commonly live for a century or more, even though precise ages are hard to determine, and no permanent sample plots are older than a few decades. For up to a century or more after a community-wide disturbance there will be change due to succession. But for a climax forest of species capable of perpetuating themselves *in situ* this is, in fact, a question it is impossible to answer with any degree of certainty because of the long time-scale involved. An individual tree only needs on average to have one single successful seedling for floristic composition to be maintained. This gives time for a great deal of

short-term fluctuation in seedling recruitment and mortality.

In Ghana there is a broad cline in species composition from southwest to northeast over a distance of 300 km, determined by a combination of soil type and the increasingly seasonal and dry climate.[244] Forests at different places along this cline seem to 'breed true', juvenile populations being most similar to the adjacent adult populations. At any one place there are big differences between species in their population dynamics. The numbers of juveniles give no indication of recruitment of adults. The two extremes are represented by *Celtis mildbraedii* and *Strombosia glaucescens*.[245] Most *Celtis* are over 0.3 m diameter, smaller plants are very rare but extremely persistent. *Strombosia*, by contrast, has very few trees over 0.3 m in diameter and numerous smaller plants which have a rapid turnover. Both species seem to be maintaining themselves but by differing means.

Similar differences were found amongst seedlings on Kolombangara Island in the Solomons where in most years *Calophyllum neo-ebudicum* and *C. peekelii* recruit often extensive populations of 300–2400 seedlings ha^{-1} of which up to three-quarters die within a year. In the same forest *Dillenia salomonensis* and *Parinari papuana* have populations of only 30–170 seedlings ha^{-1} with rare recruitment and little mortality.[246] The relative abundance of the different species in the seedling bank is different from the relative abundance of trees in the canopy above them. This is commonly the case.

In the Okomu forest, Nigeria, E.W. Jones found that the largest trees, including strongly light-demanding *Entandrophragma* spp., were scarcely regenerating.[247] On further investigation it was found the area had once been agricultural land but had been abandoned during tribal wars about 200 years previously. The regrowth forest came to be dominated by light-demanding species, and by the time of Jones' study these were being replaced by more shade-tolerant ones in the absence of further massive disturbance. The publication of these discoveries caused something of a sensation, because this was the first strong evidence that tropical rain forests are not necessarily ancient and immutable.

Permanent sample plots are essential for the monitoring of change of composition with time. At Sungai Menyala, Malaya, a 2-ha plot has been under observation since 1947.[248] Here the topmost trees are mostly light hardwood dipterocarps, climax species with fairly light-demanding seedlings. The middle-sized trees contain more heavy hardwood dipterocarps, species with more shade-tolerant seedlings. The whole plot originally contained 1075 trees over 0.1 m diameter. During the 38 years of observation 697 of these had died (64.8 per cent), which is a considerable turnover. Substantial canopy damage was noted after windstorms in 1948 and 1958 and during the 1970s. The overall effect at Sungai Menyala has been for the amount of disturbance experienced to maintain the same proportion of the two ecological species classes (light and heavy hardwoods) in the different levels within the canopy.

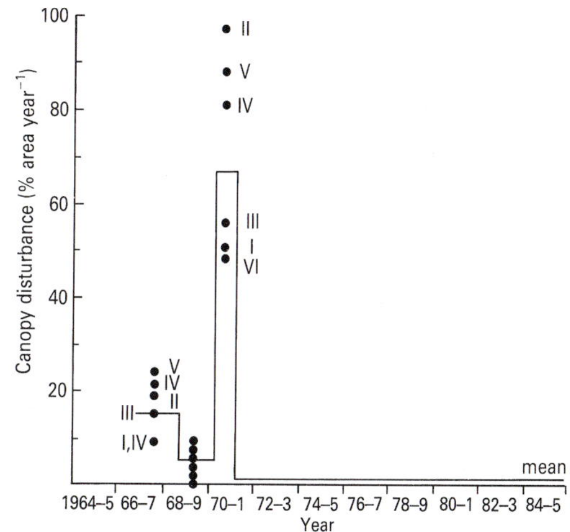

Fig. 7.44. Forest dynamics of the lowland rain forests on Kolombangara, Solomon Islands, 1964–86. Canopy disturbance to six forest communities (I–VI) of different floristic composition during different time periods. (Whitmore 1989*b*, fig. 5.)

Canopy disturbance varied from time to time and different forest floristic types suffered different disturbance.

Another long-term investigation has analysed dynamics on 22 plots of total area of 13.7 ha over periods up to 21.3 years on Kolombangara Island. Big differences were found between periods and from place to place in the amount of canopy disturbance, with a maximum 67 per cent and a minimum 0.6 per cent (Fig. 7.44). This study[249] shows that either small sample plots or short study periods look at the flux in a forest canopy through only a tiny 'window' and may give extremely misleading results, not applicable to the whole forest. In lowland rain forests the average mortality monitored from permanent sample plots is 1–2 per cent of trees $year^{-1}$.[250] The Kolombangara observations show that this average figure might conceal more than it reveals. In Sarawak the remarkable discovery has been made that rare species turn over faster than common ones.[251]

7.11 FOREST DYNAMICS—CHAPTER SUMMARY

1. The study of forest dynamics is currently very active and findings in tropical rain forests have close parallels in other types of forest.

2. There is a range of different forest microclimates. In particular, light varies in spectral composition and in quantity both in time and in space. The influence of canopy gaps extends beyond their edges and the microclimate within them is not uniform (Fig. 7.5).

3. Tree species range from light-demanding to shade-tolerant in the amount of solar radiation required for regeneration. Two classes can be recognized: pioneer and climax species. The essential differences are that pioneer species germinate and establish only in full light in a gap after its creation, so their seedlings are not found below a canopy. Climax species by contrast usually germinate and establish below a canopy. Each species class is characterized by a whole syndrome of ancillary features (Table 7.3) which are adaptive to the ecological niche it occupies.

4. Within each class there are species with different characteristics but variation is continuous. Pioneers can usefully be divided by usual height attained at maturity (Table 7.4) and climax species by the degree of seedling shade-tolerance (Table 7.5). The most light-demanding climax species resemble pioneers except for their ability to germinate and establish below a canopy.

5. Pioneer species colonize big canopy gaps. Below them climax species establish. As the pioneers die, creating small gaps, the climax species grow up to succeed them.

6. All rain forest soils investigated contain a seed bank, mostly composed of pioneer species. Pioneer species have orthodox seeds, capable of dormancy.

7. Tropical pioneer tree species have been shown to have seed germination triggered either by light in which the far-red wavelength exceeds red (Fig. 7.13), or by elevated temperature. Both conditions are found in gaps but not below a closed canopy (Fig. 7.2, Table 7.2).

8. Many climax species have recalcitrant seeds which cannot be stored and which germinate immediately. They form seedling banks on the forest floor and await a canopy gap to start upward growth.

9. Species differ in the microsite conditions in which their seedlings establish most successfully. The co-existence of numerous species in tropical rain forest depends on many facets of their regeneration, including microsite specialization.

10. A forest is most species-rich when at an intermediate state of recovery from disturbance, or when disturbance is at an intermediate intensity or frequency, because it then contains both pioneer and climax species.

11. Light-demanding species have higher unit leaf rate than shade-tolerants (Table 7.7). Seedling shade-tolerance includes the ability to utilize intermittent sunflecks and having a low respiration rate relative to photosynthesis (Table 7.8). Light-demanders show more flexible responses to changing conditions. In particular their growth is similar to shade-tolerants in tiny gaps but increases more rapidly in larger gaps

(Fig. 7.19) and this is the basis of their abundance in heavily disturbed forests.

12. Species change their response to light through life (Fig. 7.20).

13. Silviculture, the manipulation of the forest canopy to favour certain species, exploits the different gap requirements for regeneration, i.e. differences in shade tolerance. Polycyclic systems repeatedly remove trees as they mature leaving all adolescent trees to swell future yields. Monocyclic systems remove all marketable trees at a single operation (so create bigger canopy gaps) and rely on seedlings for the new crop. Polycyclic systems thus tend to favour shade-tolerants and monocyclic systems to favour light-demanding species. With good control over damage polycyclic systems yield more timber.

14. The ecology of most rain forests is sufficiently well-known for viable silvicultural systems to be devised, whereby the forest can be maintained as a continual source of timber. You can both 'have your cake and eat it' provided sufficient care is taken to work within the biological limits of the forest. Where silviculture fails it is usually because the rules are not enforced. Dipterocarp rain forests have easy silviculture because they contain numerous light-demanding commercial species which grow fast after canopy opening.

15. Nowadays the canopy opening caused by timber extraction is usually the only silvicultural operation that can be afforded; formerly pre- or post-felling treatments were applied.

16. Big woody climbers are always a problem to silviculture, and sometimes small weedy pioneer trees too, because they are strongly stimulated by canopy opening.

17. It is important to minimize damage to the forest during timber extraction, especially to the forest floor. Careful planning of extraction tracks, directional felling and pulling out logs by cable substantially reduce damage and costs (Fig. 7.29). As a result of international concern such low impact logging is now, at last, attracting interest.

18. Growth rates and longevity are difficult to estimate because rain forest trees lack annual rings. The fastest growing juveniles are most likely to reach maturity. Biomass increment can be increased by silvicultural treatment but is less in natural forest than in plantations. Height growth rates of some pioneers are spectacular.

19. Most parts of the tropics have evidence of large-scale, community-wide disturbance whose frequency and extent differs. There are various sorts of big disturbance. Extensive fires have occurred in rain forests recently in Asia and in the past in South America. Cyclones and other kinds of windstorm also cause catastrophic disturbance. Much evidence now exists of many kinds of community-wide past disturbance by man.

20. The greater the disturbance of a climax forest the longer it will take to recover. Climax species are slow to recolonize land that has been totally cleared unless seed trees remain.

21. The rain forest canopy is in a state of continual flux and different tree species have different recruitment and death rates. It is difficult to know whether a climax forest maintains the same species composition over several forest growth cycles because sample plots are too small or too short-lived to sample more than a tiny part of the spatial and temporal flux.

8

Nutrients and their cycles

This chapter investigates the effects of human usage of tropical rain forests on the ecosystem's capital stock of plant mineral nutrients and on their cycles. First the nutrient cycling aspects of shifting agriculture are described, then nutrients and their cycles in primary rain forest. Based on this we discuss the effect on nutrients of different intensities of forest exploitation and the limits of sustainable utilization. The chapter concludes with discussions of the role nutrients might play in determining the distinctive structure and physiognomy of the heath forest and upper montane rain forest formations.

8.1. SHIFTING AGRICULTURE

The trees and other vegetation are felled to lie uniformly; they form a fuel bed which is allowed to dry for six or more weeks ... The fields are burned a few weeks before the end of the dry season ... Fuel breaks are prepared ... to prevent the escape of fire ... At about midday when conditions are hottest and driest [torch bearers] begin to light the fuel bed starting on the ridges. As the fuel burns it causes an indraft, and by the time the fire reaches the lower edges of the fields a strong wind is blowing into the fire ... of 40 to 65 km per hour. Small intense firestorm whirlwinds ... spiral masses of flame up to 100 m tall, developing intense winds at their centres.[252]

Fire is an essential tool of the peasant rain forest farmer. The essence of shifting agriculture (sometimes called swidden agriculture) is to fell a patch of forest, allow it to dry to the point where it will burn well, and then to set it on fire. The plant mineral nutrients are thereby mobilized and become available to plants in the ash. One or two fast-maturing crops of staple food species are grown (Fig. 8.1). Yields then fall and the patch is abandoned to allow secondary forest to grow. Longer-lived species, such as chilli (*Capsicum annuum*) and fruit trees, and some root crops such as cassava (*Manihot esculenta*) are planted with the staples and continue to yield in the first years of the fallow period. Besides fruit and root crops the bush fallow, as it is often called, provides firewood, medicines, and building materials. After a minimum of 7 to 10 years the cycle can be repeated.

There are many variants. Shifting agriculture was invented independently in all parts of the tropical world[253] and has proved sustainable over many centuries. In Asia dry land cultivars of rice are the staple crop; in New Guinea and Melanesia, sweet potatoes or taro (*Ipomoea batatas; Alocasia macrorrhizos, Colocasia esculenta*) are grown in Africa and in the New World; maize, beans, and squash (*Zea mays, Phaseolus* spp., *Cucurbita* spp.) are commonly

Fig. 8.1. Shifting agriculture in Venezuela. A small clearing made by Amerindian farmers, the crops beginning to grow.

planted in mixture. For the later years, in Nigeria it is common to plant oil palm and the two fruit trees *Irvingia gabonensis* and *Pentaclethra macrophylla* (African oil bean)[254] and in southeast Asia bananas (*Musa*) are important.

Crop yields diminish because the soil becomes exhausted (as discussed below) and also because of a build-up of pests, diseases, and weeds. In Sarawak the most pernicious weeds are grasses (including alang alang, cogon or lallang, *Imperata cylindrica*) and sedges, mainly *Cyperus* and *Scleria*, because they spread rapidly from seeds and also from even tiny fragments of rhizome if these are overlooked when the crop is weeded.[255]

The forest regrows partly from coppice shoots and partly from seeds. Farmers have learned that certain species indicate that regrowth will be vigorous, and that if cultivation is continued for too long other species colonize which grow less strongly, or rhizomatous weeds, such as *Imperata*, get too strong a hold to be eliminated by the fallow. Farmers, therefore, know when the field should be abandoned to allow successful regrowth, with the possibility of returning for another cycle of cultivation in a few years' time. It is easier to fell and burn secondary forest than virgin jungle and repeated rotation through an area is often preferred to continual movement into new areas.

It is now realized that shifting agriculture, as traditionally practised, is a sustainable low-input form of cultivation which can continue indefinitely on the infertile soils underlying most tropical rain forest (Table 8.1), provided the carrying capacity of the land is not exceeded. Moreover, many shifting cultivators have great skill and sophistication.[256] The old idea of colonial days that shifting agriculture was a wholly bad practice has been repeatedly refuted, but still dies hard amongst some contemporary politicians. Shifting agriculture involves the rotation of fields rather than crops, and by this means breaks the build-up of pests, diseases, and weeds in a continuously wet climate where there is no cold or markedly dry season.

Shifting agriculture has the limitation that it can usually only support 10–20 persons km^{-2}, though occasionally more (Table 10.1) because at any one time only *c.* 10 per cent of the area is under cultivation. It breaks down if either the bush fallow period is excessively shortened or if the period of cultivation is extended for too long, either of which is likely to occur if population increases and a land shortage develops.

There is, however, another mode of shifting agriculture which is totally destructive (Fig. 8.2).

Table 8.1 Soils of the humid tropics and their extent grouped by main features

	Percentage of area
Old infertile loamy and clayey soils (oxisols and ultisols)	63
Relatively more fertile, less weathered soils:	
of locally less leached conditions (alfisols and vertisols)	4
on mainly alluvial lowlands (fluvents and aquepts)	12
on volcanic ash (andisols)	1
on steep slopes (tropepts and lithic soils)	11
Infertile sands (spodosols and psamments)	7
Infertile peats (histosols)	2

Based on Sanchez (1976); Vitousek and Sanford (1986) Tropical soils are far more diverse than was once believed. Although over half the tropics have infertile soils, only a small percentage are extremely infertile and, contrary to popular perception, there are also big areas of relatively fertile soils. There are two major schemes of names and classification. Those of *Soil Taxonomy* as used here, originated in 1975 by the United States Department of Agriculture and updated. For equivalents in the FAO/UNESCO *Soil Map of the World* system (1974 and updated) and further information on tropical soils consult Burnham in Whitmore (1984*a*), Richter and Babbar (1991), and Baillie in Richards P.W. (1996).

Farmers fell and burn the forest and grow crops on the released nutrients for several years in succession, continuing until coppicing potential and the soil seed bank are exhausted, pernicious weeds invade, and soil nutrients are seriously depleted. They then move on to a new patch of virgin forest. This is happening, for example, in parts of western Amazonia, practised by peasants from the high Andean plateau who have moved down the eastern slope of the Andes, along new government-sponsored roads to escape overcrowding, and have no prior experience of forest-farming. In these countries the act of forest clearance gives title to the land. After growing a few crops they sell-out to a pastoralist who attempts to raise cattle. Thus there is a moving front of cultivation with poor pasture behind invading Amazonia from its western foothills. In mainland southeast Asia the Hmong people have practised migratory destructive shifting agriculture for centuries.[257]

How shifting agriculture works[258]

The above-ground biomass of a forest contains plant mineral nutrients and these are mobilized when it is burned. Nitrogen and sulphur are volatilized in smoke, but the other nutrients mainly remain in the ash and are available to plants. Nutrient cations replace H^+ so raising the pH, and this reduction in acidity is perhaps the single most important change caused by burning. There is a reduction to about three-quarters in soil organic matter, progressive decomposition of which releases more nutrient cations but also reduces cation exchange capacity. The soil becomes more acid through loss of cations by crop-uptake, leaching, surface erosion, and aluminium becomes soluble as Al^{3+}. This ion is toxic to plants, and crop plants are especially sensitive. The important nutrient, phosphorus, combines with the soluble aluminium and with iron to form insoluble compounds and becomes unavailable to the crop plants, for which it is often critically deficient. This whole ensemble of changes in the soil leads to a decline in crop yields.

The native woody species of the bush fallow are less sensitive than crops to acid soils and to soluble aluminium. They also have deeper roots than herbaceous crops. It is likely that some are able to unlock phosphorus and utilize it, perhaps using the mycorrhiza with which their roots are infected. Thus the regrowth forest is able to thrive even though crops cannot. Nutrients are progressively restored to the above-ground biomass as the regrowth forest develops, as shown in Fig. 8.3. Soil organic matter builds up and improves the soil structure. Secondary forest continues to increase in biomass for many decades, but nutrients accu-

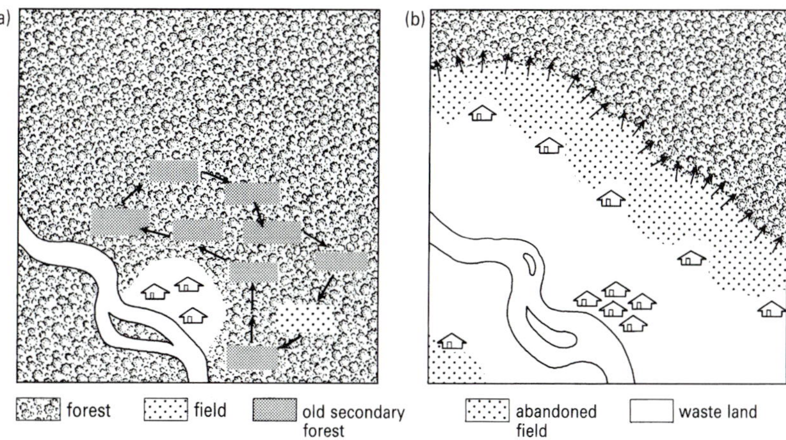

Fig. 8.2. (a) Sustainable (cyclic) and (b) unsustainable (invasory) shifting agriculture. (Rijksen 1978, fig. 146.)

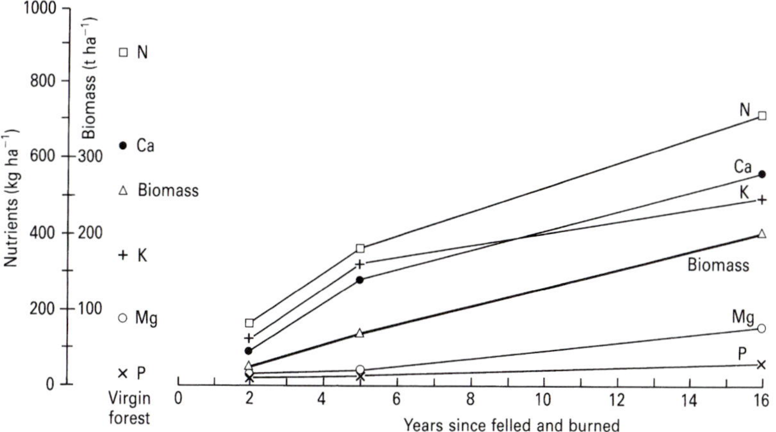

Fig. 8.3. Restoration of nutrients and biomass in forest regrowing after felling and burning. Tropical semi-evergreen rain forest, Colombia. (Data of Fölster *et al.* 1976; Jaffré 1985 gives further examples.)

mulate most rapidly over the first decade or so and by 8–10 years of age may approach their former levels. This is because the concentration of mineral nutrients in the parts of a tree is leaves > twigs > branches > trunks, and most of the leaves and twigs are restored in the first few years. Later on the main growth is as the addition of boles and branches.

Some farmers improve the efficiency of the bush fallow. In northeast India the alder tree *Alnus nepalensis* is planted.[259] This has the added bonus of strongly coppicing, so quickly regrows in the fallow. In the highlands of New Guinea *Casuarina oligodon* is planted for use as firewood. Both *Alnus* and *Casuarina* fix atmospheric nitrogen via root nodules inhabited by the fungus *Frankia*. In Nigeria *Acioa barteri*, *Anthonotha macrophylla*, and *Dialium guineense* are all planted to improve the nutrient status of the fallow. In northeast India natural fallow rapidly builds up potassium in the herbaceous climber *Mikania micrantha*. Later in the succession the same role is played by the bamboo *Dendrocalamus hamiltonianus*.

8.2. NUTRIENT POOLS AND CYCLES IN PRIMARY RAIN FOREST[260]

The study of plant mineral nutrients in tropical rain forest ecosystems, both above and below ground, has been slow to develop. There are

practical problems in the assay of nutrients in the different compartments, leaves, twigs, wood, and bark. The above-ground biomass of mature phase primary lowland forest is commonly about 300–500 tonne ha^{-1} and a sophisticated system of subsampling must be devised. A considerable fraction of the root biomass is in fine roots which are very difficult to extract (p. 57). Different standard chemical methods of assaying soil nutrients are not directly comparable. There are now enough studies, despite these difficulties, for the glimmerings of a coherent picture to be visible.

Nutrient capital

The division of the ecosystem nutrient capital between above- and below-ground compartments is shown for three lowland and two lower montane forests in Fig. 8.4. Early research led to the belief that nearly all the mineral nutrients in tropical rain forests are in the above-ground biomass and, despite much evidence to the contrary, this view is still sometimes expressed. It is certainly the case for the soluble cations K^+, Ca^{2+}, and Mg^{2+} in the Brazilian forest as shown in Fig. 8.4, but the others have a substantial fraction of their nutrients in the rooting zone 0–30 cm.[261] Note however that the significance of the recorded levels of nitrogen and phosphorus in the soil is difficult to evaluate because much is probably in forms unavailable to the plants.

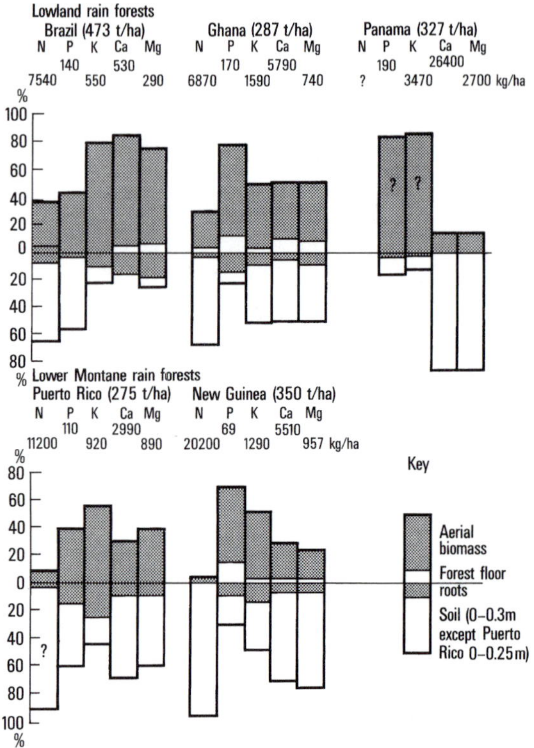

Fig. 8.4. Distribution of inorganic nutrients above and below ground in various tropical rain forests. Biomass in parentheses. (Whitmore 1984*a*, fig. 10.2.) Note that the popular belief that most of the nutrients of a tropical rain forest are in the biomass is seldom true.

Nutrient cycles

The main flows of nutrients are shown in Fig. 8.5, using the example of a lower montane rain forest in New Guinea. There is input from rainfall of nutrients in solution as well as by particulate aerosols. It was noted above that forest burning causes loss of nitrogen and sulphur in smoke. Some cations are also lost as aerosols. However, what one shifting farmer loses another one nearby gains when the air is washed clean by rain. Some rain reaches the ground by trunk flow but, except after heavy storms, 99 per cent of water reaching the ground penetrates the canopy as throughfall. As it does so nutrients are leached from the leaves, so the concentration in throughfall is greater than in rainfall. In the New Guinea example (Fig. 8.5), the soluble cations K^+ and Mg^{2+} were enhanced about 9 times, Ca^{2+} and phosphorus 5 times, and nitrogen 4.6 times. The enhancement of nitrogen (unusually high in this forest) was believed to be due to leaching from nitrogen-fixing epiphyllous algae (p. 53).

Besides nutrients in throughfall and stemflow, the forest floor also receives them from litterfall. This is conventionally divided into two components: fine and coarse litter. Fine litter is mainly leaves but also includes flowers, fruits, and fine twigs. It is relatively easy to set up a system of

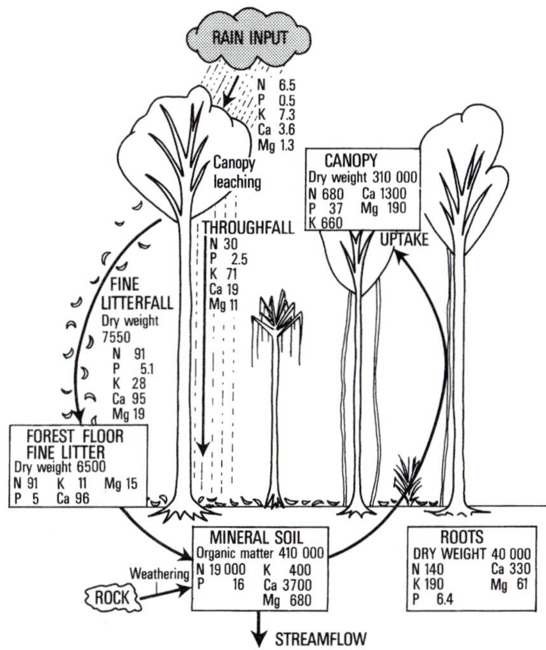

Fig. 8.5. Simplified diagram of inorganic nutrient cycling in a rain forest. Figures (in kg ha^{-1} or kg ha^{-1} $year^{-1}$) are for the lower montane rain forest at Kerigomna, New Guinea. (After Edwards 1982 in Whitmore 1984*a*, fig. 10.1.)

traps, for example fine nylon-mesh trays, and to obtain a statistically satisfactory record of fine litterfall over a year or two of study. Coarse litter, which is branches, limbs, and falling tree trunks, is very heterogeneous in both space and time and it is much more difficult to get a meaningful record, although in amount coarse litter may exceed fine litter.

After the litter has decomposed (see below) the nutrients pass into the mineral soil and may be taken up by the roots or leached out into streams. There may also be nutrient input into the soil from the breakdown of soil minerals. These may either be part of the rock on which the soil lies or in some places may be volcanic ash which has been carried in after eruptions, perhaps from many hundreds of kilometres away.

Most of the plant roots are in the top 0.1–0.3 m of the soil (p. 51). This layer of roots is important for the uptake of nutrients which enter the soil surface. Its disruption, for example, by a careless logging operation (pp. 135–6), thus has serious consequences. Most tree species have mycorrhizas and these enhance nutrient uptake, especially phosphorus, though to what extent remains unquantified.

All rain forests receive small amounts of nutrients in rainfall. Some nitrogen is converted to nitrates in thunderstorms. For other nutrients, the amounts received are highest near the sea and very low a long way inland, for example, in central and upper Amazonia.[262] There is an important distinction between rain forests on deep soils which receive nutrients solely in rainfall and others with soil parent material within the rooting zone which also receive them from that material. In the former, nutrient cycles are almost closed, and recycling is very important. In the latter the cycles are more open. It has also been shown that the amounts of most nutrients lost in streamwater vary with the lithology of the catchment. Losses are lower from infertile soils (Fig. 8.6) especially from those of Amazonia. There is also less annual nutrient loss from forests in seasonal climates. Phosphorus does not follow this pattern, because it always becomes immobilized, nor does nitrogen, which is very difficult to analyse due to its biological conversion to and from the gaseous state.

Table 8.2 shows the much higher concentrations of certain major nutrients in young shallow soils in Sabah, *c.* 0.4 m thick, contrasted to old deep soils in Malaya, 2–20 m thick, where the soil parent material is the same but is no longer within rooting depth.

Many of the world's rain forests are on rugged, hilly terrain where the soil is shallow and continually being rejuvenated by creep and landslip. Unless the parent material is very low in nutrients (as are the sedimentary soils of parts of Sarawak), these ecosystems have a continual input of nutrients from decomposing rock.[263] By contrast, rain forests on flatter terrain or older deeper soils have no such input. Even if there is addition of nutrients at the bottom of the soil

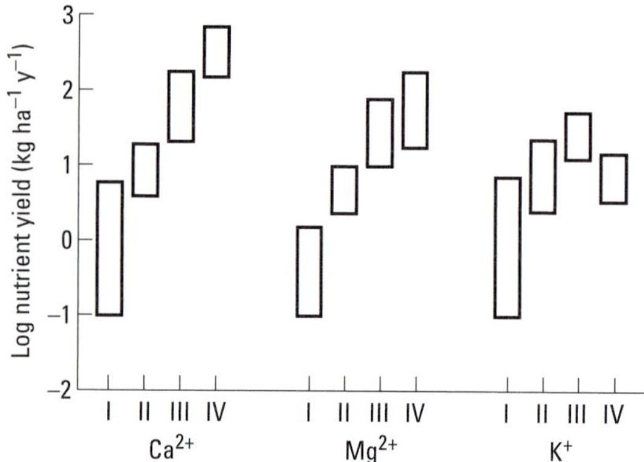

Fig. 8.6. Nutrient losses in drainage water from soils of different fertility under lowland rain forest. (Bruijnzeel 1991, fig. 2d.)

I, very infertile soils (spodosols, highly depleted oxisols).
II, infertile soils (oxisols, ultisols).
III, fertile soils (inceptisols, andisols).
IV, fertile soils (mollisols, vertisols).

For all three nutrient cations the nutrient cycle becomes increasingly more open (i.e. losses become greater) with increasing soil fertility. The very infertile soils (I) are mainly Amazonian; the most fertile (IV) are over limestone which accounts for their high losses of Ca^{2+} and low of K^{+}.

profile, in old, deep soils it is beyond rooting depth, outside and uncoupled from the ecosystem. The most extensive very deep soils occur in central Amazonia, where the forests lie on a huge sedimentary Tertiary plain.

Comparisons of quantities of nutrients in fine litterfall and in leaves still attached to the trees show that lowland rain forests cycle little phosphorus in their litterfall. Montane rain forests, by contrast, cycle little nitrogen. The implication is that phosphorus is a limiting nutrient in lowland ecosystems (as is well-known to tropical farmers), but that in the mountains nitrogen is more strongly limiting.[264]

Decomposers

Litter on the forest floor must decompose before its nutrients become available for plants. There have now been many studies made on the rates of fine litter fall[265] and its decomposition. Table 8.3 gives a sample. In the lowlands decomposition occurs in 4–12 months,[266] similar to the rates reported for other climates. The common supposition that litter disappears from the floor

Table 8.2 Plant mineral nutrients within the soil

Parent material	Nutrient concentration (m equiv 100 g^{-1})				
	Ca	Mg	K	Na	Total
Young soils, 0.4 m thick, Sabah (sampled 0.1–0.4 m)					
Sandstone and shale	0.16	0.09	0.05	0.06	0.36
Limestone	52.35	1.27	0.26	0.13	54.01
Serpentine	36.35	11.40	0.02	0.42	48.19
Basalt	5.67	8.53	0.06	0.42	14.66
Old soils, 2–20 m thick, Malaya (sampled 0.8–1 m)					
Shale	0.19	0.08	0.20	0.24	0.82
Limestone	0.42	0.11	0.24	0.22	0.99
Serpentine	0.30	0.38	0.33	0.16	1.21
Basalt	0.20	0.04	0.12	0.14	0.72

From Burnham in Proctor (1989)
The amounts of nutrients in the young, shallow soils differ between parent materials (e.g. high Ca over limestone) and are mostly greater than in the old, deep soils where parent material has little or no influence.

Table 8.3 Fine litter: fall, average amount, and disappearance in various rain forests

Forest formation and place	(a) Fine litterfall (tonne ha^{-1} $year^{-1}$)	(b) Forest floor (tonne ha^{-1})	(c) Decay factor, *k* (a)/(b)
Lowland evergreen rain forest:			
Mulu, Sarawak: ridge	7.7	5.9	1.3
valley alluvium	9.4	5.5	1.7
Pasoh, Malaya	10.6	3.2	3.3
Penang, Malaya	7.5	4.9	1.5
Manaus, Brazil	7.6	7.2	1.1
Lowland semi-evergreen rain forest:			
Barro Colorado, Panama	13.3	11.2	1.2
Kade, Ghana	9.7	4.9	2.0
Heath forest:			
Mulu, Sarawak	8.1	6.1	1.2
Forest over limestone:			
Mulu, Sarawak	10.4	7.1	1.5
Freshwater swamp forest:			
Tasek Bera, Malaya	9.2	4.8	1.9
Montane rain forests:			
New Guinea, 4 close sites (*c.* 2500 m)	6.2–6.6	4.2–6.6	1.0–1.5
Colombia (1630 m)	10.1	16.5	0.6
Costa Rica, Volcan Barva†			
1000 m	6.6	4.2	1.6
2000 m	5.8	5.2	1.1
2600 m	5.3	6.3	0.8

Mainly from Whitmore (1984*a*) table 10.8; Anderson and Swift in Sutton *et al.* (1983) table 1
†Heaney and Proctor (1989)

of tropical rain forests uniquely rapidly is not borne out by these figures. There is more variability in the amount of litter on the forest floor than in the rate of fall, which implies that differences lie mainly in the rate of decomposition. The scanty data collated in Table 8.3 suggest that decomposition is slower in montane forests (see p. 171 below).

In most soils the main decomposers are litter-feeding invertebrates. Termites are a major component of this so-called soil macrofauna,[267] but in montane forests earthworms replace them. On Gunung Mulu, Sarawak, no termites were found above 1860 m. Both termites and earthworms comminute litter as well as contributing to its decomposition. There are five families of lower termites. All have protozoan symbionts in their guts which decompose cellulose. The single higher termite family, Termitidae, dominate many rain forest soils and have bacterial gut symbionts which perform the same role. One of its subfamilies, Macrotermitinae, is widespread in rain forests, and uses mainly fresh plant material to cultivate subterranean fungus combs or 'gardens'. The fungus breaks down both cellulose and lignin and the termites feed on the food bodies that it produces. Macrotermitinae are especially frequent in more seasonal and drier climates, because the fungus combs can digest fresh, dry food-materials not utilizable by other termite groups. In Sarawak they were shown to be commoner in logged forest. Many are serious pests, attacking food- and tree-crops as well as timber buildings. Alterations in the decomposer

community which increase their abundance may have undesirable consequences.

8.3. PRACTICAL IMPLICATIONS

Humankind makes use of tropical rain forests or their soils in many ways as will be explored in Chapter 10. An important requirement for sustainable utilization without degradation is to work within the ecosystem nutrient budgets and cycles, and in the case of forest utilization not to disrupt these by excessive damage. Knowledge has developed far enough to indicate some of the major constraints.

Utilization of forests or forest soils involves removal of products and their nutrient content. The nutrient capital will run down if removals exceed inputs. Thus there may be substantial differences between forest ecosystems with closed and open nutrient cycles.

Timber removal

Human impacts on rain forests vary in severity. The selective removal of the boles of one or a few trees per hectare for timber leaves most of the nutrient capital behind. This is because nutrients are most concentrated in the branches, twigs, and leaves. With increasing volume of timber removed, the amounts of nutrients removed also increase. Loss from the ecosystem is reduced if bark is left in the forest and only the wood extracted. Evapotranspiration of the forest is reduced when trees are felled, so more rain water percolates through the soil. There is also a flush of soluble nutrients as destroyed vegetation decomposes, and there are fewer plants to take up nutrients. The overall result is extra losses from the ecosystem by leaching into stream water. Table 8.4 shows the budget for one of the few forests where leaching has been measured. In this Malayan dipterocarp forest it will take 20 to 60 years for the input in precipitation to restore above-ground ecosystem nutrients to their former amounts. The soil is too deep for there to be input from decomposing rock. The few other studies of nutrients removed in harvested timber show similar results to these in Malaya.[268] In most rain forests there are considerable amounts of nutrients in the below-ground part of the ecosystem (Fig. 8.4). Moreover, we have no knowledge yet of whether the trees need all of their nutrient content, or would grow as well with less.[269] Thus timber extraction at the usual present day levels of 80 tonne ha^{-1} or less, under a poly-

Table 8.4 Approximate above-ground nutrient budget for commercially logged lowland dipterocarp rain forest, Bukit Berembun, Malaya

	Nutrient concentrations (kg ha^{-1})			
	K	Ca	Mg	Total N
Outputs				
Removed with timber (60 m^3 ha^{-1})*	45	200	20	70
Extra leaching	75	30	15	?
Total outputs	**120**	**230**	**35**	**70+**
Annual inputs†				
Rainfall and aerosols	**5.7**	**3.8**	**0.7**	**11.3**
Years to recover	20	60	50	5–10

Based on Bruijnzeel (1992) table 2

* Taken as either 40 per cent of commercial stocking or 10 per cent of biomass (all stems > 5 cm diameter at 600 kg m^{-3} air dry density)

† It is assumed that this forest has no nutrient input from decomposing rock

cyclic system on a cutting cycle of 30 years or more, is unlikely to create a 'nutrient desert' except on particularly infertile sites, which include Malesian peat swamp forests (section 2.3) and the *terra firme* forests of the central Amazon. The current extremely heavy logging (100 or occasionally 200 m^3 ha^{-1}) of the huge dipterocarp forests of eastern Borneo (and formerly of the southern Philippines too) also give cause for serious concern (Fig. 7.24).

The most serious effect on forest nutrients of logging is likely to result from damage to the soil surface because this causes disruption to the root layer, to seeds in the soil, to the roots and stems that provide sucker shoots, and to the decomposer community. Infiltration of precipitation will be reduced. All these sorts of damage will impair recovery of the ecosystem after logging, including nutrient uptake into the biomass. Where the top-soil is scraped offor excessively compacted, the effects are strongest. On the Windsor Tableland in north Queensland, snig tracks showed losses of soil organic matter and plant nutrients which persisted for at least 4 years.[270]

Chips

Clear felling of inland rain forest and total utilization of all boles, limbs, and branches, including their bark, as a source of wood chips for paper manufacture is still uncommon. Exploitation at Bajo Calima on the Pacific lowlands of Colombia, and Gogol on the north coast of Papua New Guinea began in the 1970s and in Indonesia and at Sipitang in southwest Sabah in the 1980s.

Careful study at Bajo Calima has shown complete recovery of the former amount of biomass by extremely vigorous and dense forest regeneration.[271] Here, extraction is by aerial cable to well-spaced roads and there is no damage to the forest floor, which after felling resembles (on an extremely coarse scale) the stubble left on a cereal field after the crop has been harvested. The secret of success probably lies in the mode of extraction adopted. No studies have been made, however, of biomass nutrients.

The original forest at Gogol was rich in pioneers and these regrow vigorously.[272]

In Indonesia, mainly so far in Sumatra and Kalimantan, logged lowland and lower montane forest is being cleared and pulped and the land converted to industrial tree plantations (Hutan Tanaman Industri, HTI), mainly of *Acacia mangium* (Fig. 8.7). For the paper mill of Indah Kiat in Riau province peat swamp forest (pp. 24) is also being logged, although peat over 1–3 m deep is very acid, extremely low in nutrients, and totally unsuitable for plantations.

Fig. 8.7. Paper factory, Riau, Sumatra.
This factory will eventually use plantation grown wood. In its early years it is being supplied by wood salvaged during clearance of logged lowland rain forest. In the foreground are piles of half grown *Shorea* from cleared forest awaiting pulping. This enterprise and Aracruz in Brazil are, at over a million tons a year each, the world's largest pulp mills. Indonesia also has other huge mills already in production or planned.

At Sipitang the forest had been logged, it then burned in the Great Fire of Borneo (p. 148) and, after salvage harvesting for pulpwood, plantations of *Acacia mangium* are being established.

Plantation forestry

This also makes great demands on the ecosystem nutrient capital, because a high fraction of the above-ground biomass is removed as harvest at frequent intervals. This is shown in Table 8.5 for plantations of *Agathis dammara* and *Pinus merkusii* in central Java, grown on a 40- and 25-year rotation, respectively. It can be seen that harvesting of stems alone causes similar nutrient depletion, but when branches are also taken there is higher depletion at the *Agathis* site. When the whole biomass is removed depletion is higher still and at the *Agathis* site more is removed than the total inputs from both the atmosphere and rock weathering over the whole 40-year rotation. The soils here are andisols, derived from recent volcanic deposits, and unusually fertile (Table 8.1). On the far more extensive oxisols and ultisols soil nutrient capital is much less. One, or perhaps a few, successive, plantations may give reasonable economic yields. Nevertheless, there is no such thing as a 'free lunch' and the repeated harvesting of timber will inevitably sooner or later deplete ecosystem nutrients as more are removed than are added in rainfall. It is very likely that soil physical properties will also deteriorate. Studies on the nutrient capital in plantations are urgently needed, especially under second and later rotations.[273] The early

Table 8.5 Nutrient depletion by plantation trees, central Java

	Nutrient concentration (kg ha^{-1})							
	Ca		Mg		K		P	
Agathis dammara over a 40-year rotation								
Inputs:								
Atmosphere	395		160		385		35	
Weathering rock	3060		1565		1575		200	
Total inputs	3455		1725		1960		235	
Outputs:								
Stemwood harvest	659	19%	153	8%	437	22%	107	45%
Stem, branch and bark harvest*	1679	48%	329	19%	854	43%	204	87%
Total harvest of trees and undergrowth†	3556	103%	786	45%	1894	97%	372	158%
Pinus merkusii over a 25-year rotation								
Inputs:								
Atmosphere	245		70		180		70	
Weathering rock	1870		965		1015		82	
Total inputs	2115		1035		1195		152	
Outputs:								
Stemwood harvest	413	19%	97	1%	219	18%	73	48%
Stem and branch harvest	672	32%	127	12%	274	23%	83	55%
Total tree harvest†	1558	74%	266	26%	666	56%	134	88%

Data from Bruijnzeel (1984) table 4; Bruijnzeel and Wiersum (1985) table 4
* Bark used for fuel
† Twigs, leaves, and undergrowth used for animal bedding

stages of a plantation are critical. During clearance, burning and planting the volume and nutrient concentration of water percolating through the soil is increased and losses via leaching and erosion are high. At Sipitang, Sabah, a 9-year paired catchment experiment has investigated the difference between clearance with minimal disturbance (manual log removal, no burning) and normal practice (tractor extraction, burning before planting).[274] The results are dramatic. Minimum disturbance has resulted in 70 per cent greater growth of *Acacia mangium* during the first 3.5 years, only 4 per cent soil compaction on the parts affected by vehicles (against 24 per cent), and half the erosion and nutrient loss in streamflow. Plantation establishment costs were higher without burning, but weed growth was less and weeding cheaper, so total costs were similar. The results of this classic experiment, vividly demonstrating the benefits of low-impact plantation establishment, have widescale relevance.

Besides potential problems of nutrient depletion, all plantation forests face the build-up of pests and diseases. These are especially troublesome in perhumid lowland tropical climates and were beginning to be serious at the extensive plantations at Jari in the lower Amazon by the mid-1980s, some 20 years after the venture started.[275] To leave belts of the original forest along streams is good practice. These may reduce erosion and leaching into streams and act as barriers to the spread of pests, diseases, and fire. They may harbour insectivorous birds which glean plantation pests, thus acting as biological control agents. If such belts are large enough and joined to each other and to blocks of relict forest they may have a useful role to play in the conservation of the plants and animals of the original forest (pp. 233–4).

Agriculture

Conversion of rain forest to agriculture is even more extreme than conversion to plantation forests. Nutrient cycles are completely disrupted, especially uptake by plants and from decomposition. A study at Yurimaguas in the Peruvian Amazon found that the biomass of decomposers dropped from 54 to 3 gm^{-1} in the conversion from rain forest to arable agriculture.[276]

Traditional agriculture of the shifting kind makes use of natural ecosystem processes, as described in section 8.1. There is scope for further enhancing the efficiency of the bush fallow period by planting species that are good 'nutrient pumps' from soil to biomass. It was found in the Philippines that *Trema orientalis* restores phosphorus and *Melastoma* cf. *polyanthum*, restores potassium to the above ground biomass.[277]

Continuous cultivation of short-lived crops is difficult to sustain without inputs of nutrients, though easier on more fertile soils. Experiments on ultisols at Yurimaguas showed that continuous cultivation of rice was not possible even with the addition of fertilizer, but yields could be maintained if crops were rotated, provided that fertilizers were added (Fig. 8.8). In Nigeria crop yields have proved sustainable for 10 years or more on small farms on alfisols (which are intrinsically more fertile than ultisols Table 8.1): rotating maize, cowpea, cassava (*Zea mays, Vigna unguiculata, Manihot esculenta*), and a cover crop, with minimum tillage and the judicious addition of fertilizer.[278]

Oxisols and ultisols are the most widely occurring lowland tropical soils and amongst the least fertile (Table 8.1). It has been suggested for the Brazilian Amazon that conversion to agriculture should be confined to alfisols and to the alluvial soils that are found along the major rivers. The loss of alluvial, flood-plain swamp forests would, however, have serious consequences for the fish on which many people depend for a livelihood, as described on pp. 212–13.

Agroforestry

Mixed cultivation of trees with food crops is another possibility for maintaining yields (Fig. 8.9). This has been practised since time immemorial by peasant farmers, as for example the 'home gardens' of Central America, Java (Fig. 10.2), the Philippines, and Sumatra.

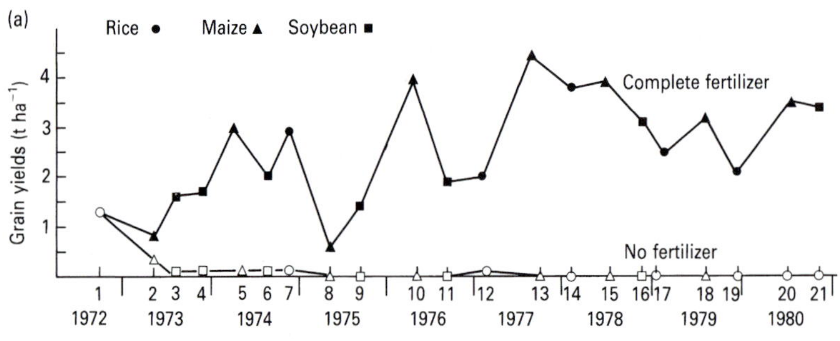

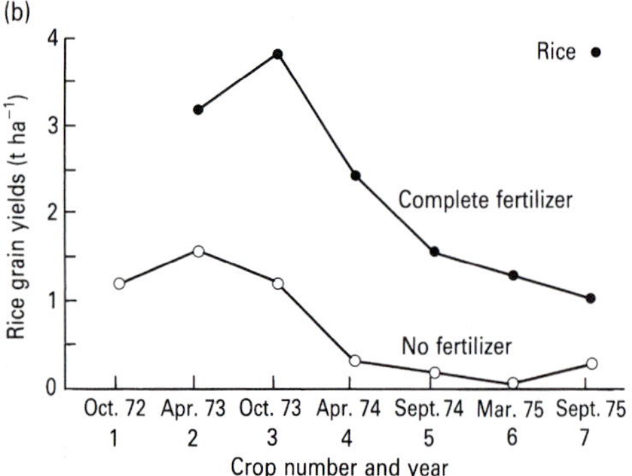

Fig. 8.8. Agriculture on nutrient-poor lowland rain forest ultisols at Yurimaguas, Peru. (a) Yield is sustainable when crops are grown in rotation, but not (b) if rice is grown continuously. (After Jordan 1987, figs. 6.10a, 6.11.)

Japanese immigrants living at Tomé Açu south of Belém in the Brazilian Amazon have developed over the last few decades a particularly intricate farming system which maintains continuous high production on infertile soils.[279] Fast-maturing food crops such as rice, maize, and beans are interplanted with the slow-maturing vines of pepper (*Piper nigrum*) and passion fruit (*Passiflora*) which are trained up tree crops such as cocoa, rubber, coconut, and *Astrocaryum* palm (grown for its edible apex, 'heart of palm'). Vanilla orchid (*Vanilla*), also a vine, is sometimes included. Some farmers specialize in chickens, whose waste, along with all other organic residues, is returned to the fields. Rotation and succession prevent build-up of pests and disease. The cultural discipline traditional to Japanese societies is important for the operation of this close-knit agricultural system.

In the 1980s mixed cultivation was 'discovered' by the aid agencies, who had become disillusioned with rain forest silviculture and tree plantations as vehicles for Third World development, and the term agroforestry was coined. The trees provide browse for domestic animals, fuelwood and lumber for buildings, and sometimes fruit as well. Agroforestry makes use of nutrient cycling by trees, as does shifting agriculture. Trees act as pumps, bringing nutrients into the superficial layers of the soil where shallow-rooted herbaceous crops can utilize them. Part of the nutrients comes via litterfall, and branches may be lopped and used as mulch to enhance this pathway. Part also comes from the decomposition of the fine roots of the trees. These are mycorrhizal and, in the case of Leguminosae and a few others, also have nitrogen-fixing nodules, so are particularly important in enhancing available phosphorus and nitrogen.

One agricultural system developed by agroforesters is so-called alley-cropping,[280] where trees are planted as hedges with belts of food crops grown in the intervening lanes or alleys. There is still considerable scope for improvement of agroforestry. For example, tree species are needed which have deep roots and will act as good nutrient pumps on infertile oxisols and ultisols.

Fig. 8.9. Cultivation of crops in mixture in the Atlantic rain forest zone of Brazil. Here pineapple; pepper, growing up the leguminous tree *Erythrina poeppigiana*, which provides shade for cocoa; and the palm *Bactris gasipaes* are all being grown.

8.4. HEATH FOREST—A FRAGILE ECOSYSTEM[281]

Heath forest is perhaps the most distinctive and easily recognized of all the lowland rain forest formations. The soils are typical podzols. Below a superficial layer of peat or mor humus is a layer, sometimes several metres thick, of bleached white silica sand overlying a hard blackish humus- or iron-pan. The forest itself has striking structure and physiognomy as described on pp. 20–4 and shown in Figs. 2.12, 2.13.

Heath forest sites cannot sustain agriculture. In Borneo this forest formation is called kerangas, land that will not grow rice. In Brazil the Rio Negro, which drains a huge area of heath forest, differs conspicuously from the rivers running through other kinds of forest in the virtual absence of riverine farming settlements. Until recently it was believed that heath forest soils were more infertile than others and that extreme oligotrophy (nutrient shortage) was the reason many heath forest plants have small very leathery leaves ('scleromorphic microphylls') and why crops cannot be grown on heath forest soils. Outside the tropics it has been known for many years that sclerophylly is associated with nutrient poverty, for example the shrubby forests around Sydney in eastern Australia and the shrublands that develop on many north temperate peat bogs.[282] The correlation between this physiognomic feature (one of the most striking attributes of heath forest), crop failure, and nutrient shortage seemed so rational that no one noticed that data on soil nutrients had not been collected. Furthermore, the Bornean heath forests have numerous pitcher plants (*Nepenthes*) and ant plants, groups that both have special means of obtaining nutrients. It seemed intuitively obvious that oligotrophy 'explained' the distinctive nature of heath forests; but intuition can be dangerously misleading in science. In 1958 P.W. Richards collected samples of heath forest soils and some adjacent ones from Brunei. The chemical analyses showed no great differences. This was so surprising that he said in his publication that he thought the samples must have become contaminated. The same results were obtained 20 years later in Kalimantan and at Mulu, Sarawak. All three data sets show that the soils under virgin heath forest are not systematically poorer in plant mineral nutrients than those under other forest formations close by. The same applies at the important San Carlos research area in southern Venezuela in the headwaters of the Rio Negro where in fact all formations occur on exceedingly infertile soils.[283] Virgin heath forest soils do, however, differ from others in being

unusually acidic, with pH less than 4.0, and in having low buffering capacity due to low concentrations of iron and aluminium sesquioxides. Polyphenols are abundant in heath forest leaves and litter, and these may be toxic or inhibit nutrient uptake when they leach into the soil.

Once heath forest is felled and burned the soil very quickly degenerates. The surface humus layer is either eroded, burned, or oxidized. The small amount of clay in the soil washes down the profile to leave almost pure silica sand, which unlike the clay and humus has no electrical charge to which nutrient ions can attach.[284] The soil may become even more acid. The extreme acidity is probably toxic to roots of crop plants and possibly other plants as well.[285] Without vegetation cover or a surface humus layer the white sand gets extremely hot in the sun. It now seems that this rapid and easy soil degradation is the reason why agriculture is impossible and why if heath forest is felled and burned it is replaced by bare sand with patches of shrubs or scattered trees, which are called padangs in the Eastern tropics and campinas in Amazonia. Heath forests are a genuinely fragile ecosystem and are easily and irreversibly degraded by human disturbance. The removal of a very few trees per hectare for timber has proved a non-degrading form of utilization of heath forests on the east coast of Malaya. Under this light selective logging regime the forest canopy is scarcely opened and the soil surface only slightly disrupted. Here, humans are working within the rather narrow limits that the fragile heath forest ecosystem can sustain.

Heath forests occur in perhumid tropical climates where dry seasons are absent or short and sporadic. It came as a revelation when in 1969 E.F. Brunig demonstrated for Sarawak, that, if instead of taking the rainfall for calendar months it was taken for successive 30-day periods through the year, and water balance was calculated, then heath forest sites can be seen to suffer periodic water shortage; this is especially seen for those near the coast which are under the influence of sea breezes which are somewhat desiccating. This is because the sandy heath forest soils are freely draining, have a low capacity to retain water, and are often shallow over the impermeable iron or humus pan. Similar periodic water shortage has been shown to occur even at San Carlos where mean annual rainfall is 3565 mm, and over 400 mm in all months except December and January when it is 150 mm. This is a very wet climate even by humid tropical standards. Nevertheless, dry spells of 3–6 days are not rare. Study has shown that during such dry spells the water-table falls. At San Carlos the heath forest occurs on low domelike hills and decreases in stature towards the summit, where the water-table drops furthest.[286]

It now seems probable that the very striking characteristics of heath forest are most likely to be adaptations to survive drought periods, even if these are infrequent. The transpiration of water is a mechanism whereby leaves are kept cool, below the lethal temperature at which protoplasm denatures (i.e. their cells are literally cooked). A whole set of the physiognomic features of heath forest either minimize heat load or optimize cooling. This applies to highly reflective shiny or pale coloured leaves. Small leaves (which are so characteristic of this forest) have a greater air flow over them than large ones. Many species have the leaves held vertically or nearly so. At San Carlos 55 per cent of species have this adaptation and an excess of leaf temperature over the air of only 1.8–5.4 °C. Experiments on the leaves of a number of species at Bako, Sarawak, found that the eight heath forest species investigated were no different in their ability to resist desiccation than those of adjacent evergreen rain forest. Other adaptations reduce water loss from the heath forest as a whole. These are the clustering of leaves, the clustering of leafy twigs into dense subcrowns, and perhaps also the very uniform forest canopy top surface which by its smoothness reduces turbulent mixing of air.

There still remains much research to conduct on heath forest to substantiate these indications. For example, stature varies from only a few to over 30 m tall and the extent to which this is

determined by water relations or by soil mineral nutrients is not known. Nor, at the extreme, is it known what determines the occurrence of natural open shrublands or grasslands. There is evidence at Mulu, Sarawak, that although the heath forest has about the same amount of litterfall as other formations nutrient concentrations in the litter are lower, which suggests the plants are retaining more of the nutrient capital, and this is most marked for nitrogen.[287]

The heath forest formation is ripe for further studies on its water and nutrient relations to substantiate the fragmentary studies just described.

8.5. THE UPPER MONTANE FOREST ENIGMA

Besides the studies on heath forest discussed above there have also been considerable efforts to relate the special structural and physiognomic features of the upper montane rain forest formation to attributes of its habitat. As with heath forest the debate has centred on whether nutrient shortage or periodic drought are important.[288] Upper montane forest presents an additional problem that continues to puzzle scientists, namely that the montane forest zones (Fig. 2.7) all occur at higher elevations on big mountains than on small ones, the so-called *Massenerhebung* effect.

A general description of upper montane rain forest and its occurrence was given on pp. 18–20. In structure and physiognomy there are strong resemblances to heath forest and also with the open stunted form of peat swamp forest (p. 24). In Borneo and Malaya a few species occur in more than one of these formations. At its lower edge upper montane rain forest occurs only on ridge crests, interdigitated with the lower montane formation which penetrates upwards along valleys and slopes. The very lowest patches exist on raised knolls with the intervening cols occupied by lower montane forest. Crests and knolls are freely draining sites which receive water from rain but lose it by lateral movement in the soil. This outward movement of ground water carries soluble nutrients with it and, moreover, the soil may dry out in spells of dry weather. Thus the sites of upper montane forest are both oligotrophic and drought-prone. As with heath forest it seems likely that either or both of these factors may determine the occurrence of upper montane forest, and which factor prevails on any given mountain can only be resolved by investigation. At one extreme the summit of Kolombangara, 1662 m, a small island in the Solomons, has excessively impoverished soils, with continual nutrient leaching in a very wet climate (estimated rainfall 8250 mm). There is persistent cloud above 800 m. Oligotrophy is undoubtedly more important than periodic drought. At the other extreme the high, extinct, or quiescent volcanoes of west Java have young and relatively fertile volcanic soils, but drought occurs in all or most years and it is likely that periodic water stress is the more important factor.

The boundary between lower and upper montane rain forest is usually sharp with only a narrow blending zone (ecotone). This boundary often occurs at the bottom of the zone at which cloud habitually develops.[289] Within the cloud the climate is extremely wet with moisture combed from the air by the trees, so-called fog-stripping. The leaf litter on the forest floor is sodden and anaerobic; decomposition is inhibited. Litter accumulates and eventually turns into peat. Bog moss (*Sphagnum*) frequently grows in these waterlogged places and accentuates peat accumulation. Because there is very little decay, nutrients in the litter remain locked up in organic form and unavailable to plants. There are hints (p. 162) that nitrogen may be limiting.[290] Where peat occurs certain upper montane tree species are favoured (including conifers and Myrtaceae) which are believed to facilitate peat development by having slowly decomposing litter. If this does occur, and the data from montane forests in Table 8.3 suggest it does, then the process of peat accumulation is self-reinforcing which will tend to sharpen the boundary with adjacent forests. The level at

which the montane cloud-cap forms, and hence at which upper montane forest develops, depends on local and regional weather patterns. It is characteristically higher on large mountains so this is likely to be part of the cause of the *Massenerhebung* effect which, however, still evades complete explanation. Detailed studies are needed to discover which environmental factors are most important in determining the forest zonation on any particular mountain. With good calibration, hydrological simulations can replace long-term series of observations. Drought may be rare but, as with heath forest, upper montane species must be adapted to survive even rare droughts.[291]

Montane forest zones provide excellent opportunities for the study of plant-environment interactions, especially in water relations and mineral nutrition. One of the difficulties in understanding the key causes of zonation, the complete change in structure and physiognomy, and a concomitant reduction in forest biomass, is that different species grow at different elevations. A potentially powerful but yet not fully explored avenue of research is to investigate those few species that grow in several different montane forest formations, or to work on a small mountain, such as Rakata Island (735 m altitude) in the Krakatau archipelago between Java and Sumatra, which has the same flora from sea-level to summit yet whose forests become stunted upwards.

8.6. NUTRIENTS AND THEIR CYCLES—CHAPTER SUMMARY

1. Shifting agriculture has been invented independently in all parts of the tropics and has proved sustainable since time immemorial. It is a low input system, suitable for infertile rain forest soils. Crops are grown for one or two years and then secondary forest, 'bush fallow', allowed to grow for 8–10 years or more.

2. Shifting agriculture works by using the capacity of trees to grow on acid infertile soils and bring nutrients from the soil into the biomass. During the bush fallow period nutrients re-accumulate in the vegetation (Fig. 8.3), partly from the soil, partly from rain. Crops cannot be grown for long because the soil becomes impoverished and more acid, and pests, diseases, and weeds increase.

3. Shifting agriculture breaks down if cultivation continues too long or the bush fallow is too short. This happens, for example, when population increases too much or when it is practised by unskilled farmers.

4. In many tropical rain forests, most nutrients are fairly evenly divided between above-and below-ground parts of the ecosystem (Fig. 8.4), not nearly all above ground as formerly believed.

5. Nutrients are added to all forests in rain, both in solution and as aerosols. Forests on young, shallow soils also receive nutrients from decomposing rock, but on old, deep soils the soil parent material is beyond the reach of roots, there is no nutrient addition, and these soils are commonly less fertile (Table 8.2).

6. Nutrients cycle through the ecosystem (Fig. 8.5). Rain, as canopy throughfall, and litterfall are the main pathways. There are greater nutrient losses in drainage water from more fertile soils, poor soils have 'tighter' cycles (Fig. 8.6). Fine litter decomposes in a year or less (Table 8.3). Soil invertebrates are important litter decomposers. In the lowlands termites predominate.

7. Sustainable utilization depends on working within the limits of ecosystem nutrient cycles. Shifting agriculture and selective removal of trees for timber do not cause serious permanent depletion (Fig. 8.3, Table 8.4). More complete biomass utilization for wood chips or in plantation forestry (Table 8.5) will deplete ecosystem nutrients unless balanced by rain and soil inputs. These practices may also disrupt the forest floor, and hence the capacity for litter breakdown and forest regeneration.

8. Permanent agriculture destroys forest ecosystem processes and requires addition of fertilizers (Fig. 8.8). The mixed cultivation of trees and crops, agroforestry (Fig. 8.9), uses the

nutrient cycling capacity of trees for agricultural purposes.

9. Podzols under heath forest are subject to periodic drought and are very acid and high in polyphenols. One or more of these factors probably accounts for their very distinctive structure and physiognomy. Virgin heath forest soils are not more oligotrophic than others, but quickly degrade after forest clearance to more or less pure silica sand, and that is what makes them impossible for agriculture.

10. Upper montane forest also has highly distinctive physiognomy and structure which also may be due to oligotrophy or periodic drought, or perhaps to toxic polyphenols in the soil that reduce water uptake.

9

Species richness

One of the most famous features of tropical rain forests is their extreme species richness which has impressed scientists ever since the earliest explorations (Chapter 1). Some species are common, but most are rare: as has been described above (p. 32), in the richest rain forests every second tree on a hectare or so is a different species. Even disregarding the trees, some rain forests are richer in species than any other vegetation on earth; and the richest communities outside the humid tropics have about the same total number of species on small plots as there are trees of 0.1 m in diameter and greater in many rain forests. The way such very large numbers of species have evolved and are packed together has been the driving force for endless speculation, constrained to varying degrees by observation, and commonly involving massive extrapolation. It is one of the most basic and fascinating aspects of these forests and one that energizes much of tropical biology. Given the complexity and diversity of tropical rain forests, generalization can be dangerous. Species richness has numerous components, many of which have been discussed in preceding chapters. Here all the different factors that collectively contribute to species richness are brought together. Concerning animals, the ways numerous species co-exist, and limits to the carrying capacity of the forest have already been considered in Chapter 4. In this chapter we concentrate on plants.[292] Some of the factors described apply equally to animals.

Historical plant geography

Extreme floristic richness involves the co-occurrence of a large number of species drawn from many genera and families. In the ultimate analysis, a region contains only those species that immigration and evolution, plus survival, have enabled to be present.

In Chapter 6 two aspects of historical biogeography were discussed which have importance for patterns of species richness today, namely evolution on the fragments of Laurasia and Gondwanaland and, more recently, fluctuations in climate. For example, a factor contributing to the floristic richness of the Eastern tropics is that taxa of both Gondwanan and Laurasian origin are intermingled. The poor rain forest flora of Africa is believed to be due to stronger desiccation and consequent heavier extinction than has occurred in the other tropical regions.

The African rain forests today have four zones with higher species richness (Fig. 6.17). These are thought to be places that were least

affected by the drier, more seasonal climates of successive Pleistocene Glacial epochs. In tropical America also a series of areas of high endemism and species richness have been detected for many different groups of animals and those families of flowering plants that have been analysed (Fig. 6.16). The areas of endemism and richness are more or less coincident for different groups. As in Africa, they are believed to be Pleistocene refugia to which rain forests and their component biota were confined at Glacial maxima. In South America the physical environment between refugia bears geo-scientific evidence of past seasonality in the climate, which strengthens the argument that rain forests became like islands in a sea of seasonal forest. The Eastern tropics are species-rich, with high local endemism. Unlike Africa and America there are no zones of relative poverty. It was shown in Chapter 6 that this is probably because the relatively species-poor regions have been drowned by rising sea-levels since the end of the last Glaciation.

In Africa tree species numbers on small plots are highest in the refugia. There are too few counts yet from the Amazon basin for a pattern to be detected. It can be predicted that there too tree species richness is greatest in refugia.

Temperate forests also have patterns of varying floristic richness which are explicable in terms of past climatic history, as discussed on pp. 105–6.

Species niches in the forest

Given a rich regional flora, forests are particularly favourable for the co-existence of many species in the same community, because they provide many different niches.

The forest canopy

The forest provides a wide array of different internal microclimates, both horizontally and vertically (section 7.1). There are tree species with different degrees of shade tolerance. The trees themselves are the framework of a wide series of habitats for climbers and epiphytes (Fig. 9.1).

The forest growth cycle

This provides another set of niches, as described in Chapters 2 and 7. Big gaps are colonized by well-dispersed, light-demanding tree species, which are eventually replaced by shade-bearers. There are many differences between species in fruiting frequency, dispersal efficiency, germination, seedling survival and establishment, and light requirements of juveniles. The same processes operate in all forests but forests have different degrees of complexity in canopy structure and differ in the number of species that occupy the many facets of what may be termed the 'regeneration niche'.

Site

The forest growth cycle interacts with differences between tree species in response to soil, drainage, and slope (section 2.5).

Thus, in any area we can expect to find a mosaic of patches of different species resulting from the operation of the forest growth cycle, superimposed on a patchiness resulting from the varied response of different species to site factors. The forest will be still more intricate if there are several species equally well suited to these various niches. Around the trees, the herbs and dependent synusiae colonize and compete in a similar complex manner.

Forest formations

At a broader scale of site, different forest formations occupy different habitats and many species are confined to one or only a few formations.

Interactions between plants and animals

In Chapter 5 the various syndromes of characters of flower and fruit were discussed which have evolved to attract particular pollinating or dispersing animals, and this diversity of food sources is one of the factors underlying animal species richness (section 4.2). It seems that one-to-one specialization between a single plant and

Fig. 9.1. Luxuriant riverine rain forest near Rio de Janeiro with a wealth of species. (von Martius 1840 XIII.) The spiny palm is *Bactris*. The epiphytes are mostly aroids: top left *Philodendron* sp. (long stem), below it *Anthurium solitarium* (huge rosette), on right hand tree *A. harrisii* (lanceolate leaves), and *A. pentaphyllum* (palmately compound leaves). Other epiphytes are bromeliads (spiny leaf margins) and orchids (seen flowering). On the ground are two more aroids *Philodendron martianum* (below, huge rosette) and *P. speciosum* (cordate leaves, centre).

animal species as a factor of species richness only exists in a few cases, such as figs (Fig. 5.14). Guilds of insects specialized to feed on (and where necessary detoxify) particular families or similar families of plants (which are sometimes called plant webs) is a looser and commoner form of co-evolution and plays a more substantial role in the packing together of numerous sympatric species (section 5.4).

In the humid tropics flowering, fruiting, and germination are possible throughout the year. Species dependent on animals for pollination or dispersal may have a better chance of success if they do not compete with others for the services of animals. This can lead to phenological separation by staggered flowering or fruiting times; examples were given in Figs. 5.7, 5.8. West Malesian Dipterocarpaceae are interesting: staggered flowering times have evolved which maximize the chance of any individual species successfully cross-pollinating, but the different species then all reach fruit maturity simultaneously which satiates predators and maximizes the chance of successful germination.

Browsing pressure ('pest pressure') of herbivores (p. 82) may be one factor that sometimes prevents any single species from attaining dominance, and acts to maintain species richness. In a similar manner dense seedling populations below a parent tree are often thinned out by disease or herbivory (section 7.3) and this also therefore contributes to the prevention of single species dominance.

Evolution

Most botanists believe that genera and families originate by a continuation of the processes of evolution by natural selection, which lead to the evolution of species. It can be argued that the flora of a region continues to evolve new species, which therefore accumulate with time. Highly stable areas come to develop a rich flora because there is little extinction. In the humid tropics the accumulation of species is accentuated because survival is not limited by a cold or dry season acting as a 'climatic sieve' and so extinction is even lower than in stable parts of other biomes.

The most species-rich plots so far enumerated are in Ecuador and Peru with 307 and 283 species of trees 0.1 m in diameter or over amongst the 693 and 580 stems on one hectare (Fig. 2.27). Here, as in all species-rich rain forests, most species are rare, and every second tree is a different species. Several plots in Borneo are almost as rich. These forests may be near the upper limit possible for sustainable tree species richness. Over the medium and long time-scale, species need to outbreed to counteract the effect of accumulating deleterious mutations and in order to express the genetic variability on which natural selection then operates. In this way the species keeps highly adapted to competition and climate as these change. For outbreeding several individuals need to exist close enough to cross-pollinate. It would be interesting in these hyper-rich forests to examine the species-area curve to see if species numbers continue to rise beyond one hectare, and also to study the distance of effective cross-pollination.

Evolution may eventually lead to the co-occurrence of series of sympatric species. In the temperate forests of east Asia and north America this is well shown by *Acer* and *Quercus*. There are many examples in lowland tropical rain forests (see p. 86).

Long series of sympatric species are found in many parts of the world. Not all are forest trees and the evolutionary origin is not always the same. In southern Africa there are swarms of several hundred *Erica* species in the Cape heathlands and of many dozens of *Mesembryanthemum* in Botswana. In mid-northern latitudes *Crataegus, Hieracium*, and *Rubus* (hawthorn, hawkweed, blackberry) are examples. The high mountain forests of Malesia, especially New Guinea, have many sympatric orchids (*Bulbophyllum, Dendrobium*) and Ericaceae (*Rhododendron, Vaccinium*). There are 16 sympatric pitcher plant species (*Nepenthes*) on Mt. Kinabalu in Borneo.

An important difference of tropical rain forests from others is the occurrence of locally endemic species (p. 33). This is one component of their species richness on the extensive scale. It means that in different places a particular niche may be occupied by different species which never compete because they never meet. It has the consequence that species are likely to become extinct when a rain forest is reduced in extent, more so than in other forest biomes.

Study of breeding systems in rain forest plants has shown that these are as diverse as in other biomes. No particular breeding system predominates to act as the driving force behind speciation.

Conclusions and summary

There are thus many different components to species richness, and many different causes. It is necessary in any discussion to specify the scale, whether local or regional, as discussed on p. 32. It is not easy to discover whether there are species of identical ecology, i.e. which occupy the same niche. Many species of *Shorea* growing together in Pasoh forest, Malaya, have not been demonstrated to have different site requirements and it seems likely that the manifold roles of chance are important there (and elsewhere) in determining forest composition.

In summary, the main reasons why some tropical rain forests are extremely rich in species results from firstly, a long stable climatic history without episodes of extinction, in an equable environment, and in which there is no 'climatic sieve' to eliminate some species. Secondly, a forest canopy provides large numbers of spatial

and temporal niches; the forest growth cycle is important here. Thirdly, richness results from interactions with animals, mainly as pollinators, dispersers, or pests. Some of these factors underly species richness in other biomes also.

The American forest ecologist R.H. Whittaker[293] can be allowed the final word on this subject. From the analysis of this chapter we may agree with his statement that 'the study of diversity has revealed divergent relationships in different groups and places and these are subject much less to prediction than to observation and evolutionary interpretation'.

The overall effect of all of humankind's many different impacts on tropical rain forests is to diminish the numerous dimensions of species richness. Not only does man destroy species, he also simplifies the ecosystems the remaining species inhabit. It is to this serious contemporary subject that we now turn in the next chapter.

10

Tropical rain forests yesterday and today

Human impact on tropical rain forests has changed in pace and scale through time. In this chapter we consider the earlier phases, up to the end of European colonial empires. This includes a look at minor forest products, which have historically been more important than timber. Then we describe the dramatic changes since World War II as timber exploitation has become of such major importance, population pressure has substantially increased, and global awareness of human impact has snowballed. The book concludes in Chapter 11 with the picture today and a prognosis for the future.

10.1 INDIGENOUS CULTURES

Humans have lived in the rain forests for millennia. The archaeological record for Africa and the Niah cave, Sarawak, goes back about 40 000 years, and for Amazonia about 5000.

Today hunter-gatherer societies live in all three rain-forest regions. In central Africa are the Pygmies, in the Amazonian forests many small groups of Amerindians, a few still with little or no contact with the outside world, and in the Eastern rain forests tribes such as the Onge and Jarawa of the South Andaman Islands,[294] and a few of the Penan of northern Sarawak and Brunei. These peoples live off the wild plants they collect and animals they hunt and do not cultivate crops. They have an intimate knowledge of the forest and do not destroy it, although they do alter it.

Nowadays most rain-forest dwellers are farmers. Settled agriculture long ago led to the permanent clearance of the freshwater swamp forests (p. 25) of the alluvial plains of Asia, which were replaced by irrigated rice, a highly labour-intensive crop which can support very dense human populations, e.g. 2500 km^{-2} on the Tonkin delta. But most agricultural societies occupied the dryland forests. For their staple foods they practised shifting agriculture, which was developed in all parts of the tropics (Fig. 10.1). This was described in Chapter 8, where it was shown that, with care and skill, it is a sustainable mode of cultivation, well adapted to infertile soils. Shifting agriculture can usually only support 10–20 persons km^{-2} (Table 10.1). These farmers also rely on forest in various ways. Secondary forest, the bush fallow of the swidden cycle, provides many useful plants, and indeed is sometimes enriched with them (Table 10.2), some being planted to attract game animals. Settled agriculturists plant orchards of mixed species of trees, and also many climbers and herbs, for timber, food, cordage and medicines (Figs. 8.9, 10.2). Some are wild species, others only known in cultivation (Table 10.3). Bantor Kalong, a village in west Java, was found in the 1970s to have 425 plant species in cultivation.

Fig. 10.1. Shifting agricultural in the Atlantic coast rain forest of Brazil near Rio de Janeiro in the early 19th century. The huge tree on the left is a *Ficus*. (von Martius 1840, plate XVI.)

Farmers also enrich the primary forest by planting or tending useful species, to create what are sometimes called agroforests. In southern Sumatra there are thousands of hectares of damar forests rich in *Shorea javanica*, tapped for its resin, some of which are believed to be 900 years old.[295] In Borneo forests have been enriched with other *Shorea* spp., of the group which produce illipe nuts or with rattan.[296] Recent investigations in the Amazon have found forests rich in useful species, some growing outside their natural geographical range. For example, an inventory of 1 ha of tall uninhabited forest near the river Xingu in Brazil recorded 147 tree species of 10 cm diameter or over.

Table 10.1 Population density of shifting agriculturists in various Malesian rain forests

		Persons (km^{-2})	Fallow period (years)
Kalimantan	Kantu	16	7
Sarawak	Kenyah	11–18	20–12
	Iban	18	12
Philippines	Hanunoo	48	12
New Guinea	Tsembaga	34	15–25

Various sources
The calculations assume 50–70% of the land is cultivable

Table 10.2 Useful plants in the fields and forest fallows of Bora Indian shifting agriculture, Amazonian Peru

Stage	Planted	Spontaneous
High forest	–	Numerous species for construction, medicine, handicrafts, and food
Newly planted field, 0–3 months	–	Dry firewood from unburnt trees for hot fires
New field, 3–9 months	Corn, rice, cowpeas (*Vigna unguiculata*)	Various early successional species
Mature field, 9 months–2 years	Manioc,† some tubers,† banana,† cocona (*Solanum sessiliflorum*),* and other quick maturing crops	Vines and herbs of forest edges
Transitional field, 1–5 years. Seedlings of useful trees appear	Replanted manioc, pineapples,* peanuts,* coca, guava†, caimito (*Pouteria caimito*),* uvilla (*Pourouma cecropiifolia*),* avocado,* cashew,* barbasco (*Lonchocarpus nicou*),* peppers (*Capsicum*),* tubers; trapped game	Medicinal plants within field and on edges. Abandoned edges yield straight, tall saplings, including *Cecropia* and *Ochroma lagopus*
Transitional fruit field, 4–6 years, with abundant forest regrowth	Peach palm (*Bactris gasipaes*),* banana, uvilla, caimito, guava, annatto (*Bixa orellana*), coca, some tubers; propagules of pineapples and other crops; hunted and trapped game	Many useful soft construction woods and firewoods. Palms appear, including *Astrocaryum*§. Many vines; useful understorey aroids
Orchard fallow, 6–12 years	Peach palm, some uvilla, macambo (*Theobroma bicolor*);* hunted game	Useful plants as above; self-seeding *Inga*. Probably most productive fallow stage
Forest fallow, 12–30	Macambo, umari (*Poraqueiba sericea*),* breadfruit,*, copal (*Dacryodes* sp.)	Self-seeding macambo and umari. Some hardwoods becoming harvestable, e.g., cumala. Many large palms: huicungo (*Astrocaryum huicungo*), chambira (*A. chambira*), assai (*Euterpe* sp.), ungurahui (*Jessenia bataua*)§.
Old fallow, high forest over 30 years	Umari, macambo	A few residual planted and managed trees

From table 1, Denevan *et al.* in Lugo *et al.* (1987)
* Fruits
† Carbohydrates
§ Oil
Other species mainly for medicinal or utilitarian uses

Fig. 10.11. The Atlantic Coast rain forests of Brazil have very nearly disappeared. (After Brown and Brown in Whitmore and Sayer 1992, fig. 6.2.)

In 1882 Charles Darwin had his first taste of the tropics at Salvador, 13°S. These forests have a high percentage of endemics amongst both plants and animals, and surprisingly, so far, few have become extinct.

one million km^2, have been 95 per cent cleared (Fig. 10.11). Once roads were built, the larger trees were felled for charcoal to fuel iron smelters (Brazil has little indigenous fossil fuel), and pastoralists completed the destruction. Along with the plants, most of the animals of the Atlantic Coast forests have become endangered, but so far unexpectedly little extinction has occurred (we return to this later, p. 232). Very big pastures are also eating into the Amazonian rain forests, especially on the southern margin in Brazil. The figures quoted for the total pasture area vary widely, from 6 to 11 million ha in the early 1980s, and will have increased since then (the area of Holland is 3 million ha).[330] Southeast Pará State has the most extensive pastures. A further area of about 10 000 ha has been converted to pasture north of Manaus (Fig. 10.12) in order to provide meat for the city as a supplement to the river fish it was largely dependent on. The rapid extension of the pastures is reflected in cattle numbers (Fig. 10.13). These Brazilian pastures were largely funded by fiscal incentives given by government in order to develop the region and which made cattle ranching highly profitable. The fiscal incentives were removed, partly as a result of adverse worldwide publicity on the extent of forest loss. However, an economic

Fig. 10.12. Cattle pasture north of Manaus, Brazilian Amazon. Primary lowland evergreen rain forest behind.

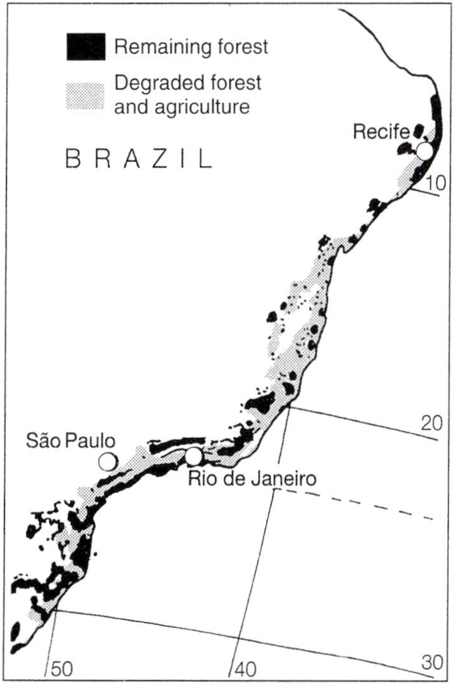

Fig. 10.11. The Atlantic Coast rain forests of Brazil have very nearly disappeared. (After Brown and Brown in Whitmore and Sayer 1992, fig. 6.2.)
In 1882 Charles Darwin had his first taste of the tropics at Salvador, 13°S. These forests have a high percentage of endemics amongst both plants and animals, and surprisingly, so far, few have become extinct.

one million km^2, have been 95 per cent cleared (Fig. 10.11). Once roads were built, the larger trees were felled for charcoal to fuel iron smelters (Brazil has little indigenous fossil fuel), and pastoralists completed the destruction. Along with the plants, most of the animals of the Atlantic Coast forests have become endangered, but so far unexpectedly little extinction has occurred (we return to this later, p. 232). Very big pastures are also eating into the Amazonian rain forests, especially on the southern margin in Brazil. The figures quoted for the total pasture area vary widely, from 6 to 11 million ha in the early 1980s, and will have increased since then (the area of Holland is 3 million ha).[330] Southeast Pará State has the most extensive pastures. A further area of about 10 000 ha has been converted to pasture north of Manaus (Fig. 10.12) in order to provide meat for the city as a supplement to the river fish it was largely dependent on. The rapid extension of the pastures is reflected in cattle numbers (Fig. 10.13). These Brazilian pastures were largely funded by fiscal incentives given by government in order to develop the region and which made cattle ranching highly profitable. The fiscal incentives were removed, partly as a result of adverse worldwide publicity on the extent of forest loss. However, an economic

Fig. 10.12. Cattle pasture north of Manaus, Brazilian Amazon. Primary lowland evergreen rain forest behind.

Table 10.4 Rain forest fruits

America:	*Annona muricata* (soursop)
	A. reticulata (bullock's heart)
	A. squamosa (sweetsop)
	Bertholletia excelsa (Brazil nut)
	Brosimum alicastrum
	Carica papaya and some relatives
	Chrysophyllum cainito (caimito)
	Eugenia uniflora (Cayenne cherry)
	Manilkara zapota (sapodilla, and whose latex is also used for chicle, chewing gum)
	Persea americana (avocado)
	Psidium cattleianum (strawberry guava)
	P. guajava (guava)
	Spondias mombin
	Theobroma grandiflora (cupuaçu)
Africa:	*Afromomum melagueta* (grains of paradise)
	Blighia sapida (akee)
	Canarium schweinfurthii
	Cola nitida
	Dacryodes klainiana
	Dialium guineense
	Hibiscus sabdariffa (roselle)
	Irvingia gabonensis
	Pentadesma butyracea (tallow tree)
Orient: (mostly arillate)	*Artocarpus altilis* (breadfruit)
	A. integer (chempedak)
	A. heterophyllus (jackfruit)
	Averrhoa carambola (starfruit)
	Durio zibethinus (durian)
	Eugenia aquea (water apple)
	E. jambos (rose apple)
	E. malaccensis (Malay apple)
	Garcinia mangostana (mangosteen)
	Lansium domesticum (duku, langsat)
	Mangifera spp. (mango)
	Nephelium lappaceum (rambutan)
	N. ramboutan-ake (pulasan)
	Salacca zalacca (salac)
	Sandoricum koetjape (sentul)
	Shorea spp. (illipe nuts, for a cocoa butter substitute)
Melanesia: (many nut trees)	*Barringtonia magnifica*
	Canarium indicum (ngali)
	Pandanus sp.
	Pometia pinnata
	Terminalia spp.

This is a list of just some of the commoner regional fruits; several of them are now cultivated throughout the tropics.

Fig. 10.7. Limes, pupuña (peach palm) and cupuaçu fruits (*Citrus aurantifolia*, *Bactris gasipaes*, *Theobroma grandiflora*) on sale at a roadside stall, near Manaus, central Amazonian Brazil.

Rattans in international trade are currently worth *c.* £1500 million per year; 150 000 tonne are traded annually. They still come almost entirely from natural rain forest and have traditionally been one of the most important trade commodities for forest dwellers, passing through many middlemen to the great entrepôts of Ujung Pandang and Singapore. More recently, social controls on their exploitation have broken down.[321] Following the advent of logging roads in modern times and the consequent ease of access, rattans have been heavily over-exploited (Fig. 10.9). Export of unprocessed rattans has been progressively banned by the nations where they grow, and substantial local manufacturing industries have developed, creating jobs and adding export value. Trade in natural rattans is likely to collapse when the last virgin forests have been opened up.[322] Since the early 1980s rattan cultivation on a large scale has begun[323] and there are now many thousand hectares reaching maturity across the Malay archipelago.

There have been attempts by campaigning groups in recent years to turn the clock back, sometimes claiming forests have a greater cash

Table 10.4 Rain forest fruits

America:	*Annona muricata* (soursop)
	A. reticulata (bullock's heart)
	A. squamosa (sweetsop)
	Bertholletia excelsa (Brazil nut)
	Brosimum alicastrum
	Carica papaya and some relatives
	Chrysophyllum cainito (caimito)
	Eugenia uniflora (Cayenne cherry)
	Manilkara zapota (sapodilla, and whose latex is also used for chicle, chewing gum)
	Persea americana (avocado)
	Psidium cattleianum (strawberry guava)
	P. guajava (guava)
	Spondias mombin
	Theobroma grandiflora (cupuaçu)
Africa:	*Afromomum melagueta* (grains of paradise)
	Blighia sapida (akee)
	Canarium schweinfurthii
	Cola nitida
	Dacryodes klainiana
	Dialium guineense
	Hibiscus sabdariffa (roselle)
	Irvingia gabonensis
	Pentadesma butyracea (tallow tree)
Orient: (mostly arillate)	*Artocarpus altilis* (breadfruit)
	A. integer (chempedak)
	A. heterophyllus (jackfruit)
	Averrhoa carambola (starfruit)
	Durio zibethinus (durian)
	Eugenia aquea (water apple)
	E. jambos (rose apple)
	E. malaccensis (Malay apple)
	Garcinia mangostana (mangosteen)
	Lansium domesticum (duku, langsat)
	Mangifera spp. (mango)
	Nephelium lappaceum (rambutan)
	N. ramboutan-ake (pulasan)
	Salacca zalacca (salac)
	Sandoricum koetjape (sentul)
	Shorea spp. (illipe nuts, for a cocoa butter substitute)
Melanesia: (many nut trees)	*Barringtonia magnifica*
	Canarium indicum (ngali)
	Pandanus sp.
	Pometia pinnata
	Terminalia spp.

This is a list of just some of the commoner regional fruits; several of them are now cultivated throughout the tropics.

Fig. 10.7. Limes, pupuña (peach palm) and cupuaçu fruits (*Citrus aurantifolia*, *Bactris gasipaes*, *Theobroma grandiflora*) on sale at a roadside stall, near Manaus, central Amazonian Brazil.

Rattans in international trade are currently worth *c.* £1500 million per year; 150 000 tonne are traded annually. They still come almost entirely from natural rain forest and have traditionally been one of the most important trade commodities for forest dwellers, passing through many middlemen to the great entrepôts of Ujung Pandang and Singapore. More recently, social controls on their exploitation have broken down.[321] Following the advent of logging roads in modern times and the consequent ease of access, rattans have been heavily over-exploited (Fig. 10.9). Export of unprocessed rattans has been progressively banned by the nations where they grow, and substantial local manufacturing industries have developed, creating jobs and adding export value. Trade in natural rattans is likely to collapse when the last virgin forests have been opened up.[322] Since the early 1980s rattan cultivation on a large scale has begun[323] and there are now many thousand hectares reaching maturity across the Malay archipelago.

There have been attempts by campaigning groups in recent years to turn the clock back, sometimes claiming forests have a greater cash

Fig. 10.8. Rattans drying after treatment at a factory in Singapore.

These are the stout species rotan manau (*Calamus manan*), known in the trade as canes, and used for chair legs. Singapore was the main entrepôt through which rattans from all over the Malay archipelago were traded. Some were made into furniture, others were processed and exported to Hong Kong or Europe. Now producing nations have their own factories.

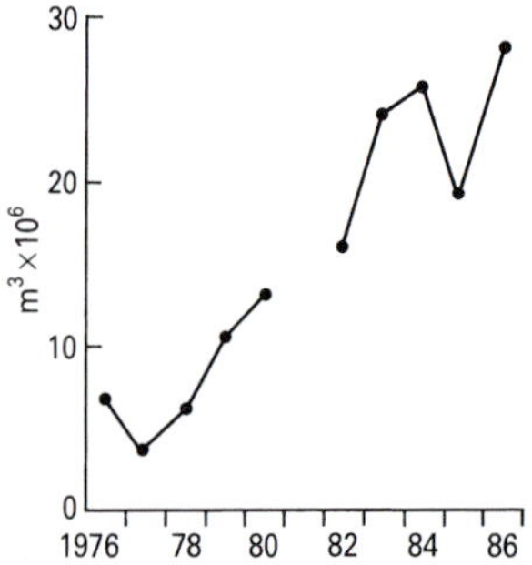

Fig. 10.9. Philippine rattan production 1976–86. (Philippines Bureau of Forest Development Statistics.)

As forests were opened up for logging, rattan production quadrupled to an unsustainable rate.

value for minor forest products than for timber.[324] A review of 24 studies found that the median annual value per hectare of sustainably produced, marketable non-timber forest products was \$50 $year^{-1}$.[325] As a natural rain forest grows commercial timber at 1–2 m^3 $year^{-1}$ ha^{-1} or more, and this is worth over \$100 m^{-3}, sustainable production of timber is of greater value by a factor of at least two to four. What really matters about minor forest products is their importance as components, either for consumption or trade, of the livelihood of rural people who live in or near the forest.[326]

10.4 TROPICAL RAIN FORESTS IN MODERN TIMES

The past half-century, especially the period since the mid-1960s, has seen rapidly increasing human impact on tropical rain forests. The technology has been mastered to build roads in, and to fell and remove giant trees from, remote forests. Formerly debilitating diseases have been controlled by newly developed drugs. At the same time, global communications have been revolutionized. People can travel half way round the world in a day, and the information superhighway using satellites and fibre-optic cable puts them in instant and constant contact. The global economy is increasingly closely integrated and well-informed. Modern cheap air travel has led to the rise of the new phenomenon of ecotourism. Holidays can be spent in rain forest, and eyes opened to their wonders and to what is happening. Scientists are more than ever before members of a single worldwide community. Many industrial-nation biologists and foresters now have a rain-forest project in their portfolio; so rain-forest science is no longer isolated and we increasingly perceive how these forests fit

into a global pattern, often at one end of a spectrum of features or responses, for example, in species richness and nutrient cycling.

Nowhere have modern technologies caused a greater transformation than in Southeast Asia, amongst the so-called 'tiger economies' of the Pacific rim. Thailand, Malaysia and Indonesia, formerly living by agriculture and primary production, are using forest wealth to help fund their transformation into industrial economies. The tiger economies of Korea and Taiwan have been major importers of rain-forest timber.

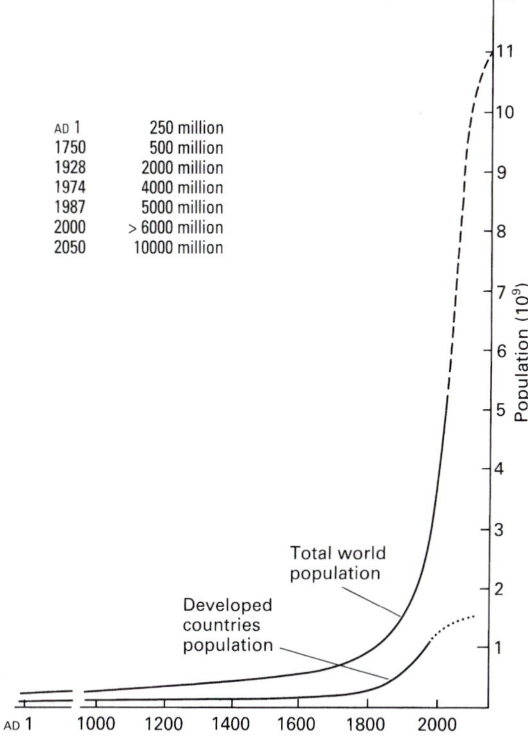

Fig. 10.10. Human population numbers. (World Bank figures cited in Anon. 1985.)

The world population doubled between 1650 and 1850. The most recent doubling took only 44 years. Population by mid-1997 was 5.8×10^9. The prediction is that the balance between falling fertility and increasing longevity will result in 10×10^9 being reached by 2050, and a peak of *c.* 11×10^9 around 2070–80, then followed by a slow decline (Lutz *et al.* 1997).

Human numbers are increasing dramatically (Fig. 10.10). The increase is mostly in the tropics and subtropics, where the demographic transition to small families typical of industrialized communities is only now patchily beginning to take place. The impact of all these people is exacerbated by greater economic aspirations and consequent greater demands per person on the environment. Tropical rain forests are one of the last great forest frontiers to be rolled back by humankind.[327] Deforestation of the Mediterranean basin forests began 8000 years ago.[328] Massive deforestation began in China 5000 years ago. The temperate forests of Europe have been reduced to only a few per cent of their post-Glacial natural extent over the past two millennia. England was in fact less forested at Domesday (1086 AD) than it is now. Later, the temperate forests of North and South America and Australasia were hard hit by European immigrants. Timber extraction on its own does not destroy rain forests, but is sometimes the first step. Destruction of rain forests to create more agricultural land is taking place much more rapidly than these past episodes of destruction. Of the world's cultivable land, 950×10^6 ha are already in use and another 550×10^6 ha are available, much of it in the tropics where the demand is greatest and also will be particularly acute in the three next decades to 2030 AD.[329]

10.5 FOREST CONVERSION TO OTHER LAND USAGES

Agriculture

Agriculture is the main purpose for which rain forests are cleared. There are several major kinds of agriculture and their impact varies from place to place. Important detail is lost by pantropical generalization.

Conversion of tropical rain forests to pasture is especially widespread in the neotropics, where there is a long tradition of cattle husbandry. The Atlantic Coast rain forests, which once covered

Fig. 10.11. The Atlantic Coast rain forests of Brazil have very nearly disappeared. (After Brown and Brown in Whitmore and Sayer 1992, fig. 6.2.)
In 1882 Charles Darwin had his first taste of the tropics at Salvador, 13°S. These forests have a high percentage of endemics amongst both plants and animals, and surprisingly, so far, few have become extinct.

one million km^2, have been 95 per cent cleared (Fig. 10.11). Once roads were built, the larger trees were felled for charcoal to fuel iron smelters (Brazil has little indigenous fossil fuel), and pastoralists completed the destruction. Along with the plants, most of the animals of the Atlantic Coast forests have become endangered, but so far unexpectedly little extinction has occurred (we return to this later, p. 232). Very big pastures are also eating into the Amazonian rain forests, especially on the southern margin in Brazil. The figures quoted for the total pasture area vary widely, from 6 to 11 million ha in the early 1980s, and will have increased since then (the area of Holland is 3 million ha).[330] Southeast Pará State has the most extensive pastures. A further area of about 10 000 ha has been converted to pasture north of Manaus (Fig. 10.12) in order to provide meat for the city as a supplement to the river fish it was largely dependent on. The rapid extension of the pastures is reflected in cattle numbers (Fig. 10.13). These Brazilian pastures were largely funded by fiscal incentives given by government in order to develop the region and which made cattle ranching highly profitable. The fiscal incentives were removed, partly as a result of adverse worldwide publicity on the extent of forest loss. However, an economic

Fig. 10.12. Cattle pasture north of Manaus, Brazilian Amazon. Primary lowland evergreen rain forest behind.

edible protein can be extracted from leaves,[315] and lignin is utilized for the manufacture of plastics, ion-exchange resins, soil stabilizers, rubber reinforcers, fertilizers, vanillin, tanning agents, stabilizers for asphalt emulsions, and dispersants for oil-well drilling and for ceramic processing. Cellulose can be utilized for rayon and plastics and as a raw material for hydrolysis to sugar which, using yeasts, can be turned into alcohol and edible protein. Complex molecules, such as steroids, are conveniently obtained from plant sources. These so-called genetic resources are by no means yet fully known or exploited.

Crop plants which have developed major significance in international trade and whose origin is in rain forests are: cocoa[316] and para rubber (*Theobroma cacao*, *Hevea brasiliensis*) from the neotropics; banana, citrus fruits, coconut[317] and sugarcane (*Musa*, *Citrus*, *Cocos nucifera*, *Saccharum*) from the Eastern tropics; and oil palm (*Elaeis guineensis*) from Africa.

In addition, there are numerous fruits of regional importance, some now grown throughout the tropics (see Table 10.4, Fig. 10.7) and others whose potential for wider cultivation is newly attracting interest from development agencies. Some of these species have scarcely altered from their forest ancestors. Others, for example *Durio zibethinus*, *Garcinia mangostana* and *Carica papaya* (durian, mangosteen and papaya), are unknown in the wild, as is the case of many of humankind's more important cultivated plants and domesticated animals too.

Rain forests provide innumerable species that are used locally in traditional medicine. Rather few have gained international importance.[318] In this latter category, America is the source of a muscle relaxant used in major surgery, from the roots of a climber (*Chondrodendron tomentosum*) which is one of the constituents of the arrow-poison curare, and of quinine (from the bark of *Cinchona*). Cocaine also originates from South America; it is prepared from the leaves of *Erythroxylum coca* (and was once introduced as a hedge plant to Asia but soon eliminated). Tubers of the yam (*Dioscorea*) are gathered from the wild in Central America, India and China to provide diosgenin, the steroid used as the precursor molecule from which oral contraceptives and cortisone are made; but soyabean oil has recently become an important alternative source and total synthesis is also common nowadays. Reserpine, a cardiac glycoside, which reduces high blood pressure and is also used in treatment of mental illness, is extracted from the roots of both African and Asian species of *Rauvolfia*, a shrub. Two *Calophyllum* species from Sarawak have shown some promise in the treatment of AIDS, though their potency depends on where they are growing. This illustrates the potential of tropical rain forests to meet new human requirements as they arise. However, the claim sometimes made that rain forests contain enormous numbers of drugs just awaiting exploitation does not survive critical examination.[319] Reality is more complex, and there are serious difficulties in developing an economic case for biodiversity conservation based on undiscovered pharmaceuticals. Firstly, drug companies need today to consider whether screening jungle plants will yield better drugs than computer modelling of molecules and their synthesis. Collecting and screening, then purifying and testing them, takes many years and has low success and high costs. Many drugs have already been developed from the 25 000 species used worldwide in traditional medicine, and the future success rate is likely to be less. Many leads have been followed through, exploring, for example, the alkaloids of the Apocynaceae. Today there is commendable attention to sharing the profits of success with local communities but systems being proposed to protect native property rights are based on Western legal principles and can undermine patterns of community ownership.

The single most important minor forest product is rattan, the stems of climbing palms, abundant throughout the Eastern tropics, second only in value to timber (Fig. 10.8).

> When you sit in a rattan chair at a rattan table on a rattan carpet with a rattan glass-holder, behind a rattan blind, with a rattan holdall, think of the hands through which the rattan came...[320]

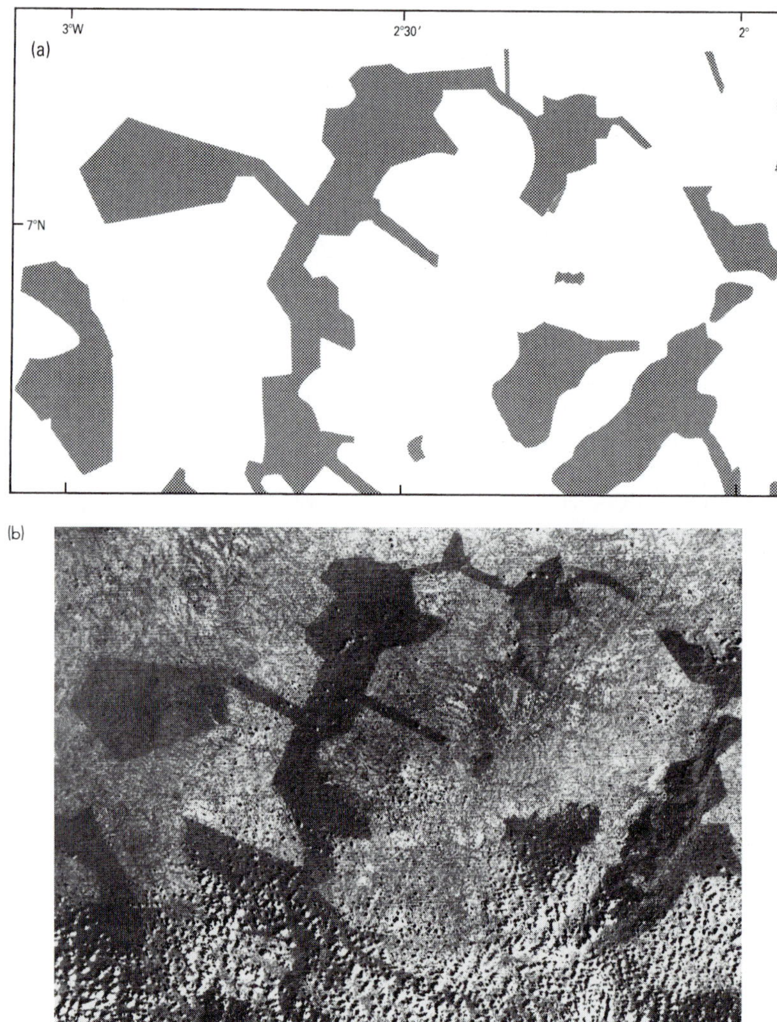

Fig. 10.6. Rain forest remnant patches in northwest Ghana. (a) Forest reserve map, 1955. (b) LANDSAT 2 images, 1973. Sunyani is just north of the scene. On the LANDSAT image the forests appear dark, like islands, in a pale-coloured sea of densely farmed land. Note that the shape and size of the reserves remained essentially the same over this 18-year period. This was despite heavy population pressure. The satellite scene was recorded at about 09.30 and in the south small white clouds are casting dark shadows to their west. Relict forest patches such as these provide important service functions as water sources and in ameliorating local climate.

Tropical rain forests contain many wild-fruit trees, some of which are the ancestors and relatives of cultivated species, and many are species of medicinal importance or potential.[314] Furthermore, when fossil fuels are eventually depleted, humankind is likely to make greater use of plants as sources of complex organic molecules, often as raw materials for manipulation. For example, the trunk of *Copaifera langsdorfii* (pau oleo), a leguminous tree of the Amazon, produces an inflammable oil at a rate of 20 litres per 6 months, which is tapped and used locally instead of kerosene. Forests are a biochemical storehouse, scarcely yet exploited. Already,

Fig. 10.5. *Agathis dammara* tapped for the resin in its bark which is traded as Manila Copal, Sulawesi. (Whitmore 1980, fig. 6.)

The trunk has been damaged by excessively large tapping scars.

led to increased and destructive tapping so that many trees were killed in the *Agathis* forests of central Indonesia (Fig. 10.5). The colonial administration of the Dutch East Indies introduced regulations that attempted to prevent this and other over-exploitation.[310]

Forest Departments were established and Forest Reserves created. In the British Empire, the officers were called Forest Conservators. The survival today of 81 ha of partly primary rain forest at Bukit Timah on Singapore, a tiny City State of 2.7 million people on 62 600 ha, and of 1 million ha of an original 1.2 million ha of the Queensland rain forests, is testimony to the ethos of the colonial era.[311] Fig. 10.6 shows how the only remaining tropical moist forests in Ghana survive as Forest Reserves created by the former colonial government, and set today in a sea of dense agricultural settlement.

A consequence of European impact on the humid tropics, whose full extent is only now coming to be discovered, was the total disruption of, and in places the extirpation of, whole societies of indigenous people. The Amerindian population of the New World rain forests collapsed as tribes succumbed to introduced respiratory diseases and measles. Today the Amazon basin has only about 2.5 per cent of its former indigenous population.[312] Those who remain are mostly hunter-gatherers and fishermen. There is far less shifting agriculture practised there today than formerly.

10.3 MINOR FOREST PRODUCTS

Minor, or non-timber, forest products, the terms used for everything the forest provides in addition to timber, are a cornucopia of useful goods for mankind. Many are used only locally, a few enter international trade. Some were exploited formerly but have been replaced by factory products, such as plastics.[313] Others await discovery. Forest Departments issue licences and collect royalties on minor forest products, though many are now obsolete. There are specific rates, for example, for leaves collected for food wrappers (Malaya, *Macaranga gigantea*), and for edible birds' nests (Borneo, produced by cave-dwelling swiftlets), and turtle eggs.

The commercial value of minor forest products has been eclipsed in recent years by timber. In 1938 the relative importance of timber and minor forest products traded from Indonesia was 55 to 45 per cent. Since then the giant modern markets for commodity timbers (section 10.6) and the machinery to extract them have been developed. The revenue from timber has rocketed, and for most minor forest products it has diminished as modern manufactures have superseded many of them.

of population increase, and following the building of roads, vegetables are grown for sale in nearby urban markets.

10.2 THE COLONIAL ERA

European exploration of the world and later settlement unwittingly led to the decrease or even the extirpation of some forest-dwelling peoples. It had other profound effects throughout the humid tropics.

European interest in tropical rain forests was at first focused on them as the source of spices, especially important before refrigeration, to mask the flavour of preserved meat. The New World was discovered during the search for a better route to the Spice Islands of the East Indies, present-day Maluku (the Moluccas). It is no exaggeration to say that nutmeg, cloves, cinnamon, and pepper (*Myristica fragrans*, *Syzygium aromaticum*, *Cinnamomum* spp., *Piper nigrum*) have shaped the history of the world. Trade for spices was soon followed by settlement, to consolidate, control, and increase the supply. The Dutch, who monopolized the trade in nutmeg[307] from the mid-17th to the early 19th century, restricted its cultivation by force to the Banda group of tiny islands in Maluku, destroying the trees on other islands; but they were thwarted by pigeons which, as specialized frugivores, dispersed the seeds.[308] Still today the primitive cultivars of cloves are difficult to locate because the Dutch only permitted cultivation on Ambon, and plantations elsewhere were destroyed.

The trading posts were followed by full colonial domination. With it the European powers began the movement of useful plants between the different tropical regions. Captain Cook and other voyagers brought back descriptions from Polynesia of the tree from which 'bread itself is gathered as a fruit', and Lieutenant William Bligh was conveying breadfruit (*Artocarpus altilis*) from Tahiti to the Caribbean on the Bounty when his crew mutinied on 28 April 1789. Botanic Gardens were established as centres for plant introduction and trial, for example, Peradeniya in Sri Lanka, Singapore, Bogor in Java, Rio de Janeiro, and St. Vincent in the Caribbean (where the original breadfruit tree planted after Bligh's second voyage in 1793 still stands). Weeds were inadvertently transported, often in ships' ballast. The first New World weeds to reach Asia arrived via the Spanish galleon route between Acapulco in Mexico and Manila in the Philippines.[309] Their spread through the Malay archipelago can be traced outwards from Manila. Until the age of European expansion no cultivated plant was pantropical. Today even traditional societies extensively cultivate species which originated in another continent, bananas (*Musa*, from Asia), and cassava, maize, and sweet potatoes (*Manihot esculenta*, *Zea mays*, *Ipomoea batatas*, from America) are major examples (see also Table 10.3). In the highlands of Papua New Guinea a population explosion followed the introduction to Asia from South America of sweet potato by the Spaniards; previously no high-yielding staple food occurred which could be grown at such high altitude.

The Industrial Revolution increased demand for tropical products. By the late 19th century plantations had been established, for example, in Asia to produce para rubber, an Amazonian species (*Hevea brasiliensis*), plus the indigenous gutta-percha (*Palaquium gutta*) (cf. p. 131) and gambir (*Uncaria gambir*) for tannin, as well as food crops, beverages (tea, coffee, cocoa *Camellia sinensis*, *Coffea* spp., *Theobroma cacao*), and spices. The rain forests also provided industrial raw materials, gathered by forest dwellers and traded through middlemen to the great entrepôts, such as Manaus and Belém on the Amazon, Makassar and Singapore. Over-exploitation sometimes occurred. Where the market was strong, social controls on occasion broke down. For example, the conifers *Agathis* produce Manila copal, a resin used in varnishes and formerly also in linoleum. There had always been a small trade as the resin is also used in batik cloth manufacture and for torches and boat caulking. The high European demand

that the outside modern world is impinging on these societies. In West Africa,[301] the patches of high forest found by colonizing Europeans were the no-man's-lands in a complex human landscape, and had survived because the area was contested. An example is the Okomu forest of Nigeria (p. 153). Colonial powers took ownership of most of the land and boundaries were superimposed either with (Ghana, Fig. 10.6) or without (Ivory Coast) consultation.[302] There are no owners in the Western sense. For example, at Gola, Sierra Leone, indigenous guerillas, smugglers, diamond-diggers and pit-sawyers are as much owners as anyone else; the various human societies have negotiated for themselves a niche in a complex and labile sociological landscape. This has had major implications recently as international conservationists have arrived, intent to superimpose Western ownership patterns.

In some places, settled agricultural societies have disappeared and the land has reverted to forest. This has happened in Sri Lanka, and in Cambodia at Angkor, where the ruins of irrigation systems and buildings (Fig. 10.4) occur within tall species-rich forest. In Middle America, traces of Aztec and Mayan agriculture, whose history can be traced back 3000 years, persist as forests rich in fruit trees and mahogany (*Swietenia macrophylla*) in regions where the first Spanish explorers recorded open country; and where also forests have regrown on land the European settlers cleared for cattle.[303] In the Mayan lowlands the useful trees occur as tall groves, set in a low deciduous forest, called 'pet kot' and of 20–2400 m^2 in area, on deeper soil and sometimes surrounded by low stone walls.[304] The recent discovery (p. 146) that some 12 per cent of the Brazilian Amazon rain forests have signs of former human occupancy is one of the major surprises of recent years. Old secondary forest is extensive in Borneo, New Guinea and the Solomons with past farming as one possible explanation. Fruit trees growing deep in the jungle, such as durian (*Durio zibethinus*) in western Malesia,[305] should always alert one to possible past disturbance. The myth of the primeval rain forest dies hard.

Fig. 10.4. A strangling fig growing over a Hindu stupa at the huge temple complex of Angkor Wat, Cambodia, which was built during the 10th–12th centuries and abandoned in the 15th century when the Khmer civilization collapsed. It was reclaimed from the jungle in the late 19th century by then French rulers of Indo-China.

Shifting agriculture is diverse. The various tribal groups in northern Thailand, for instance, practise different forms.[306] The Hmong prefer to farm primary forests for up to 5 or even 10–15 years, leaving behind a grassy fallow that takes decades to recover. The Karen and Lawa in the same area farm for the more usual 2–3 years, followed by a woody 'bush fallow'. In this area, as elsewhere, there is continual adaptation to changing circumstances. Today under pressure

Table 10.3 Cultivated plants at Kampong Melor, Malaya

Staple food plants:
New World origin
Manihot esculenta
Zea mays
Fruit trees:
(a) Cultivated, also found wild in rain forest
Artocarpus integer
Baccaurea griffithii
B. motleyana
Bouea macrophylla
Durio malaccensis
Garcinia atroviridis
Nephelium lappaceum
N. ramboutan-ake
Pangium edule
Parkia speciosa
Pithecellobium (Archidendron) jiringa
Sandoricum koetjape (sentul)
(b) Cultivated, wild ancestors of same species in adjacent rain forest
3 *Eugenia* spp.
Lansium domesticum
Mangifera sp.
Sandoricum koetjape (sertapi)
(c) Cultivated, indigenous to the tropical Far East, but wild ancestors unknown
Areca catechu
Artocarpus heterophyllus
Cocos nucifera
Durio zibethinus
Garcinia mangostana
(d) Introduced, all from the New World
Annona muricata
Averrhoa bilimbi
Carica papaya

Data compiled by the author on a short visit on 2 July 1971

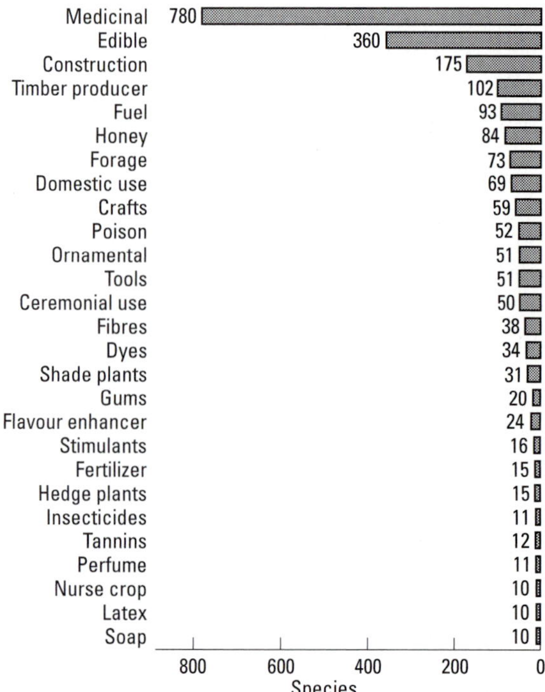

Fig. 10.3. Useful plants from the tropical rain forests of Mexico. (Toledo *et al.* 1995 fig. 4.)

The 1330 useful species, 1052 from primary forests, yield 3173 products.

wild sago palm *Eugeissona utilis*, staple food of the Sarawak nomadic Penan; and of minor forest products, mainly rattans and resins, traded far down-river for kerosene, cloth and steel tools, from the remote Apo Kayan region of northeast Kalimantan.[298]

The hunter-gatherers live in symbiosis with settled farmers with whom they form the first link in long trade chains. In Southeast Asia, trade in jungle produce with China can be dated back two millennia.[299] Borneo, for example, bartered hornbill ivory, bezoar stones, gaharu (eaglewood, diseased heartwood of *Aquilaria malaccensis* used for incense),[300] kingfisher feathers, turtle eggs, damar (resin from *Agathis*, Burseraceae and Dipterocarpaceae), camphor (from *Dryobalanops*), dragon's blood (fruit surface resin of certain rattan palms) and edible birds' nests, for ceramics, which have become heirlooms and symbols of wealth. Some of this trade persists today.

Land ownership does not follow European patterns in the sense of their being exclusive and heritable rights, and this causes problems now

Fig. 10.2. A Javanese village set amongst trees. Left to right: durian (in fruit), sugar palm (*Arenga pinnata*), coconut, two betel nut palms (*Areca catechu*), sugar palm, and kapok. (Blume 1835, plate 104.)
This is a cultivated variety of kapok, *Ceiba pentandra* var. *pentandra*, cf. Figs. 1.6, 3.31.

Three of the ten dominants were food trees (the palm babassu, *Attalea speciosa*, *Inga* and a cocoa relative) and many of the other species were also useful, e.g. Brazil nut (*Bertholletia excelsa*). It seems likely that this forest had been augmented by former human inhabitants, either shifting cultivators or hunter-gatherers, who have since died out, because another forest nearby had only 100–120 species ha^{-1}, and lacked the useful ones.[297] Finally, shifting farmers commonly leave patches of primary forest intact to retain useful plants, for example, trees for dugout canoes or house beams, and as a habitat for some of the animals they hunt for meat.

The forest landscape that these traditional societies inhabit is thus a mosaic of different sorts of vegetation, much of it under their influence. There is probably no strong conscious conservation ethic, but population density is overall low, the rich biodiversity of the natural forest survives, albeit altered, and total biodiversity is enhanced by the addition of anthropogenic elements to the landscape. Numerous plant species from this mosaic enter the local economy (Fig. 10.3). Utilization of the forest resources is under social controls which prevent over-exploitation and only a sustainable harvest is gathered. This is the case, for example, of the

Table 10.2 Useful plants in the fields and forest fallows of Bora Indian shifting agriculture, Amazonian Peru

Stage	Planted	Spontaneous
High forest	–	Numerous species for construction, medicine, handicrafts, and food
Newly planted field, 0–3 months	–	Dry firewood from unburnt trees for hot fires
New field, 3–9 months	Corn, rice, cowpeas (*Vigna unguiculata*)	Various early successional species
Mature field, 9 months–2 years	Manioc,† some tubers,† banana,† cocona (*Solanum sessiliflorum*),* and other quick maturing crops	Vines and herbs of forest edges
Transitional field, 1–5 years. Seedlings of useful trees appear	Replanted manioc, pineapples,* peanuts,* coca, guava†, caimito (*Pouteria caimito*),* uvilla (*Pourouma cecropiifolia*),* avocado,* cashew,* barbasco (*Lonchocarpus nicou*),* peppers (*Capsicum*),* tubers; trapped game	Medicinal plants within field and on edges. Abandoned edges yield straight, tall saplings, including *Cecropia* and *Ochroma lagopus*
Transitional fruit field, 4–6 years, with abundant forest regrowth	Peach palm (*Bactris gasipaes*),* banana, uvilla, caimito, guava, annatto (*Bixa orellana*), coca, some tubers; propagules of pineapples and other crops; hunted and trapped game	Many useful soft construction woods and firewoods. Palms appear, including *Astrocaryum*§. Many vines; useful understorey aroids
Orchard fallow, 6–12 years	Peach palm, some uvilla, macambo (*Theobroma bicolor*);* hunted game	Useful plants as above; self-seeding *Inga*. Probably most productive fallow stage
Forest fallow, 12–30	Macambo, umari (*Poraqueiba sericea*),* breadfruit,*, copal (*Dacryodes* sp.)	Self-seeding macambo and umari. Some hardwoods becoming harvestable, e.g., cumala. Many large palms: huicungo (*Astrocaryum huicungo*), chambira (*A. chambira*), assai (*Euterpe* sp.), ungurahui (*Jessenia bataua*)§.
Old fallow, high forest over 30 years	Umari, macambo	A few residual planted and managed trees

From table 1, Denevan *et al.* in Lugo *et al.* (1987)
* Fruits
† Carbohydrates
§ Oil
Other species mainly for medicinal or utilitarian uses

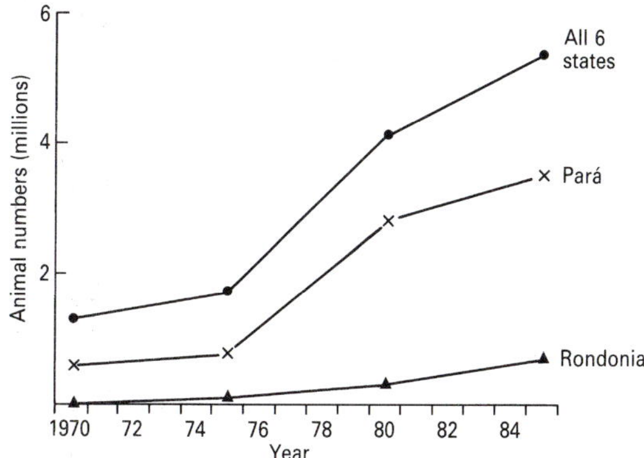

Fig. 10.13. Total cattle numbers in the Brazilian Amazon 1970–85 and in the two states with the largest herds. (From Hecht *et al.* 1988.)

model has show that, in the conditions of very high currency inflation that persisted until 1994, 'improved' land increased in value sufficiently to make it profitable to create pasture even with no fiscal incentives or credit at all. Furthermore, pastures are used for under 10 years and then abandoned because, due to rising land values, it is profitable to shorten their life by overgrazing.[331] These reasons explain why livestock raising has continued to expand. In Middle America pasture creation has substantially reduced rain-forest cover. It is the main reason for forest attrition in Costa Rica, shown in Fig. 10.14.[332]

Rain forests have also been felled and replaced by plantation tree crops, principally rubber and oil palm (*Hevea brasiliensis*, *Elaeis guineensis*) with cocoa (*Theobroma cacao*) taking third place. This form of conversion is having a major impact in Malaysia (Figs. 10.15, 10.16), Indonesia (Fig. 10.17), and

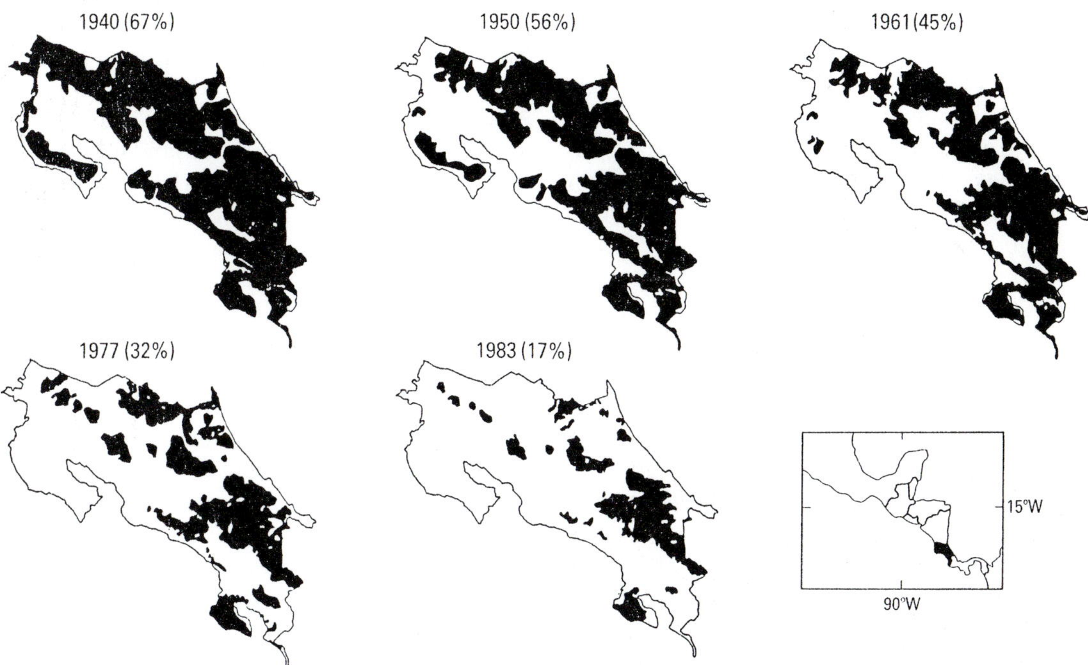

Fig. 10.14. Loss of primary forest, Costa Rica 1940–83. (After Sader and Joyce 1988, fig. 1.) The forest formations ranged from rain forest to thorn forest in progressively drier climates. Note that besides primary forest (shown black) much degraded and regrowth forest also occurred, e.g. for 1983 the map shows the 17 per cent of primary forest cover but 33.5 per cent of the country had forest of some kind. By 1983 only rugged montane forests remained relatively undisturbed.

Papua New Guinea. The crops are either grown on large plantations or smallholdings and are for export; they produce greater annual wealth than forest, in hard currency with which the nation can buy industrial goods. Indonesia and Malaysia have for a long time been the major producers of rubber, and now also of palm oil, which has replaced coconut oil in importance, and in which they have eclipsed West Africa.

Plantations of trees grown for paper or for timber, which is in effect a kind of long-term agriculture, are becoming increasingly widespread and important throughout the humid tropics, as an alternative source of revenue as the virgin rain forests become depleted (Figs. 8.7, 10.16, Table 10.11). The tropical rain-forest climate is ideal for tree growth, which continues (except in dry periods) for nine or more months of the year and far exceeds what can be attained in other parts of the world (Fig. 7.37). Throughout the tropics, *Gmelina arborea* (yemane, native to seasonal forests of continental Southeast Asia) has grown well and is widely planted for sawlogs or chips. In Sabah and the Philippines much *Paraserianthes (Albizia) falcataria* has been planted, also for sawlogs or chips. The most important species for chips and pulpwood in the Eastern tropics is now *Acacia mangium* and its close relatives, pioneer trees of Queensland and adjacent Malesia.[333] These show a very high rate of successful establishment on extremely poor soils, and with skilful tending grow very fast to yield at 8–10 years age a crop grown at a rate of 45 m^3 ha^{-1} or more (Fig. 11.7). *Eucalyptus grandis*, *E. urophylla*, their hybrid, and *E. deglupta* are also widely grown and can attain similar high rates. The major pulpwood plantation at Jari in the lower Amazon (Fig. 10.18) is now *c.* 100 000 ha in extent and mainly planted to hybrid *Eucalyptus* with *Gmelina* on the best sites.[334] *Pinus caribaea* and relatives mainly from Middle America, given extensive trial in the 1970s, have not been successful in the humid tropics; they only thrive in the seasonal tropical climates from which they originated.

Agriculture for food production, sometimes mixed with cash crops, has also made inroads into rain forest and has taken several forms. In Peninsular Malaysia, landless and unemployed people have been resettled by a quasi governmental agency, FELDA,[335] in big blocks of small holdings, created out of lowland rain forest on flattish land (Fig. 10.16). Indonesia, under its scheme of transmigration, has moved more than 4 million peasant farmers, hundreds of thousands of persons annually, from densely populated Java, Madura, and Bali to other islands, commonly on to land carved out of the rain forest.[336] However, the population of Java, Madura, and Bali increases by about 3 million annually. One consequence of transmigration in Indonesia is to bring together people with very different cultural backgrounds, and this sometimes creates friction.

Brazil in the 1960s and early 1970s built several strategic roads through the Amazon forests and moved peasant farmers to smallholdings along them from the overcrowded arid northeast, notably along the Trans-Amazonica highway. The slogan was 'a land without men for men without land'. The result is that today there are strips of farming stretching part way across southern Amazonia. The Andean nations of the western Amazon have also built roads down into the forest (Fig. 10.19).

The final mode of agriculture may be called unplanned, because government plays no direct role and it is usually illegal. This is the felling of rain forest and subsequent settlement by land-hungry peasants. The unsustainable kind of shifting agriculture these people often practise, and its impacts on the western Andean fringe of Amazonia, were mentioned on p. 158. Roads built for access by logging companies often provide the arteries along which such invasion of the forest occurs. For example, Buginese from Sulawesi moved westwards across the Makassar Strait into east Kalimantan and set up farms, with pepper as a cash crop, in previously uninhabited dipterocarp forests, once logging roads gave access.[337] The intention of the

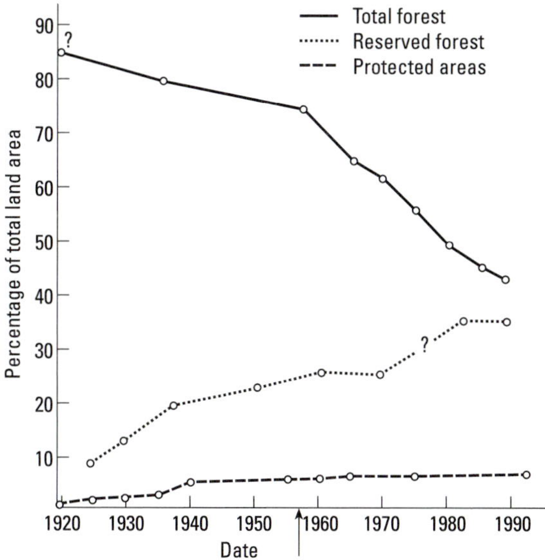

Fig. 10.15. Rain forest cover, reserved forest and protected areas, Peninsular Malaysia. (Aiken 1994, fig. 1.)

The nation attained Independence in 1957.

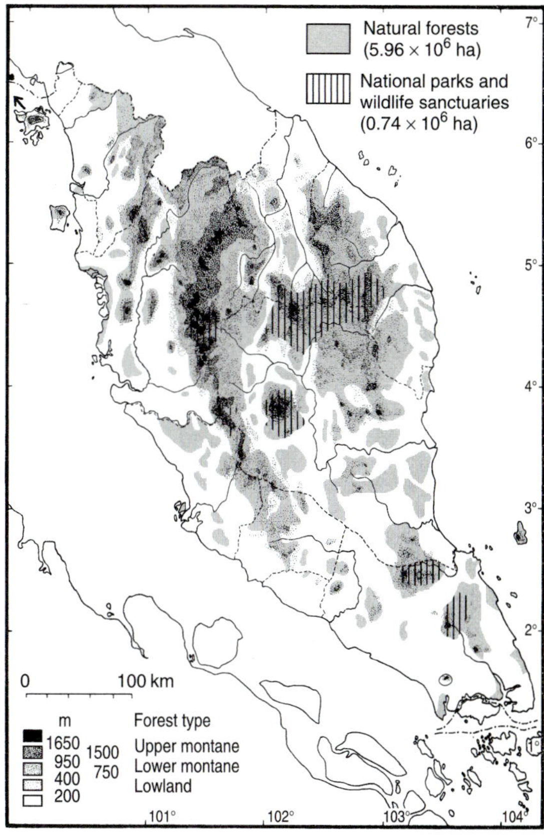

Fig. 10.16. Forest cover of Peninsular Malaysia in 1995. (After Thang 1995.)

Most remaining forests are only on hilly and mountainous land. Much is lower and upper montane rain forest and contains no dipterocarps or other commercial timber. The fauna is different and less species-rich than the lowlands and lacks most of the charismatic Sunda Shelf animals (primates, cats, elephant, gaur, etc.).

Indonesian government had been to make these forests part of the permanent forest estate to provide a continual source of timber. In the Philippines most production forests have been similarly invaded and destroyed. A ban on commercial logging there, progressively enforced from the early 1990s, has hampered access to the few remaining blocks of virgin lowland forest. On Palawan, where most of this remaining forest lies, an attempt is being made by a European Union aid project to stabilize the agricultural frontier by increasing farmers' productivity and controlling the amounts of minor forest products harvested, working through and improving local democratic institutions. But with population increasing at 8 per cent per year, due equally to natural increase and immigration, pressure on the forest is intense. In Nigeria, subsistence farmers have now nearly totally destroyed the rain forest: less than 10 per cent remains. In Ivory Coast, the last remaining large lowland rain forest, the Taï National Park (a World Heritage area, Table 10.15), is under intense threat from encroaching farmers, many of them refugees.

A road, the BR-364, funded by the World Bank, was built into Rondônia State in southwest Amazonian Brazil during the 1970s. It was planned to move in farmers to settle the better soils. In fact, landless peasants invaded from southern Brazil and overran the region. Population increased from 10 000 in 1960 to over a million by 1985 and has continued since. Forest loss accelerated from 1.2×10^5 ha in

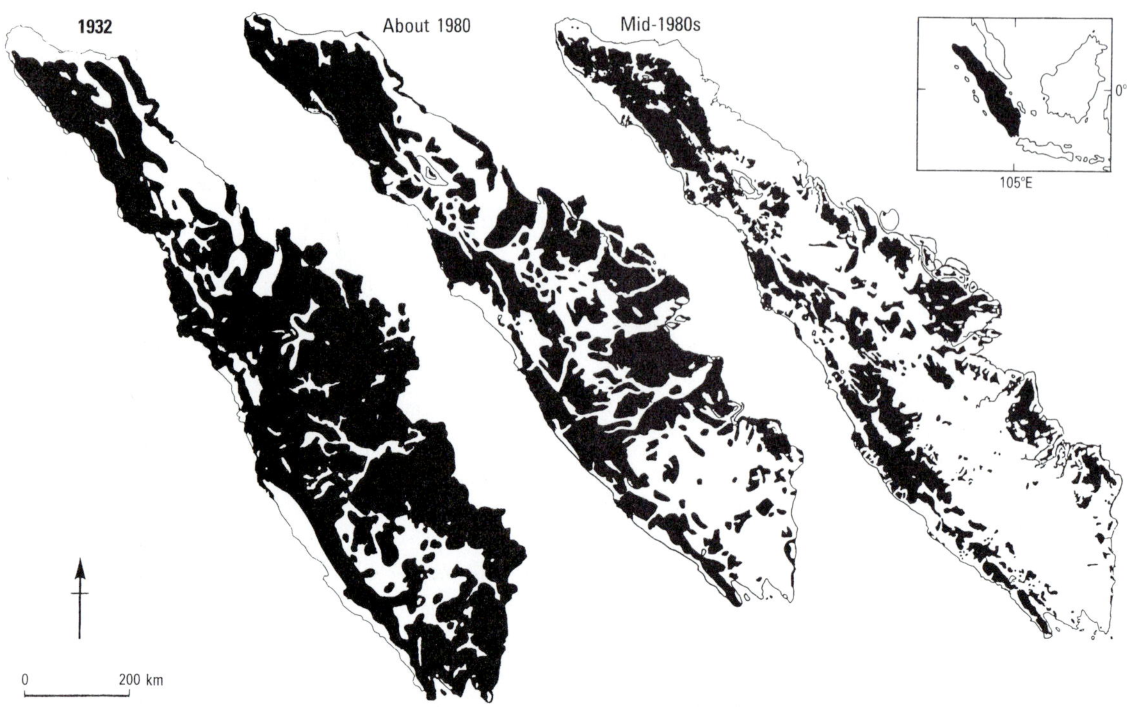

Fig. 10.17. Loss of primary forest, Sumatra. Shown black is primary forest and forest that is logged but not substantially degraded. (Whitten *et al.* 1984, fig. 1.20 and data of Whitmore 1984*b* and Laumonier *et al.* 1983, 1986*a*,*b*.)

Loss is principally in the lowlands and is due to planned conversion to agricultural, including transmigration, or to plantation forestry. In addition there has been illegal clearance by farmers after logging. Once logging roads gave access, spontaneous invasion occurred. In the three decades from 1970 humankind essentially liquidated the lowland rain forest from this, the fourth largest island on earth, 1700 km long, 475 000 km^2 in area.

Fig. 10.18. Lowland semi-evergreen rain forest being cleared for tree plantations, the wood used to fuel a power station. Jari, Amazonian Brazil.

Fig. 10.19. Access road for oil exploration built from the base of the Andes into the Amazonian rain forest, Ecuador.

Throughout the humid tropics farmers and hunters quickly penetrate previously inaccessible rain forests as soon as such roads are built.

1975 to 16×10^5 ha in 1985 (Fig. 10.20). Parts of the land were so infertile that agriculture soon failed and a low scrub replaced what had been pristine high forests populated at a low, sustainable density by Amerindians only a few years before. In the 1980s this road was extended westwards into Acre State. The tappers of wild rubber in Acre united in opposition to the felling of their forests for pasture and were harassed by certain of the developers. Many hundreds of tappers were murdered, including, in late 1988, their spokesman. The outside world at last awoke to what was happening and stood aghast. This sad episode certainly increased international sensitivity to the progressive rape of Amazonia. One response, supported by international campaigning groups, has been to create Extractive Reserves, within which traditional collection of rubber, Brazil nuts and other produce is protected. In 1996 Rondônia created 1×10^6 ha of such reserves. The challenge is to demonstrate that they are an economically viable form of land use now that the forests are so much more accessible than formerly.

Pressure of increasing population and associated landlessness, poverty and inequality, drives all these human migrations. Increasing population leads also to an increase in the area under the sustainable kind of shifting agriculture. It too will break down when expansion is no longer possible and increasing population forces cultivation beyond the carrying capacity of the site (section 8.1). Commonly, however, farmers show flexibility, and come to rely on shifting agriculture to provide only a part of their food (Sarawak) or else modify it towards cash cropping (northern Thailand (p. 185)).

Shifting agriculture in its multifarious forms has been identified as the single most important reason for loss of tropical forests by FAO. In the early 1980s it was estimated to be responsible for 35 per cent of all New World deforestation (with conversion to pasture second), and in Africa and Asia 70 per cent and 49 per cent, respectively.[338]

Dams[339]

Many forest people live along rivers; they farm in valleys, and depend largely on fish for protein. Thus, although dams flood only a small fraction of the landscape, they have a disproportionately heavy impact, inundating riparian forests and fertile alluvial soils, destroying aquatic ecosystems, and displacing whole villages. The resettlement of the people ousted by the reservoirs has nearly always led to great hardship, and has been one of the most contentious aspects of dam building.

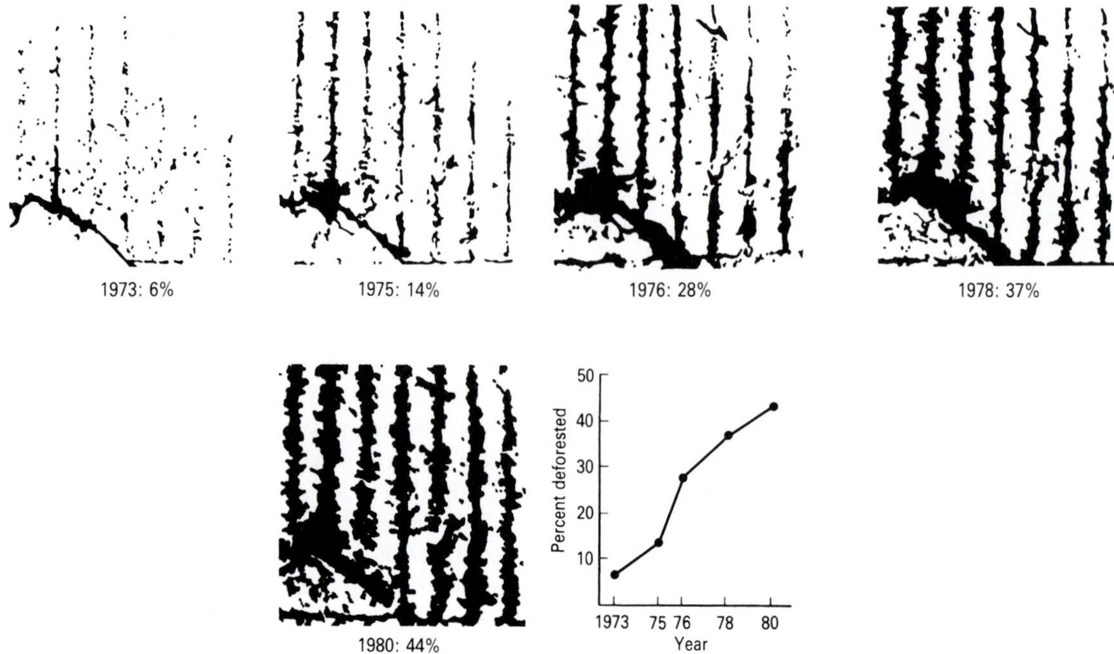

Fig. 10. 20. Progressive clearance (shown black) of forest for farms in southwest Amazonia 1973–80 along roads 5 km apart. The main national highway, BR-364 is in the southwest. The area is the quarter degree square whose northwest corner lies at 11°15′S, 61°30′W. (Fearnside 1986, fig. 5.)

In Brazil, a country with very little fossil fuel, the Tocantins river, a major southern tributary of the lower Amazon, has had one large hydro-electric dam constructed south of Tucuruí which has flooded 216 000 ha. Other south-bank tributaries have further suitable sites, where these northwards flowing rivers run as cataracts down off the hard rocks of the Brazilian shield on to the great Tertiary central Amazonian sedimentary plain. Vigorous protests in the late 1980s by Amerindians and others led to the World Bank withdrawing its offer of finance and the postponement or cancellation of further planned dams, including six on the Xingu.

In Sarawak, one hydro-electric dam was completed in 1985 and a second is under construction. The Batang Ai dam flooded 8700 ha and displaced some 2800 traditional shifting cultivators from 26 longhouses. The Bakun dam, in construction in the mid 1990s, will flood 69 500 ha (an area larger than Singapore), and displace 9400 people. Both schemes disrupt the longhouse societies they displace. The hope is that, given fair compensation, the farmers will display their historic flexibility in adapting to new circumstances.[340]

Mines

Mines are also totally destructive of rain forest and restoration of forest cover after their operation may be unsuccessful or neglected. Although the mine itself may only occupy a small area, there is sometimes much wider-scale forest disruption. An example is Carajas in southeastern Amazonian Brazil.[341] Here, vast rich deposits of iron ore and other minerals have been exploited since the mid-1980s. The mine itself is to be reforested but the Greater Carajas project involves massive development of a large region centred on a 900-km railway built to the Atlantic coast. The ores are largely exported to

Europe, and the scheme has received substantial funds from both the European Union and the World Bank. Some of the iron ore is smelted at furnaces situated along the railway, using charcoal made out of rain-forest trees.[342] It is planned eventually to produce charcoal from vast *Eucalyptus* plantations. Based on yields of eucalypts at the Jari plantations further north, 700 000 ha will be needed, seven times the area of Jari. However, because of the high cost and technical difficulties of establishing plantations, charcoal is likely to come from the natural forest for as long as any exists. Felling for charcoal and for agriculture is currently laying waste a broad corridor of forest along the railway and the livelihood of the people who inhabited it.

Exploration for and extraction of oil has led to damage to the rain forest; for example, in the eastern Andean foothills in South America. Traces cut for seismic prospecting, roads for moving equipment, and pipelines all create access routes into previously inpenetrable forest. Hunters and sometimes farmers follow. Drill sites are now being built that are accessible only by helicopter. This reduces but does not eliminate impact.

10.6 LOGGING AND THE TIMBER TRADE

Replacement of forests by agriculture totally destroys them. If farmland is abandoned it is likely to take several centuries before all signs of forest succession have disappeared, and species-rich, structurally complex primary forest restored (section 7.9). By contrast, selective removal of timber leaves a forest rich in primary species of trees, epiphytes and climbers (Fig. 10.21), which has only to return to structural maturity. The time that takes will depend on the intensity of exploitation (Figs. 7.22, 7.23, 7.24, 10.22) but certainly after a century only the very practised eye will be able to detect that there has been human disturbance.

It is conventional to divide wood utilization into fuelwood, which includes charcoal, and other uses, so-called industrial wood. Table 10.5 shows global annual wood use and its various components. About two-fifths of global use is as fuelwood of which two-thirds is from tropical broadleaf species (i.e. hardwoods).[343] This reflects the importance of domestic fuel in the tropics, and because the greatest populations are in the seasonal tropics the main source is tropical seasonal forests.[344] Industrial wood is supplied 70 per cent by conifers which, because of their long fibres, are still pre-eminent for paper manufacture (though new technology increasingly permits use of short-fibred wood), and provide the main constructional timber of the northern industrial nations. Only 11 per cent of total annual use of industrial wood is supplied from the tropics, and this mainly comes from tropical rain forests, not seasonal forests.[345] With the great attention given today to the exploitation of rain forests it is commonly forgotten that they are small contributors to the global timber trade by comparison with boreal conifer and north temperate forests (Table 10.6).

The exploitation of rain forests for timber has progressively expanded (Fig. 10.23). Southeast Asia[346] has become the increasingly dominant

Table 10.5 World wood production (million m^3)

1 Production by end use and region (1991)	World	Tropics
Total production	4410	1650 (37%)
Fuelwood*	1830	1300 (71%)
Industrial wood	2580	346 (13%)
Pulp and paper	399	22 (5%)

2 Production by timber class and source (1989)	Conifer (softwood)	Broadleaf (hardwood) tropical	Broadleaf (hardwood) temperate
Fuelwood*	16%	65%	19%
Industrial wood	70%	11%	19%

* Fuelwood includes charcoal
From (1) ITTO (1993*a*), (2) FAO (1989)

Fig. 10.21. Lowland evergreen rain forest on Vanikolo, Santa Cruz Islands, Melanesia, 8 months after the timber (mainly *Agathis macrophylla*) was extracted. Note regrowth of the big-leaved pioneer *Macaranga aleuritoides* on the right. The climate is extremely wet, hence the abundance of climbers and epiphytes.

Logging destroyed all the virgin stands (and source of seed) of *Agathis* but there was vigorous regeneration from seedlings. No silvicultural system was in operation. This was a 'cut and clear-out' operation, and could loosely be referred to as a clear-felling for *Agathis*, but as can be seen the matrix of the forest survived (except along tracks as in the centre of this photograph), as did the very rich flora (it is unlikely that any species was extirpated). Clear-felling of tropical rain forest for timber has a quite different result from clear-felling in Europe or North America because much less is removed. Where, as here, the term is loosely applied to a monocyclic felling operation, it has misleading connotations which can cause confusion for the tyro.

Fig. 10.22. Lowland dipterocarp forest soon after logging. Kalabakan, Sabah.
The extraction tracks are visible, but the matrix of the forest remains intact, although more ragged than that of virgin forest.

source (Table 10.7, Fig. 10.24). West Africa has become less and less important. Nigeria is now a net importer and, with Ivory Coast and Sierra Leone, faces total disappearance of its rain forests. In the past few years tropical American exploitation has begun to increase.

Throughout the Eastern rain forests there has been a trend to advance from the export of logs to sawnwood, veneer, plywood, mouldings, furniture and fittings. Log exports from the region dropped from 60 per cent of total production in 1980 to 13 per cent in 1994–6. The same boom cycle followed by bust has been followed by individual countries in succession[347] (Fig. 10.25). The Philippines and Thailand are now timber importers. In Sarawak, heavy exploitation of dipterocarp rain forest began in the mid-1970s. Late in 1989, and after international criticism, the government invited a mission from ITTO,[348] since when annual production has been reduced, but is still much above the estimated sustainable level (Fig. 10.26).

The total export market is divided roughly three ways: the main market for logs is Japan, which took 48 per cent of all tropical logs traded in 1992; Europe imports sawnwood and plywood; and North America imports plywood.

Table 10.6 Top ten timber exporters 1989 (10^9 £)

Canada	11.5 (*c*. 20% world total)
USA	7.6
Sweden	5.4
Finland	5.3
West Germany	3.8
USSR	2.4
Indonesia	**2.3 (*c* 3.6% world total)**
France	2.1
Malaysia	**1.9**
Austria	1.8
All developed countries: (82% world total)	

From ITTO (1993*b*), the figures are for exports of all forest products
The international tropical timber trade accounted for only c. 10 per cent of the total volume of all logs and sawnwood exported globally during 1987–96, while tropical plywood exports now account for two-thirds of global export volume (Pleydell & Johnson 1997).

Now, as the Southeast Asian nations are becoming progressively exhausted and timber supplies are dwindling, a disturbing predatory pattern is evolving. Asian logging companies are moving in to Africa, and even more to tropical America, to extract logs to feed the timber processing industries back at home. Cameroon,

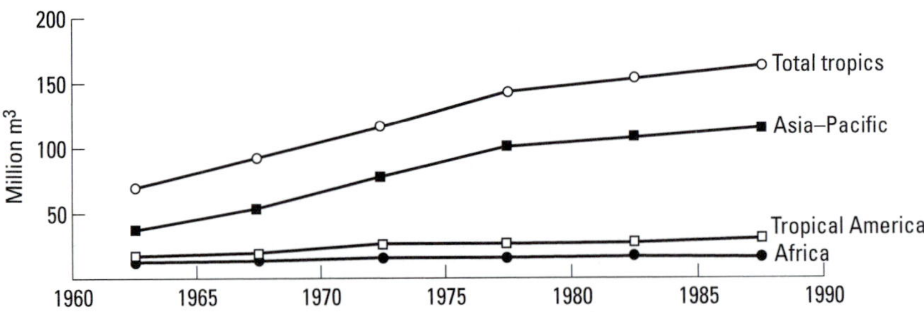

Fig. 10.23. Tropical hardwood production (saw logs and veneer logs). (After FAO 1993, fig. 20.)

Table 10.7 Annual logging 1981–90 in all natural tropical forests

	Area logged (10^6 ha)	% Natural forest	% Logged first time	Harvest intensity (m^3 ha^{-1})
America	2.58	0.3	89	8
Asia	2.15	0.7	82	33
Africa	0.91	0.2	74	13
Total	5.64	0.3	84	18

From FAO data analysed in Whitmore (1997) table 1.3

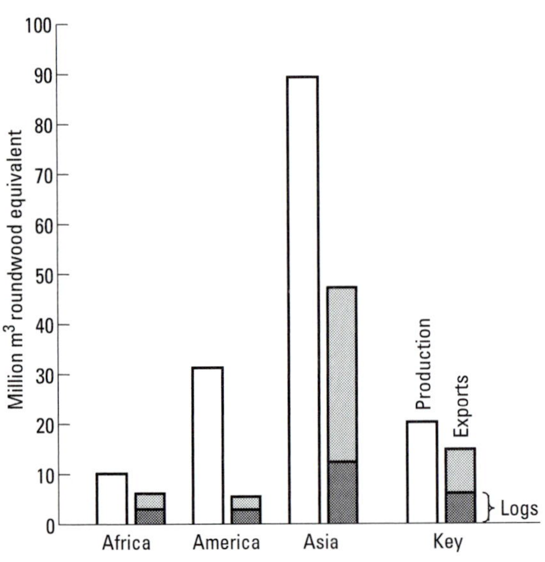

Fig. 10.24. Annual tropical timber production and exports 1994–96 by continent. (Data of Johnson 1997.)

Asia dominated production and total exports; although it also dominated log exports these were a much smaller fraction than elsewhere. Tropical America had the smallest proportion of exports to production because there are large internal markets. Indonesia, Malaysia and Brazil accounted for *c.* 70 per cent of total production. Non-ITTO countries are excluded, but their production was tiny.

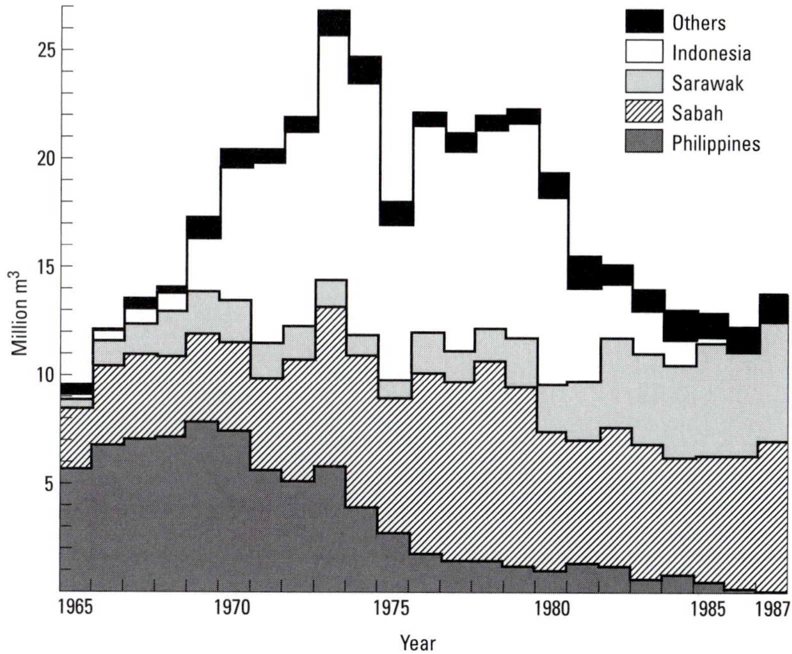

Fig. 10.25. The changing source of rain-forest logs imported by Japan 1965–87. (Nectoux and Kuroda 1989, fig. 10.)

The supply from Philippines, Sabah, Indonesia and Sarawak expanded in sequence as each country in succession banned log exports. After 1987 the supply from Sarawak increased further (Fig. 10.26), together with Papua New Guinea and Solomon Islands. The Philippines has now banned commercial logging: it took 30 years to virtually liquidate its lowland rain forest and is now a net importer of timber. In Sabah production is down from *c.* 10×10^6 to *c.* 3×10^6 m^3 $year^{-1}$ and very little unlogged forest remains. Indonesia will exhaust her primary forest early next century.

Gabon, Belize, Suriname,[349] Guyana, and Brazil all now have Asian-owned or funded concessions. Their forests have much less commercial timber than Southeast Asia per hectare, commonly under 10 m^3 as opposed to 50 m^3 or more, so the economics of harvesting are totally different. Moreover, having logged carelessly and sometimes wantonly in the Asia–Pacific region, where there are reasonably strong regulations, the auguries of this new South–South partnership for the future of these newly opened-up rain forests are not encouraging. By 2010 it is predicted that the total tropical log harvest will have declined by 65 per cent, with the gap within the tropics between supply and demand filled from plantations, by better utilization, and by imports.[350]

Rain forests have a few high-value so-called 'cabinet' woods, worth \$1000 m^{-3} or more. These are timbers with a specialized market. Ebony, satinwood and mahogany[351] are examples, together with teak from tropical seasonal forests. But most rain forest timbers are of low value, \$200–300 m^{-3}, which in real terms has stayed the same for half a century. These so-called 'commodity timbers' are interchangeable with one another and with timbers with similar properties from temperate climates, for example, plantation-grown *Pinus radiata* and North American mixed hardwoods. It is the rain

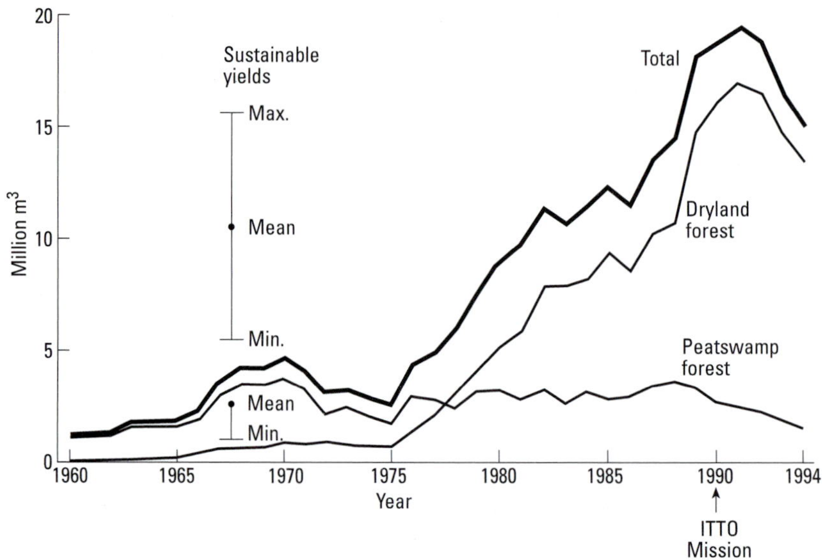

Fig. 10.26. Recorded log removals, Sarawak. (Brunig 1996, fig. 5.1, from Forest Dept. reports.) Dryland evergreen rain forest exploitation expanded rapidly after 1975 and soon overtook peat swamp forests as the main source of logs. In Sarawak, as elsewhere, part of the cut comes from land being cleared for other uses, the rest from areas designated as forest reserves, the so-called Permanent Forest Estate. It is these reserves which, if managed properly (section 7.6), provide the sustainable yield.

forest commodity timbers whose exploitation has burgeoned over the past half century. If attempts are made to raise their price, or their supply is reduced, the market simply moves to another source.

Dipterocarp rain forests have special attractions for the logging industry. The harvest intensity of commercial timber is much higher than in other tropical rain forests (Table 10.7). In addition, dipterocarps have the unique, commercially-useful attribute that they can be grouped for sale into just a few end-use classes (Table 10.8). Thus, even though the Southeast Asian dipterocarp rain forests are very diverse and rich in species, a steady supply of a few grades of timber is maintained and this suits the users. An attempt to boost the sale of West African timbers by grouping them largely failed because the properties of the individual species were too diverse. South American timbers too have defied grouping into end-use classes, with the result that continuity of supply cannot be assured and a big international market has not developed. Moreover, many Amazonian species have heavy, dark, siliceous timber (Table 10.9), which does not meet modern requirements and for which there is little international demand except for flooring. The main markets are internal and less discriminating, for example, within Brazil,[352] where the share of Amazonian timbers in the national annual wood harvest increased from 14 to 44 per cent in the decade 1976–86; they were extracted from dryland rain forests which had been made accessible by newly built roads. Now, from Paragominas municipality in northern Pará State, for every beef truck leaving for southern Brazil, dozens more leave carrying wood. Residual trees are in future likely to be felled for charcoal, with the Carajas-San Luis iron ore railway only 200 km distant. The first charcoal for the smelters built along this railway left Paragominas late in 1988.

Light hardwoods form only a low percentage of many Amazonian forests. In particular

Table 10.8 Numbers of timber species in the top five end-use classes and their share of total log production

	Number of species	% Total log production	Production* ($m^3 \times 10^6$)
Southeast Asia†	195	59	50
Africa	7	45	10
South America	13	40	31

From Erfurth and Rusche (1978) table 4
* In 1995 (Johnson 1996)
† A few species occur in more than one class
Only in Southeast Asia can numerous species be sold together as a few timber groups and that region has far higher production.

Table 10.9 Timber density classes in Pará State, Brazilian Amazonia, amongst trees over 0.25 m in diameter with over 2 m^3 timber ha^{-1}

		Timber density class (%)*		
Region (area and % sampled)	Volume ($m^3\ ha^{-1}$)	Light (≤ 500)	Medium (500–700)	Heavy (≥ 700)
Xingu-Tocantins (1.7 million ha, 0.01%)	154	3	18	79
Tocantins-Guama/Capim (3.1 million ha, 0.01%)	136	2	15	83
Tapajos-Xingu (1.5 million ha, 0.03%)	127	1	37	62

Data of FAO. Simplified from Whitmore and Silva (1990)
* Air dry density in kg m^{-3}

Swietenia macrophylla, mahogany, is absent from the central part of the basin, occurring solely in a broad southern belt. Amazonian mahogany[353] has entered international trade since the strategic roads were built in the early 1970s. The pastures of Pará State occur partly within the mahogany belt, and many fenceposts and farm gates there are built from it. The export market has been divided about equally between USA and Great Britain, though in the mid-1990s British imports declined. In late 1995, Brazil declared a morotorium on new concessions for mahogany exploitation, in response to fears that Amerindians living in the mahogany forests had been abused.

Chips

A more intensive utilization of rain forest than as a timber resource is the production of wood chips. The few forests where this is happening were discussed in section 8.3. Indonesia has plans for extensive industrial tree plantations[354] and associated pulp mills, all of which in the early years will exploit not only the forest being cleared for planting (Fig. 8.7) but also other forests within range of log transport, and so represent a considerable threat to the rain forest. Wood chips are also obtained from some Asian mangrove forests and used in Japan as the raw material for rayon.

Fuelwood

Fuelwood production from rain forests has until recently been relatively unimportant, with the notable exception of the virtual destruction of the Brazil Atlantic Coast rain forests for industrial charcoal.[355] There are major plans for the Amazon forests: fuelwood for power stations and to produce charcoal for iron-smelting along the Carajas railway, which has already commenced.

10.7 ANIMALS IN LOGGED FOREST[356]

Research since the early 1980s by A.D. Johns at three locations in lowland rain forest, at Sungai Tekam in the Jengka area of Malaya, at the Danum valley, Sabah, and near Tefé, Brazil, has shown that at those forests bird and mammal populations were surprisingly robust in the face of a single logging operation. Similar discoveries are now being made in other rain forests. Species relative abundances can change markedly but total species numbers and the relative frequency of species in the different feeding guilds show little change.

To consider Sungai Tekam in detail, logging occurred on part of a big area of forest. Of 56 'large' mammal species 9 per cent occurred exclusively in primary forest. The canopy-dwelling primates (Fig. 10.27) and squirrels all returned soon after logging. By 12 years after logging, all but five of the 193 bird species had been rediscovered, though their population densities were lower and there were changes in the proportions in different feeding guilds, with specialized insectivores hard hit. At the Danum valley all but 10 of 223 bird species were found in logged forest. This second survey took place in an isolated block of forest which was totally logged. It showed that animals found after logging are indeed the original population. Other studies of birds before, during and after logging in Queensland and French Guyanan rain forests gave similar results, but with less complete recovery in the latter.[357]

These studies indicate that most animal species are adaptable to the changed conditions

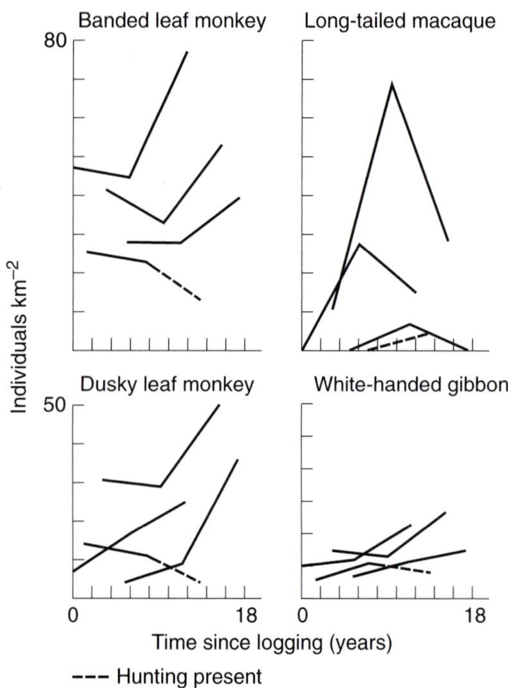

Fig. 10.27. Primate populations 18 years after logging. Sungai Tekam, Malaya. (Johns and Johns 1995, fig. 1.)

The long-tailed macaque, which strongly increased then decreased as the forest recovered, is the species common in disturbed forest near habitations. The increase after several years of both leaf monkeys is believed to be due to the development of very dense and leafy low regrowth forest.

of logged forest. During logging the animals retreat to pockets which remain untouched, and keep quiet (no calling is heard). The restocking of the forest after logging is not by immigrants from outside but from these animals moving back out of these pockets. At Sungai Tekam primates ceased to breed during the logging period, as one year's births were absent when the forest was recensused. Both species of leaf monkey divided into smaller subgroups in response to changed food availability. The gibbons ate more leaves and less fruits, and rested more. The data suggest that species utilizing patchy food resources are pre-adapted to survive logging, unless their resources are reduced too far.

Furthermore, terrestrial browsing species, such as deer, elephant, and tapir, prefer to feed in logged forest because of the greater abundance of browse near the ground.

At Kibale, Uganda, elephants are attacted to gaps created by logging, and perpetuate a low herbaceous tangle. In order to get tree regeneration it is necessary to reduce logging intensity.[358]

There is as yet little information on how invertebrate animals survive in logged forest. Primates in particular, and many birds too, are versatile and able to adapt. By contrast, many insects are very specialized feeders (section 5.3); they may suffer as the proportion of secondary and light-demanding primary-forest species increases.[359]

Thus, the conservation value of logged forest for vertebrates has been clearly demonstrated where, moreover, no special measures were taken to protect them. There are three factors of importance. Firstly, it is necessary to ensure enough food just after felling, for example, of the primary forest fruits required by some specialist frugivores. Secondly, adequate refuges must be maintained, for example, riparian strips of unlogged forest. Thirdly, it is essential to ensure all species recolonize and stabilize before the next logging event. Under a polycyclic system repeat logging can occur after about 25 years (section 7.6). By that time the national timber supply is likely to be less and more species to be commercial,[360] so patches of primary forest that provided refuges at the first logging may be eliminated. Thus, although research so far gives grounds for some optimism, in fact the long-term prognosis for species survival in repeatedly logged production forest is not good unless logging practises are modified in explicit realization of the essential contribution production forest must make to biodiversity conservation (p. 227).

10.8 RATES OF DISAPPEARANCE

Loss of area

It is estimated that in 1990 the total remaining area of all tropical forests was 1756 million ha, with 52, 18 and 30 per cent in America, Asia and Africa, respectively. The area of tropical moist forests (after deduction of dry and very dry forests) was 1510 million ha, 86 per cent. Within moist forests, lowland rain forest totalled 715 million ha, with roughly three-fifths, one-fifth and one-quarter in America, Africa and Asia, respectively. Seasonal forests, 591 million ha, were less extensive than rain forests, and least extensive in Asia (Table 10.10). These data were collected by FAO[361] and are the most comprehensive and reliable available. FAO also estimated (Table 10.11) that the loss rate of all natural tropical forests over the decade 1981–90 was 15.4 million ha (0.8

Table 10.10 Estimated area of tropical moist forests in 1990 as million ha (and percentage)

	Lowland rain	Lowland seasonal	Hills and mountains	Total area
America	451 (63%)	298 (50%)	122 (60%)	871 (58%)
Asia	178 (25%)	42 (7%)	46.5 (23%)	266 18%)
Africa	86.6 (12%)	251 (42%)	35.2 (17%)	373 (25%)
Global total	715	591	204	1510

FAO data analysed in Whitmore (1997) table 1.1

Table 10.11 Annual loss of tropical moist forest from 1981–90 as percentage (and million ha)

	Tropical moist forests				All natural tropical forests (%)
	Lowland rain	Lowland seasonal	Hills and mountains	Total	
America	0.4 (1.9)	0.96 (3.18)	1.2 (1.66)	0.72 (6.74)	0.75% (7.4)
Asia	1.2 (2.23)	1.4 (0.68)	0.95 (0.49)	1.1 (3.4)	1.1% (3.9)
Africa	0.51 (0.47)	0.82 (2.25)	0.75 (0.29)	0.75 (3.0)	0.72% (4.1)
Global total	0.64 (4.6)	0.94 (6.1)	0.93 (2.4)	0.9 (13.1)	0.81% (15.4)*

FAO data analysed in Whitmore (1997) table 1.2 (corrected)
* The annual loss has since dropped to 13.7×10^6 ha yr^{-1} (FAO 1997).

per cent) year^{-1}. For lowland seasonal forest the loss rate was 0.94 per cent year^{-1}, one and a half times the rate for lowland rain forest. A previous FAO assessment in 1980[362] had estimated annual loss of natural tropical forests likely to be 11.3 million ha for 1981–85. The upward revision is believed to be due to under reporting by certain countries in 1980. One unexpected discovery by FAO[363] was that the annual attrition of tropical forests has been steady at about 15 million ha year^{-1} since the 1960s and has not accelerated, as had been widely believed. Through this third of a century the location of heavy deforestation has moved around, so the global figure is not very informative.

The overall global loss of all tropical moist forests during 1981–90 was estimated to be 13.1×10^6 ha year^{-1} or 0.9 per cent. At the continental scale loss of lowland rain, lowland seasonal, and hill and mountain forests was at a rate of about 1 per cent year^{-1} or less, and nowhere over 1.5 per cent (Table 10.11). However, these average figures obscure large differences between nations. Table 10.12 shows examples from each continent of countries with high and low loss. In the Philippines, national forest cover was down to a quarter by 1990 and lowland forest had virtually disappeared. At the other extreme, French Guyana remained 91 per cent forested. Human population pressure lies behind these statistics: forest area per head in these two countries is 0.1 and 87 ha, respectively. Other countries besides the Philippines are losing forests at a rate which is causing their virtual disappearance, at least from the lowlands, for example, the Atlantic Coast of Brazil, Costa Rica, Malaya and Sumatra (Figs. 10.11, 10.14, 10.16, 10.17).

Loss of quality

Forest area is relatively easy to monitor and can give a more or less instantaneous pantropical picture because forest and other kinds of land surface cover can clearly be told apart on satellite images. But area is a crude measure of the state of the forest: its biomass, structure or biodiversity. More informative measures of forest quality are less simple to make. Two measures which can be obtained fairly easily from remotely sensed images are perimeter/area ratio, and edge/core ratio, the latter being a measure of the fraction of the forest within say 10 km of an edge. For example, the West African forests

Table 10.12 Annual loss of all natural tropical forest from 1981–90 in selected countries (thousand ha)

	Area in 1990	Annual loss since 1980	Forest area per person (ha)
Low			
French Guyana	7997	0.3 (0%)	87
Brunei	458	1.8 (0.37%)	1.7
Congo	19865	32.3 (0.16%)	10
High			
Costa Rica	1428	49.6 (2.3%)	0.5
Philippines	7831	316.0 (2.5%)	0.1
Ghana	9555	137.5 (1.2%)	0.6

FAO data analysed in Whitmore (1997) table 1.4

Table 10.13 Loss and degradation of lowland rain forest, Brazilian Amazonia (million ha)

	Total area deforested	Area in fragments* or < 1 km from an edge
by end 1978	7.8	20.8
by end 1988	23	50.8
Annual rate†		
1978–89	2.11	–
1989–90	1.38	–
1990–91	1.11	–
1991–92	1.0	–
1992–93	1.2	–
1993–94	1.5	–

From Skole and Tucker (1993)
* Area under 100 km^2
†Brazilian Space Agency (INPE) figures, *pers. comm.* N. Higuchi May 1997

have a high edge/core ratio because they are now reduced to small fragments. In southwestern Brazilian Amazonia there is a contrast in the edge/core ratio between the states of Rondônia (Fig. 10.20) and Acre, scoring 75 and 25 per cent, respectively, because the Rondônian forest is much more broken up by roads.

The loss of forest from Brazil in the 1980s (Table 10.13) attracted much international attention. Attrition was largely along the south-eastern fringe where vast areas were cleared and burned to create cattle pasture (p. 192). It declined by the early 1990s, partly because fiscal inducements for clearing were eliminated, but the subsequent rise suggests a downturn in the national economy was also implicated. Table 10.13 shows that allowance for fragmentation and the effects of forest edges, as well as for deforestation, more than triples the area of forest affected by human impact.

Another measure of the loss of forest quality is the reduction in biomass as trees are harvested. Two successive forest inventories of the Malayan commercial forests show that during the 1980s the area was reduced by 18 per cent, but the biomass by 28 per cent, as logging progressed.[364] At present, most forests are still being logged for the first time (Table 10.7). Repeat logging in less than about 25 years is likely to cause greater forest degradation, by removing biomass faster than it is accumulating.

Area logged is less easy to measure than area deforested. The distinction is important but has not always been made. Some earlier and frequently quoted very high figures for the amount and rate of tropical deforestation are in fact the sum of the two forms of impact, a crude measure of the sum of the losses of forest quantity plus quality.[365]

10.9 CONTROLLING HUMAN IMPACT

The dominant image of the age in which we live is that of the earth rising above the horizon of the moon—a beautiful, fragile sphere which provides the home and sustains the life of the entire human species.

The 1972 United Nations Conference on the Environment at Stockholm, at which this statement was made,[366] was an important milepost in attempts to control human impact on the biosphere. Late 20th century humans, with their immense technological power, are seriously out of harmony with nature and fast destroying it. The human species has come to live beyond the resources of 'spaceship earth'. Massive damage is being inflicted on global support systems, using up the natural capital of topsoil, groundwater, tropical forests, fish, and biodiversity.[367] It is now widely but not yet universally accepted that matters have to change, and that exploitation of natural resources must be made sustainable, a concept introduced by the influential Brundtland report.[368] A few optimists still maintain that new resources will always be found and that it is human nature not to respond until the last minute. But extinction of species is irreversable, and what will we bequeath to the future?[369] The control of human impact on tropical forests is part of the control of impact on the biosphere and there have been important developments since 1972.

Fig. 10. 28. Conservation poster, Papua New Guinea. (Gagné and Gressitt in Gressitt 1982.) A literal translation of the Pidgin caption is 'Preserve the dugong for the future. If you want to injure our national animal [don't], behave as if there was a taboo'.

Citizens' responses

Throughout the world citizen groups have sprung up, so-called NGOs, Non-Governmental Organizations. These have focused the concerns of ordinary people and have been powerful stimulants for governmental action. Doyen amongst them is the World Conservation Union (IUCN) and its offshoot Worldwide Fund for Nature (WWF).[370] Also prominent on an international level are Friends of the Earth, Greenpeace, and World Resources Institute. On a national level may be mentioned the Malaysian Nature Society,[371] national branches of WWF, WALHI in Indonesia, and IMAZON in Pará State, Brazil, to name just a few. It is difficult for the NGOs of industrial nations to focus protests accurately about matters within nations with rain forest. Potentially more powerful are the increasingly vocal local protests[372] (Figs. 10.28, 10.29). Citizen groups, spearheaded by the Malaysian Nature Society, have since 1970 fought off three separate threats to the country's main national park, Taman Negara (434 000 ha),[373] but have so far not achieved secure legal conservation status for additional large areas in the north and south, at

Fig. 10.29. Conservation poster, Brazil. This, the most ape-like monkey of the neotropics and the largest endemic mammal of Brazil, is one of the many endangered species of the Atlantic Coast rain forest. The Portuguese caption reads, 'It's ours, the largest in the Americas, the Muriqui monkey'.

Belum and Endau-Rompin,[374] respectively; the latter 89 000-ha area contains the last remaining Malayan breeding herd of 20–25 Sumatran rhinoceros.

There is a spectrum amongst NGOs from radical campaigners to measured judgers. The campaigners act to sensitize public and political opinion, more by dramatic appeal than factual accuracy. The latter group aim to maintain dialogue with decision makers. Both groups have a role to play, and their actions regarding tropical forests have evolved as their knowledge and sophistication has increased. Unfortunately, tropical forest science and forestry have become more technical and complex but professionals have been fearful of their reputations if seen to be involved in political campaigns and a gulf has opened between scientists and foresters and the more radical NGOs.

NGOs have initiated major inventories of forests and their constituent species, for example, the World Conservation Monitoring Centre (WCMC) produces Red Data Books on threatened and endangered species, and lists of National Parks and Equivalent Reserves. *The World Conservation Strategy* and its sequel *Caring for the Earth* a decade later have made powerful contributions to the evolving global conservation ethic.[375]

International responses

Collectively, NGOs, by lobbying governmental and other aid-giving bodies, by dialogue, and through publicity on television and in magazines, have been important catalysts for the developing and changing responses of national and multinational institutions. The numerous initiatives that have attempted to address human impact on tropical forests are summarized in Table 10.14. They have burgeoned since the mid-1980s. The second United Nations Conference on Environment and Development (UNCED) at Rio de Janeiro in 1992, 20 years after Stockholm, was another important milepost.[376] There, developing nations rightly broadened concern for forests to cover the whole earth. No binding agreement could be reached,[377] but both the Biodiversity and Climate Change Conventions, which were agreed, have important implications for forests. Since Rio there has been a plethora of committees, commissions, workshops, and dialogues and the cynic may ask what real action might one day materialize.

Governmental responses

Looking back[378] it can be seen that colonial administrations, whilst actively encouraging capitalist entrepreneurs, sometimes did show concern both for conservation objectives and for the rights of indigenous peoples. Post-colonial societies have given so much weight to the political imperative of encouraging rapid economic development that national governments have

Table 10.14 Alphabet soup: important international, multilateral initiatives and bodies now impinging on rain forest conservation

WORLD BANK		and its affiliates
FAO	1946	Food and Agriculture Organization of United Nations
IUCN	1948	International Union for Conservation of Nature
CITES	1975	Convention on International Trade in Endangered Species of Flora and Fauna
ITTO	1983, 1992	International Tropical Timber Organization
TFAP	1985, 1991	Tropical Forest Action Plan (later Programme)
WCED	1987	World Commission on Environment and Development (the Brundtland report)
WCMC	1980s	World Conservation Monitoring Centre
IUFRO	mid-1980s	International Union of Forest Research Organizations launched a developing countries programme
GEF	1991	Global Environmental Facility, providing funds to implement CBD
UNCED	1992	'Earth summit' United Nations Conference on Environment and Development, Rio de Janeiro: UN Framework Convention on Climate Change UN Convention on Biological Diversity (CBD) AGENDA 21 Chapter 11 CSD Commission on Sustainable Development (oversees implementation of Agenda 21) IPF Intergovernmental Panel on Forests 1996–97 (spawned by CSD)
CIFOR	1993	Centre for International Forestry Research, part of Consultative Group on International Agricultural Research (CGIAR)
WCFSD	1994	World Commission on Forests and Sustainable Development
G7	1990s	Pilot programme for the Brazilian Amazon of the Group of Seven industrialized nations

tended to disregard such constraints on policy. Nevertheless, since the 1970s large areas of tropical rain forest have been put into National Parks, which now cover up to 10 per cent of national land area.[379] In many countries it is now too late to create large new National Parks *de novo*, no matter how strong the scientific case, because land-use plans have been crystallized (Fig. 11.1). Emphasis now needs to shift to develop and implement management plans for what are still often only parks on paper and which may be less than ideal in configuration. Some conservation areas have been declared Biosphere Reserves, a concept invented by UNESCO in which an inviolate pristine core is surrounded by a 'buffer zone' of forest managed for sustainable production and which may also include cultural landscapes of plantations, orchards, fields, and pastures. It exemplifies well the concept of sustainable utilization with the maintenance of full diversity and species richness: humanity living in balance with nature. Other parks have become World Heritage areas (Table 10.15): in order to qualify for inclusion an area has to be of 'acknowledged universal natural or cultural value'.

Another governmental response to modern issues has been to create institutions for environmental management. These commonly lie outside the power structure and so have limited capacity for enforcement. Typically their responsibility includes Environmental Impact Assessments (EIA) of proposed developments. These are now usually required before funds are released for projects that will necessitate alteration of forests. Unfortunately, sometimes only lip service is paid to their findings. The objective of an EIA is to unravel the complex linkages in an ecosystem and see how a proposed development would affect them. It may then be necessary to modify the plans to reduce or eliminate bad consequences. An example of how alteration of a forest may have totally unexpected but serious repercussions is afforded by the

Table 10.15 World Heritage tropical rain forests

Africa:	
Cameroon	Dja Faunal Reserve, 1987
Guinea and Ivory Coast	Mt Nimba Reserves, 1981, 1982
Ivory Coast	Taï National Park, 1982
Uganda	Bwindi Impenetrable Forest National Park, 1994
	Ruwenzori Mountains National Park, 1994
Congo	Kahuzi-Biéga National Park, 1980
	Salonga National Park, 1984
	Virunga National Park, 1979
America:	
Colombia	Los Katios National Park, 1994
Costa Rica	Talamanca Range—La Amistad Reserves, 1983
Ecuador	Sangay National Park, 1983
Guatemala	Tikal National Park, 1979
Honduras	Río Platano Biosphere Reserve, 1982
Mexico	Sian Ka'an Biosphere Reserve, 1987
Panama	Darien National Park, 1981
	La Amistad International Park, 1990
Peru	Manu National Park, 1987
	Rio Abiseo National Park, 1990
Venezuela	Canaima National Park, 1994
Eastern tropics:	
Australia	Wet Tropics of Queensland, 1988
India	Sundarbans National Park, 1987
Indonesia	Ujung Kulon National Park, 1991
Sri Lanka	Sinharaja Forest Reserve, 1988
Thailand	Thung Yai—Huai Kha Kueng Wildlife Sanctuaries, 1991
(United Kingdom)	Henderson Island, 1988

The World Heritage Convention of UNESCO has been ratified by 135 countries and 128 natural sites have been inscribed on the World Heritage List. Twenty-five of these are in tropical rain forest, a third declared in the 1990s. Note that the Amazon rain forests are scarcely represented. Of the two nations with the most rain forest Brazil has no World Heritage rain forests and Indonesian forests are only minimally covered.

floodplain of the Amazon. The swamp forests which fringe the Amazon and its tributaries lie over alluvial soils, the most fertile in the basin, and occupy *c.* 10^5 km^2. It is sometimes suggested that the land should be converted to agriculture, concentrated here not on the infertile *terra firme* (dry land) oxisols and ultisols. Many Amazonian fish feed on the fruits of swamp-forest trees (p. 79), and both rural and urban populations depend on fish as a major source of protein. Loss of the swamp forests would destroy an essential link in the ecosystem chain, fish populations would collapse, and much human suffering would ensue. There would also be serious repercussions on the hydrology of the river system because of the removal of floodwater storage capacity.

10.10 CAUSES FOR CONCERN

Three topics in particular have emerged as foci for concern: loss of biodiversity, the disruption of human societies, and the role tropical forests might be playing in climatic change.

Loss of biodiversity

Unlike essentially all other scientific disciplines conservation biology is a science with a time limit, with the clock ticking faster as the human population continues to increase.[380]

Biodiversity became a buzzword of the late 1980s. As applied to a tropical rain forest it has several distinct facets. Firstly, there is the structural diversity of the forest canopy, whose reduction inevitably follows from human intervention, as the forests are progressively simplified by increasing degrees of interference, for example, for timber utilization. Secondly, it refers to species numbers of both plants and animals, which are also usually reduced because where destruction is total, by conversion of the forest to agriculture, populations disappear or are diminished.

Local endemics which have only a limited geographical range are particularly prone to extinction, and for animals destruction of habitat can lead to extinction. Reduction within species of either the numbers of individuals or of geographical range is a more insidious loss of biodiversity than extinction, sometimes evocatively called 'genetic erosion', because of the likely loss of ecotypes and other genetic variants.[381]

Various attempts have been made to estimate how many species are becoming extinct at current rates of tropical forest destruction. Such calculations are fraught with imprecision. For example, there are wildly different estimates of how many species occur in these forests (p. 67); species richness and distribution across the landscape is not accurately known, and is not the same for all forests or species; and the effects of forest fragmentation are unknown.[382] A reasoned calculation, based on species-area curves[383] is that, if the 1981–90 rate of tropical moist forest destruction (Table 10.11) had continued for another 25 years, then, by AD 2015, 5–11 per cent of its species would have been either extinct or doomed to extinction. Most of the losses would be unknown and undescribed insects. For birds, already in the decade 1978–88 in Indonesia the number that faced extinction increased from 14 to 126; and in Brazil from 29 to 121, of which 64 occurred in the largely destroyed Atlantic Coast rain forest.[384] It is also fair comment that human-induced extinction today is as great as any of the five previous extinction spasms life on earth has experienced.[385]

The discussion of minor forest products (section 10.3) showed that direct economic benefit is at stake. Contemporary human impact on rain forests and disappearance of traditional ways of life has led to a renaissance in ethnobotany, as a salvage operation to record traditional usages, and to renewed exploration for useful species.[386] This is to some extent a backlash inspired by campaigners to recent overemphasis on timber. There are several products from rain forests which have recently begun to be exploited. Amongst these are a perennial teosinte, *Zea diploperennis*, discovered in late 1978 in relict forest in Mexico,[387] which has useful disease and pest resistance genes for the improvement of maize (*Zea mays*). There are many more opportunities still awaiting development. For example, *Elaeis oleifera* is the South American congener of the African *E. guineensis*, the commercial oil palm. The Amazonian species offers a high quality oil, a lower height because of its creeping trunk, and better resistance to diseases.[388]

Impact on human societies

There are human societies throughout the tropics who live in, or are closely dependent on, the rain forest. Other communities have their roots there and make use of many jungle products. These peoples have their lives disrupted and altered by forest destruction, although many show great resilience and flexibility in the face of enforced change.[389]

The nomadic pygmy hunter-gatherers of the eastern Congo basin Ituri forest in Africa use small forest-dwelling antelope (mainly five species of duiker, *Cephalophus*) as an important source of meat. The antelope populations dimin-

ish unless there are unhunted areas. Gregariously fruiting trees of Leguminosae/Caesalpinoideae are a major food, and are not found in secondary forest, so the antelope cannot survive where large areas of primary forest disappear.[390]

In the Amazon the few scattered Amerindians who have survived the centuries of attrition from outside are coming increasingly into conflict with modernity as dams, mines, farmers, road builders, missionaries, gold prospectors, and most recently mahogany-loggers, move into their territories. Such peoples seldom have political muscle to prevent encroachment, or the skill and the contacts with mainstream society needed to seek redress.

The modern world creates manifold pressures. As a result, traditional societies closely dependent on the tropical rain forest are experiencing many changes. This is poignantly documented for the longhouse dwellers of the Baram river, Sarawak, by a visiting American, W.W. Bevis.[391] He drew a parallel with the north American Red Indians a hundred years earlier, when the modern world moved in to exploit their buffalo and seize their lands for European agriculture.

> Like most longhouses in the middle and upper Baram, Long Anap is on a flat between the river and a hill. The land traditionally belonging to the longhouse extends from the river back to the top of the ridge behind the settlement, and also across the river in a similar swath [e] up to the crest of the opposing ridge ... each longhouse recognizes three kinds of land ... First is cultivated or cleared land ... the second ... is tribal forest, again about a third of the area ... open to common use, including ... cultivation, felling trees for boats and house planks, collecting vines, and any form of gathering or hunting. The third category is a forest preserve, a protected ecosystem. Hunting and renewable resource gathering are allowed, but cutting is prohibited ... Regulations for gathering are very specific: illipe nuts are to be left on the ground for the pigs...
>
> On paper the forest preserves can be protected from logging by each tribe. The natives may go through a process ... within a certain period after a timber licence is granted, that is if they hear about the licence ... To date not a single forest reserve request has been granted...
>
> If they lose 80 per cent of their meat supply, all their free building materials and boats, their wild fruits and nuts to supplement garden vegetables, and perhaps half their cash income, their economy is not [just] affected, it is destroyed. And after the logging has passed ... their sons no longer bring home company pay cheques...
>
> Most ominously, however, they are changing to a new kind of living, from a free economy to a cash economy, in which they are as lost as I am in the jungle. The compensation will be in cash, yet they have no basis, no experience, on which to judge cash value ... Before the logging thirty dollars sounds to these people like a substantial sum; it is, when only five per cent of their economy involves cash ... [only used] for the occasional outlay for roofing and the pooled resources to buy a motor. But if they have to buy all their baskets (plastic baskets ?) ... thirty dollars will seem like nothing. It is exactly at this point that the injustice so easily occurs: a Japanese trading firm worth billions sits down, through its representatives, on the longhouse porch with Borneo natives, and offers cash for their old world to bring them into a new world they cannot imagine. The medium of forest resources, which the native knew so well, is not the language spoken ... the terms are cash, in that medium the natives are illiterate. The Japanese are not.

Despite the impact on traditional forest dwellers it is, nevertheless, necessary to strike a balance and not overlook other poor and vulnerable members of society. This is well-illustrated by the case of Guyana. This nation has 75 000 Amerindians living patchily in or near the rain forest which still covers most of the country, and 750 000 dwellers, of great ethnic diversity, who live on the coastal plain. All are Guyanese and protection of the interests of one group cannot override and ignore the needs of others. The forest resources are substantial and have the capacity to make a major contribution to the welfare of all Guyanans.

Climatic change

An issue on which discussion of rain forest destruction focuses is the changes that might occur to local, regional or global climate. This is

an immensely complicated subject, and it is beyond the scope of the present book to discuss it in full. The experts are not in full agreement but knowledge is improving all the time. Vegetation can affect climate in several different ways, via heat balance, surface roughness, the hydrological cycle (precipitation, how water penetrates and is stored within the soil, and evapotranspiration), and the production of various gases, notably carbon dioxide.

Two other issues are what impact climatic change might have on the composition and functioning of tropical moist forests and the role these forests could have in reduction of the global warming now taking place.

Heat balance

Forests look dark on aerial photographs (Fig. 10.6). This is because they absorb more of the incident solar radiation than other vegetation types, in other words their albedo is low. It has sometimes been suggested that destruction of rain forests will have dramatic effects on regional or global climate. The argument is that any replacement vegetation has a higher albedo because it absorbs less of the incident radiation, and that this alters the regional heat balance and hence the climate. The argument hinges, however, on how much the albedo differs between forest, cultivated crops, and bare land. It can be seen in Table 10.16 that, in fact, all vegetation has a fairly low albedo compared to bare earth. In reality tropical forest is either replaced by agriculture, often as tree crops, or by regrowth forest or scrubland. Nature abhors a vacuum and nowhere more so than in the humid tropics, where bare earth is revegetated within a matter of weeks or months. It follows that the effect on climate by a change to the albedo of parts of the humid tropics may not be very large.

The more heat that is absorbed the more there is available for evapotranspiration of water and to create air convection currents, so albedo by these mechanisms also influences the hydrological cycle.

Surface roughness

Forest canopies have an uneven upper canopy surface, most marked in lowland tropical rain forest with its scattered emergent trees (Table 10.16). Tree plantations and tree crops are also fairly rough compared to pasture. Surface roughness of vegetation affects climate by its influence on air currents and by the increased evapotranspiration which occurs as a result of turbulent mixing at a rough surface.

Hydrological cycle

Vegetation evaporates and transpires much of the rainfall it receives and this water vapour forms clouds which yield further rain. All vegetation has high evapotranspiration compared to bare land (Table 10.16) and in particular primary forest, logged forest, secondary forest, and tree plantations (oil palm, rubber, *Elaeis guineensis*, *Hevea brasiliensis*; plus timber species) have fairly high evapotranspiration rates; this is partly because of the albedo and surface roughness effects described above.

Several studies have been made in the tropics to compare the water yield in the drainage

Table 10.16 Some climatic properties of different surface covers

	Forest	Grass	Herbaceous crops	Bare ground
Potential evapotranspiration (mm year^{-1})	850	550–750	550–750	400–500
Surface albedo	0.12–0.13	0.16–0.2	0.2	0.35
Roughness (m)	2.1	0.026	–	–

Various sources, including Lean *et al.* in Gash *et al.* (1996). See also O'Brien (1996)

streams of paired, adjacent catchments with different vegetation cover and each of a few square kilometres area.[392] In general, the water yield increases if tree cover is removed because evapotranspiration diminishes.[393] These studies also demonstrate that the ability to absorb periodic rainfall into the soil and release it into streams in a steady flow (sometimes referred to as the 'sponge-effect') is not unique to forests, but also occurs in grassland so long as there is a sufficient volume of roots to maintain soil porosity. But where, as is usually the case, a catchment has rock outcrops, bare earth, roads, or buildings its sponge effect is reduced and outflow may become very uneven.[394] Some reduction in the sponge effect is very likely to occur following agricultural development or where a forest is dissected by logging tracks and compacted areas.

There are two cases where tropical rain forest cover probably does have a direct effect on the amount of precipitation.[395] One is that upper montane forests (which occur above the level of prevalent cloud formation, section 8.5) physically strip water from the fog which frequently pervades the canopy. In the wet season this so-called 'occult precipitation' is about 7–18 per cent of rainfall but may reach over 100 per cent during the dry season. If upper montane forest is removed, water entering the ecosystem is reduced. The second case is the Amazon. This vast forest receives its rain from the northeast trade winds blowing in from the Atlantic Ocean; rain falls; much water is evapotranspired, re-enters the atmosphere, is blown towards the Andes, and falls again as further rain. Measurement of the heavy isotopes of oxygen and hydrogen ^{18}O and D (deuterium), which are less evapotranspired, shows that inland Amazonia appears to receive about half of its rainfall from recycled water.[396]

Amazonian deforestation

Climatologists have been curious to explore the consequences of Amazonian deforestation on climate. This can be done by use of General Circulation Models which have become increasingly sophisticated and realistic, but to require progressively greater information on albedo, roughness and water relations of forest and pasture. A joint Anglo-Brazilian team ABRACOS have recently collected such data for several years.[397] Their prediction for complete deforestation is of reduced rainfall (Fig. 10.30). Partial deforestation, a more likely scenario (Table 10.13), was predicted to have similar but less extreme climatic consequences, but the impact on the remaining rain forest might be severe: Manaus already has 73 dry days $year^{-1}$ and it is drier still in the lower Amazon (Fig. 2.1).

'Greenhouse' gases

After years of uncertainty the International Panel on Climate Change (IPCC) finally concluded in 1996 that 'the balance of evidence suggests that there is a discernable human influence on global climate'.[398] Despite a great deal of annual and longer term variation the signal is now greater than the noise, the world's climate is indeed getting warmer under human influence. This is due to the increasing concentration of certain gases in the atmosphere. These gases act like the glass of a greenhouse by preventing the radiation back to space of infra-red wavelengths.[399] The main gas is carbon dioxide and this creates about half the warming, so-called radiative forcing (Table 10.17). The other major natural contributor is methane (CH_4), 15 per cent, with nitrous oxide (NO) contributing 5 per cent. In addition, man-made chlorofluorocarbons (CFC and CHFC) contribute a quarter. Tropical deforestation contributes just under a quarter of all radiative forcing. The rest comes from the burning of fossil fuels, partly in internal combustion engines (which contribute oxides of nitrogen, NO_x, as well as CO_2). The burning of tropical forest adds both CO_2 and NO_x. Regrowth of forest subsequently fixes CO_2 again. Table 10.18 gives some details. The soil below a tropical forest or plantation has a higher humus, and hence soil organic carbon, content than does the soil under agricultural crops.[400] In the lower Amazon forests (which

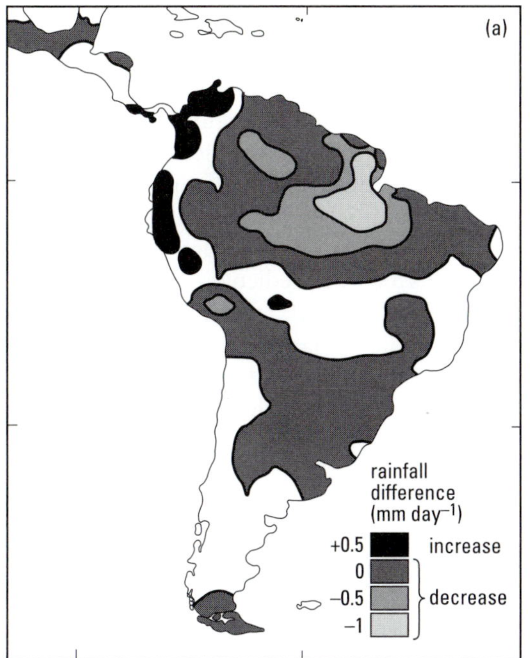

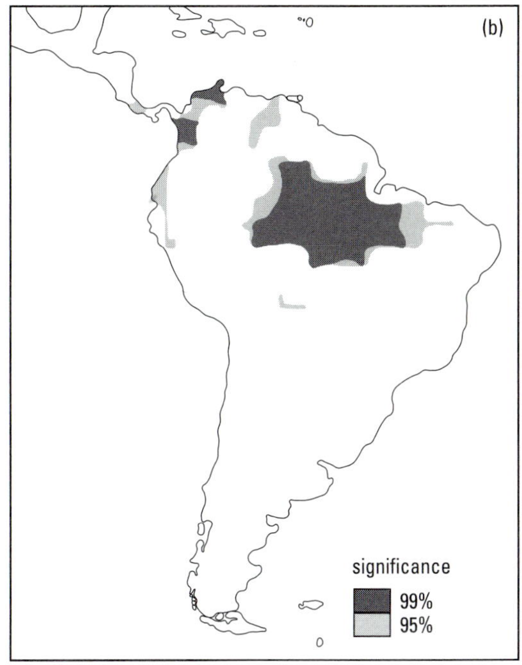

Fig. 10.30. (a) Difference in rainfall after Amazonian deforestation (mm day^{-1}) and (b) its significance. (Lean *et al.* in Gash *et al.* 1996, fig. 6*a*, *b*.)

appear extreme in this respect) there is more carbon in the soil below 1 m depth than in the above ground biomass.[401] Removal of forest cover leads to oxidation of much of the soil humus and release of CO_2 and NO_x. Cattle produce *c.* 300 l methane each per day by bacterial fermentation in the rumen[402] and cattle numbers are increasing as pasture replaces rain forest, especially in the neotropics (Fig. 10.13).

Estimations of known sources and sinks of carbon allow computation of an annual global budget. One major sink is the phytoplankton, which live in the surface waters of many oceans and which incorporate CO_2 both in organic matter by photosynthesis and in their calcareous cell walls. Some CO_2 also dissolves in the oceans as bicarbonate and the magnitude of this buffering effect is difficult to calculate. Sources exceed sinks by 1.4×10^9 tonne year^{-1} even after allowance has been made for the increase in atmospheric CO_2. There is some evidence that certain forests may be this missing carbon sink, partly the boreal forest and possibly also tropical rain forests.[403]

Global climatic warming could affect tropical forests

Consideration of a scenario of probable changes to tropical climates based on General Circulation Models[404] has shown that changes likely to be most influential to forests[405] are decreased rainfall, a stronger dry season, and greater interannual variability. For example, the relative success of different species in the forest floor seedling bank could change, and this would then influence species composition of the canopy.[406]

Mitigation of global warming

As part of the implementation of the Climate Change Convention, IPCC have considered

Table 10.17 Annual contribution of (tropical) deforestation to the greenhouse effect during the 1980s

	Deforestation		Industrial sources	Natural sources	% Radiative forcing	
					total	deforestation
Carbon dioxide (10^6 t C)	2000*	(25%)	5600	0	55	14
Methane (10^6 t CH_4)	250	(50%)	100	150	15	7.5
Nitrous oxide (10^6 t N_2O)	2.0	(20%)	1.3	7.2	6	1
CFCs and CFHCs (10^3 t CFC)	0	–	1	0	24	0
Total radiative forcing (%)					100	22.5

From Houghton (1995*a*) table 15.4
* See Table 10.18
By the 1980s most deforestation was in the tropics.

Table 10.18 1980 flux of carbon into the atmosphere (million tonne) from deforestation and subsequent regrowth of vegetation

Burning of biomass on cleared and harvested sites	+ 700	+ 39%
Decay of slash	+ 1100	+ 61%
Oxidation of harvested products	+ 400	+ 22%
Oxidation of soil humus	+ 300	+ 17%
Regrowth of vegetation + redevelopment of soil organic matter	– 700	– 39%
Totals	+ 1800	100%

From Houghton (1991) table 38.2
By 1980 deforestation was mostly tropical (Houghton 1995*b*) and resulted in a net flux into the atmosphere. By contrast, carbon released by the annual burning of savanna and agricultural waste is assumed to be balanced by subsequent regrowth, so this net flux was nil.

various steps that could be taken to mitigate the increase in atmospheric CO_2. They suggest that forests could be manipulated to sequester more carbon (Fig. 10.31); an extra 147×10^9 tonnes of biomass could be grown in the period 1995–2050, equivalent to 39×10^9 tonne of coal.[407] One move would be to insist on low-impact logging (p. 137), which reduces the amount of CO_2 released, because it decreases damage to the forest.

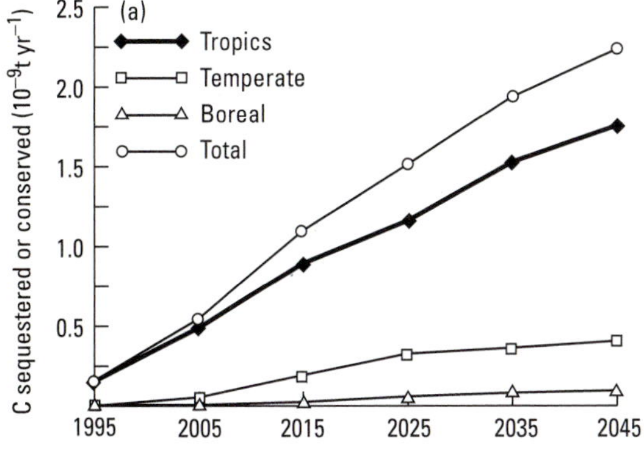

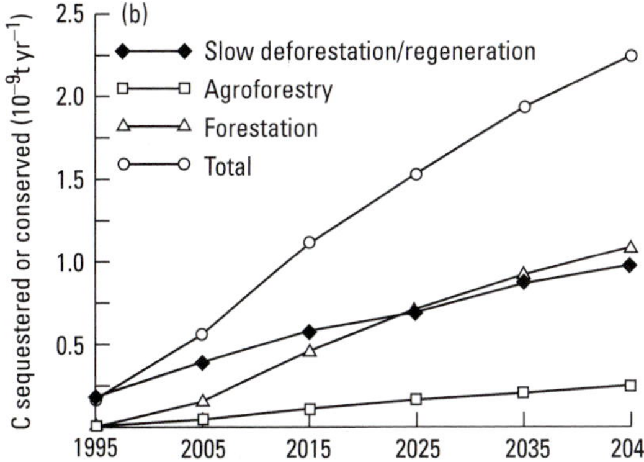

Fig. 10.31. Recommended scenario for carbon sequestration by the world's forests, (a) contribution of different biomes, (b) contribution of different activities. (Watson *et al.* 1996, fig. 24.2.)

The second report of the International Panel on Climate Change recommended various steps to be taken to counteract global warming. Trees grow fastest in the tropics and more carbon could be fixed there than elsewhere.

10.11 TROPICAL RAIN FORESTS YESTERDAY AND TODAY—CHAPTER SUMMARY

1. Human societies in many parts of the humid tropics, be they hunter-gatherers, shifting cultivators or settled farmers, have developed in close dependence on the rain forests.

2. The colonial era had a profound impact on tropical societies; some disappeared but left tell-tale traces in forest structure or species composition which are now being discovered. The spice trade shaped the history of the world. Useful plants were moved between continents and introduced to their new region via Botanic Gardens. Today, staple foods and many fruits are pantropical as a result. The Industrial Revolution increased demand for many forest products, such as rubber and resins. Only recently has trade in these minor forest products been overtaken by that in tropical hardwood timber.

3. Rattans of the Eastern tropics are by far the most valuable minor forest product. Cultivation in plantations has begun.

4. Forests can produce both timber and minor forest products sustainably. Timber is more valuable because minor forest products have a limited market and cash value.

5. Human impact on tropical rain forests has increased dramatically over the past half-century with the development of modern technology. Modern communications have ensured wide awareness of this. At the same time population has doubled (Fig. 10.10), mainly in the tropics. Rain forests are the latest in a long line of forest biomes to be heavily altered by mankind.

6. Forests are destroyed to make way for agriculture. This takes several forms and there are major differences between regions.

7. One class of agricultural activities is planned by government. Pastures for cattle are most important in the neotropics. Tree plantations producing export crops (mainly oil palm and rubber) are most extensive in the Eastern tropics. Plantations for wood production occur everywhere. Agricultural smallholdings created by government are also important especially in the East, for example transmigration in Indonesia.

8. The other class of agricultural activities is unplanned. In many countries peasant farmers are moving into rain forest, usually to practice shifting agriculture and often illegally. In some places there is an advancing frontier of destruction progressing through the forest (Fig. 8.2).

9. Forests are also destroyed for dams and mines.

10. Forests are utilized without being destroyed in various ways. In the tropics fuelwood comes mainly from seasonal forests (because most people live in seasonal climates), whilst industrial wood (timber, fibres) comes mainly from rain forests. Production of tropical hardwoods has increased greatly since 1950 and continues to rise. Asia is the dominant source (Figs. 10.23, 10.24). Nations there have successively gone through a cycle from boom to bust (Fig. 10.25) and from exporting logs to processed timber. Asian loggers have recently entered Africa and tropical America.

11. Rain forest bird and small mammal populations have shown surprising resilience to logging in the few forests studied so far, but repeated logging might be more deleterious.

12. Annual rate of loss of all tropical forests over the decade 1981–90 was 15.4×10^6 ha, 0.81 per cent (Table 10.11). This rate has been steady since the 1960s. Annual loss of tropical moist forests was 13.1×10^6 ha, 0.9 per cent, but with big differences between countries. An additional area is logged but not destroyed. The remaining forest is losing quality as it becomes increasingly fragmented (Table 10.13) and canopy height is reduced.

13. The loss of rain forests has become a major issue for burgeoning conservation movements. These have stimulated increasing concern at national and international levels. Damage to forests was a major issue at the 1992 UNCED conference at Rio de Janeiro. A plethora of bodies now discusses forest issues (Table 10.14).

14. Human interference reduces the biodiversity of rain forests by causing simplifications in structure and by loss of species. Loss of area is leading to some species becoming extinct, and many more losing genetic diversity.

15. Human life is impoverished by forest loss, especially for the rural people who live in or near the forest.

16. Interference with rain forests might affect climate by decreasing rainfall (Fig. 10.30). A change to bare land (which is unlikely over big areas) has greatest effect, to pasture less, and to a different woody vegetation least effect.

17. Tropical deforestation is responsible for about a quarter of radiative forcing, mostly due to releasing CO_2. Man-made chloro (fluoro) carbons make a similar contribution (Table 10.17). Forests, especially tropical forests, could play an enhanced role in sequestering carbon (Fig. 10.31), and hence play a role in curtailing the warming of the world's climate.

11

Tropical rain forests at the cusp of the new millennium

There is now a general awareness that careless logging of tropical rain forests, or their reduction to fragments by clearance for agriculture, damages their biology and the livelihood of rural peoples dependent on them. This perception, plus the responses needed from signatories to the Conventions on Biodiversity and Climate Change, is likely increasingly to colour the actions of the governments of rain-forest countries and the aid of donors who support them. Large sums are becoming available through the Global Environmental Facility (GEF)[408] to assist the implementation of these Conventions.

International timber prices reflect the value of timber, not the numerous other values of the forests whence it was extracted, which are damaged by logging and so should be costed in.[409] Nevertheless, there is an increasing realization, for which advocacy groups have been largely responsible, that tropical rain forests do have a multiplicity of values. Forests are no longer being regarded just as timber quarries. Some enthusiastic campaigners have indeed gone so far as to regard rain forests as sources of many things but not timber. This is to forget the importance of timber as a versatile, naturally renewable construction material which, unlike steel, concrete or aluminium, is generated without using non-renewable resources. Moreover, in contrast to these other materials, timber production sequesters CO_2 from the atmosphere instead of releasing it.

We can perhaps expect more care to be taken of tropical rain forests in the future than in the past half-century. In this final chapter we look briefly at the situation today and the challenges that lie ahead.

11.1 MANAGEMENT FOR SUSTAINABILITY

The International Timber Trade Organization, ITTO, has set the target that all tropical forests will be managed sustainably by the year 2000 for its producer country members, and they encompass over 90 per cent of tropical rain forests. ITTO has published guidelines.[410]

The sustainable production of timber from a tropical rain forest rests on meeting a few basic requirements, which in principle are very simple.

1. Long-term security of the forest. In Malaya much was lost when lowland forest successfully regenerated after logging by the Malayan Uniform silvicultural system (p. 141) was converted to agriculture (see also Fig. 8.7). Intergenerational equity

depends on long-term security and is best achieved by governments.

2. Sound operational management. This requires a strong Forest Department with adequate knowledge and the capacity to control harvesting.
3. A suitable financial environment, namely, corruption-free revenue collection and adequate ploughback of revenue into forestry.
4. Adequate knowledge. In recent years much money and effort has gone into research to increase knowledge whereas, in fact, enough is known about most forests and sustainable production would be better served by improving broader management aspects than in refining silvicultural techniques. For example, a pantropical survey commissioned by ITTO in the mid-1980s revealed that only one million ha of tropical rain forest were under sustainable management but that several countries had substantial areas of forest where it could be achieved with only a little extra effort. Moreover, most logged forests could readily be brought into sustainable production.[411]

At the level of the individual rain forest the requirements for sustainable management are equally simple and clear.

1. Demarcate the concession area, leaving blocks for conservation[412] and in which people have traditional rights.[413]
2. Determine likely growth rates and hence the annual allowable cut (aac).
3. Divide the area into annual coupes, based on the aac with a safety margin built in.
4. Harvest the timber using the proven techniques of low impact logging (see p. 137). These will include retaining belts of unlogged forest along streams and at their headwaters, felling trees only above preset diameter limits, and retaining species of local importance, such as fruit, resin, and latex producers (e.g. Brazil nut, jelutong, almaciga, *Bertholletia excelsa, Dyera costulata, Agathis* spp.).
5. Prevent illegal felling. Police legal felling operations, levy fines for infringement and as a last resort suspend or cancel the concession licence.
6. Under no circumstances allow repeated logging or re-entry of a coupe until the next felling cycle.

11.2 FOREST CERTIFICATION

In the late 1980s Friends of the Earth initiated an embargo within consumer nations on the use of timber from rain forests that were not being exploited sustainably.[414] The cessation of logging is not a realistic option, as too much money is at stake for both the nations and individuals involved. Insofar as the boycott was effective, it merely deflected timber to other, less sensitive markets. The focus was soon changed to press for the use only of timber coming from sustainably managed forests. Here there is a convergence with 'ITTO 2000', the requirement that all producer country members of that body have their forests under sustainable management by the turn of the millennium. The necessity arises to certify that a given forest is indeed sustainably managed, and by the late 1990s much thought and effort was being given to forest certification.

Several different certifying bodies have sprung up across the world. In 1992 the NGOs themselves founded the Forest Stewardship Council to give a seal of approval to individual certifiers.[415] In addition, for rain forests, Indonesia, the West African nations and the Amazonian nations are developing schemes of certification.[416] Part of the currently rapidly moving debate is whether certification could be linked to an International Standards Organization (ISO) standard.[417]

The fraction of the market judged in the mid-1990s to be sensitive to certification was tiny, at maximum 20 per cent of the European plus 10 per cent of the United States tropical timber markets. This was valued at 4 per cent of the 1991 timber revenue of rain-forest producer

countries ($430 × 10^6), and, at 2.81 × 10^6 m^3 roundwood equivalent in 1993, accounted for the annual growth of less than 1 per cent of the world's productive tropical forests.[418] So far eco-labelled tropical timber has not fetched a premium price. The expectation, however, is that although the market is small it is influential and a standard of exploitation will be set that other producers will aim to follow. Already as a result of the move to certification a few logging companies have been created to serve this niche market, one which they hope will grow larger.

Whilst the principles of sustainable forest management are clear and simple (section 11.1), thrashing out the details in order to verify sustainability is proving difficult. A great pall of semantic smoke is being belched from the chimneys of a bunch of international research agencies, bilateral aid programmes and sincere but sometimes ill-informed environmentalists. There is a sticky fog of conferences, networks, workshops, and newsletters, glued together by cheap air fares, fax, and the internet, all trying to decide between principles, criteria and indicators. The simple, basic underlying requirements have vanished from sight. Technical forestry issues are the least contentious, but it is proving to be much more difficult to come to grips with the issues relating to biodiversity and the traditional rights of local people.

11.3 THE FUTURE PATTERN OF LAND USAGE

Till mid-century most rain-forest landscapes were still like oceans of forest with scattered islands (usually lying along valleys) of anthropogenic vegetation as the signs of contemporary human activity. This pattern persists today in, for example, New Guinea, Congo, Gabon, southeast Cameroon, and much of the great South American rain forest that reaches from the Amazon basin north to the Orinoco and the shores of the Caribbean in the Guyanas. The other extreme also occurs, islands of rain forest set in an ocean of man-made vegetation, a landscape similar to parts of Europe such as Bavaria or France. All intermediate stages also exist. Countries where the rain forest has been reduced to fragments are the Philippines, Malaya (Fig. 10.16) and Sumatra (Fig. 10.17)[419] in Asia, much of West Africa (for Ghana see Fig. 10.6), and most of Middle America, from the Mexican rain forests south to Panama (for Costa Rica see Fig. 10.14). In most of these areas human population pressure is high.

These cultural landscapes have some open land as pasture or arable crops. Otherwise they are tree-covered with orchards, the secondary forest phase of shifting agriculture, tall forests rich in fruit or useful trees (sometimes called agroforests, p. 180), and sometimes also plantations of timber trees, rubber or oil palm. The relict rain forest itself will have unlogged blocks, and other areas at various ages after one or more logging events.

Today, the mosaic of land usage is increasingly the result of planning decisions by government, often strongly coloured by the interplay of political forces, and with conservation forest usually a low priority (Fig. 11.1).

The original species of the rain forest will have to survive in this landscape in which the forest fragments themselves are only partially linked together. Let us therefore look briefly at different parts of the mosaic and the role they can plan individually and together in biodiversity conservation. First, we must discuss the conservation of forest genetic resources.

Genetic resource conservation

> Breeders and genetic engineers can devise more and more ingenious ways of using available genes but they cannot create new ones. Although wild genes are a brand new resource and although their utility is growing dramatically, the lack of progress in conserving them ... casts a shadow over their future.[420]

There are two modes of genetic resource conservation with the ugly but descriptive appellations, *ex situ* and *in situ* conservation (Fig. 11.2).

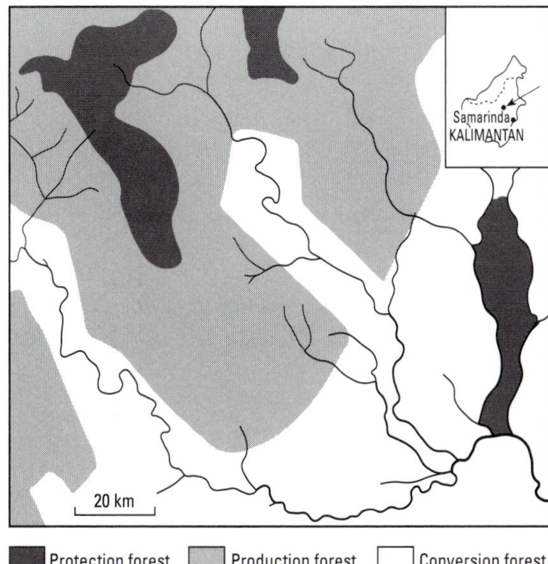

Fig. 11.1. Landuse planning, Belayan river, East Kalimantan.

This sparsely populated area 300 km inland from Samarinda was until recently largely clothed with virgin lowland dipterocarp rain forest. The land-use plans are to convert the most accessible forests along the major rivers to agriculture, and maintain most of the rest as production forest. Protection forest has lowest priority. It will not include samples of the whole landscape but will be confined to the hilliest most remote places (west) and a swamp forest (east).

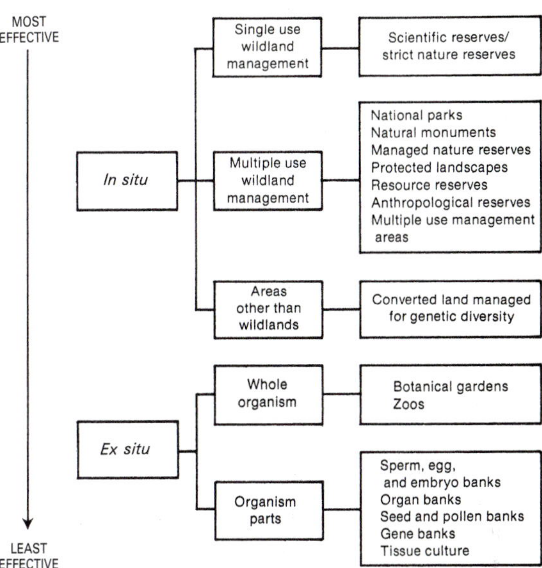

Fig. 11.2. Different ways genetic diversity can be conserved. (After Ledec and Goodland 1988, fig. 2.1.)

Conservation outside the forest

Storage off-site, or *ex situ*, can take several forms. Storage of seed in seed banks is only practicable for those species that have orthodox seed (p. 120). It has only been applied to a few major crops of the humid tropics, notably rice for which IRRI, the International Rice Research Institute, in the Philippines has 200 000 holdings (70 per cent traditional varieties, 10 per cent wild species); and to a few pioneer timber species, e.g. *Cordia alliodora* of northern tropical America and some subtropical pines (mostly *Pinus caribaea*, *P. oocarpa*). Genetic diversity can be represented in seed banks, but on-going adaptation to pests, diseases, biotic interactions, and climatic change then ceases. A few important plantation tree species have certain races ('provenances') conserved in plantations. In the case of some pines the original wild populations have now disappeared.

Arboreta and Botanic Gardens are less satisfactory for *ex situ* conservation because each species is represented by only a few individuals. The Botanic Gardens of the humid tropics were established as centres of trial and introduction but have become increasingly less relevant to the needs of agriculture and forestry. Botanic Gardens worldwide have recently awoken from a long slumber to realize that, like zoos, they may have a role to play in conservation, if only they can precisely define it. They have made a start to co-ordinate their activities and to compare stock inventories, and have established a Botanic Gardens Conservation Secretariat based at Kew and operated by IUCN.[421]

Conservation within the forest

Conservation *ex situ* can never conserve more than a tiny fraction of the species richness of tropical rain forests. Conservation of species in

place (or *in situ*) in the forest is much more important and is the best means of conservation of the numerous species which have recalcitrant seed (p. 122).

Identification of priority areas for conservation has often been attempted. Some nations are richer in species than others, and have come to be called 'mega diversity' countries, but to a considerable extent this merely reflects their size. Brazil and Indonesia, for example, always rank high in any biodiversity league table. Within a nation it is tempting to identify 'hot spots' for biodiversity conservation. For birds and mammals it may indeed be possible to pinpoint particularly rich or diverse localities[422] but other animal groups and plants are not well enough known, beyond the broad biogeographic patterns described in Chapter 6, for example, the Pleistocene refugia of the African rain forest (Fig. 6.17) are indubitably species-rich. Knowledge of species richness and diversity is biased to the places that have been studied (Fig. 11.3) and these are what will be picked out by any analysis of 'hot spots'. The only sound scientific basis is to conserve adequate habitat, spread across its geographical extent, and to sample all biogeographical regions.[423] For example, a rational plan for biodiversity conservation in Kalimantan (Fig. 11.3) would be to sample all the forest formations (Chapter 2) (and these have been accurately mapped), and the slightly seasonal climates of the east and southeast coast, as well as the perhumid climates elsewhere. The result would be a number of National Parks spread across the island. What is now known about tree species packing in rain forests (p. 32) suggests that fairly small areas can be expected to contain most of them.

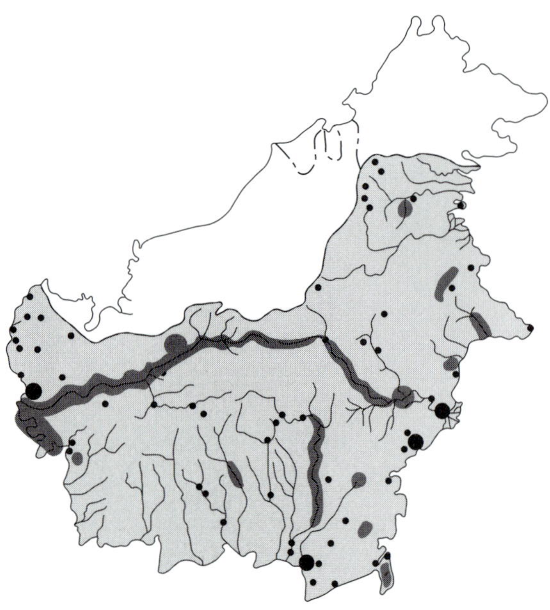

Fig. 11.3. Plant collecting localities in Kalimantan to 1990. (From *Flora Malesiana*, mapped by J. Burley.) Statements on endemism or rarity inevitably reflect this pattern.

For many rain forest species, including most plants, it is difficult to make meaningful statements about rarity and hence about conservation priority. Here too the picture is entirely dependent on the thoroughness of collecting. It is dangerous but common to rely on the abundance of herbarium material. *Macaranga lamellata* illustrates the problem. This pioneer tree species was described in 1974 from a single collection, so the temptation was to list it as rare, but a subsequent study in the early 1990s of the ecology of *Macaranga* in Sarawak and Brunei revealed it to be one of the most abundant roadside tree species. It had been overlooked because plant collectors shun the weedy trees of secondary forest. In Malaya *Macaranga amissa* and *M. curtisii* also appeared rare; in this case because they rarely flower or fruit and so are seldom collected.[424] Again, forest observation revealed their true status. These cautionary tales suggest that some of the considerable funds at present pouring into biodiversity conservation would be better spent on first-hand exploration than in recataloguing existing data or attempting to read too much into the present fragmentary knowledge.

Animal geneticists have given considerable thought to the question of how many individuals are necessary to maintain the full genetic integrity of a species in perpetuity.[425] Much has been learned from zoos. A simple but extremely crude rule-of-thumb is that a minimum popula-

Table 11.1 Density of hornbills and primates in the Krau Game Reserve, Malaya, and areas estimated to support 500 individuals

		Numbers	Area (km^2) to support 500 individuals
Kuala Lompat, 200 ha lowland rain forest			
Hornbills:	Helmeted	1	1000
	Rhinoceros	1	1000
	Southern Pied	2	500
	Black	4	250
	Bushy-crested	5	200
Whole reserve, 537 km^2, lowland, lower and upper montane rain forest			
Apes:	Siamang	1440	186
	Lar gibbon	1568	171
Monkeys:	Dusky langur	7952	34
	Banded langur	12 298	22
	Long-tailed macaque	2975	90
	Pig-tailed macaque	2800	96

From Whitmore (1984*a*) table 21.1, corrected

tion of 50 breeding adults maintains fitness in the short term, thus preserving a species 'frozen' at one instant of time. To prevent continual loss of genetic diversity ('genetic erosion') over the long term, and therefore to conserve the potential for continuing evolution, requires a big population, and a minimum of 500 breeding adults has been suggested to be necessary. This 50/500 rule is only a very rough approximation and can differ widely between species. The numbers certainly need to be increased to allow for ecotypic differentiation and (as occurs in many plants) for overlapping generations and restricted gene flow (section 5.6).

Most difficult to conserve are animals (or indeed plants too) that live at very low population density (e.g. hornbills, tapir, and top carnivores, such as jaguar and tiger), or that have large territories (e.g. gaur, elephant) (Table 11.1). If enough forest is conserved for these rarities, then most plant species and the commoner animals will be adequately represented (Table 11.2). Conservation means, in practice, caring for rare species. There is no escape from the fact that even for common species numbers drop as forest is destroyed (Table 11.3), and in many cases this will mean there is a loss of genetic diversity.

Virgin forest

There is an indisputable case for retaining parts of the rain forest inviolate, as natural reserves, kept intact for species to continue to interact between themselves and with the environment. For, however careful human intervention to extract useful products may be, there is an inevitable alteration in the relative abundance and population structure of species, and usually also the simplification of forest structure. Reserves of natural forest act as benchmarks against which change elsewhere can be monitored. Their usefulness is increased if they are surrounded by production forest, not cultivated land. The Virgin Jungle Reserves set up throughout Malaya and Sabah by the Forest Departments, as compartments of Forest Reserves not to be logged are an example.[426] However, no developing nation can afford the luxury of 'locking up' more than a tiny fraction of its

Table 11.2 Frequency of occurrence of species and genera of trees with horticultural or pharmacological potential in the rain forest of Ulu Kelantan, Malaya

		Number per 40 ha	Area (km^2) to support 500 big trees*
(a) With horticultural potential			
Anacardiaceae	*Mangifera* (machang)	6.7	30
Bombacaceae	*Coelostegia* (punggai)	0.5	400
	Durio (durian)	3.7	50
Euphorbiaceae	*Baccaurea griffithii* (tampoi)	2.3	90
	Elateriospermum tapos (perah)	33.0	6
Guttiferae	*Garcinia* (kandis, manggis)	0.9	200
Leguminosae	*Parkia* (petai, p. meranti)	8.8	20
Meliaceae	*Lansium domesticum* (langsat)	0.2	1000
	Sandoricum koetjape (sentul)	0.8	200
Moraceae	*Artocarpus integer* (bangkong; chempedak)	1.3	250
	A. lanceifolius (keledang)	18.0	10
	A. rigidus (temponek)	7.4	30
	A. scortechinii/elasticus (terap)	3.4	60
Sapindaceae	*Nephelium lappaceum* (rambutan)	5.9	30
(b) With pharmacological potential			
Apocynaceae	*Alstonia* (pulai)	0.8	200
	Dyera costulata (jelutong)	7.2	30
Moraceae	*Antiaris toxicaria* (ipoh)	0.5	400

From Whitmore (1984*a*) table 21.2, corrected

* Trees ≥ 1.2 m girth. All these species are likely to be fertile by 0.6 m girth, so the area needed to support a viable population is halved.

Survey of 26 628 trees in a 676 ha sample of 10 100 km^2 of mainly lowland evergreen rain forest

Table 11.3 Estimated loss of primates in Malaya 1957–1975 due to reduction of forest area from 84% to 51%

	Population in 1958	population in 1975	Loss	Per cent loss
Apes:				
Siamang	111 000	48 000	63 000	57
Gibbons	144 000	71 000	73 000	50
Monkeys:				
Dusky langur	305 000	155 000	150 000	49
Banded langur	962 000	554 000	408 000	42
Silver langur	6 000	4 000	2 000	33
Long-tailed macaque	414 000	318 000	87 000	23
Pig-tailed macaque	80 000	45 000	35 000	44

From Mohd. Khan in Cranbrook (1988) table 17.1

forest. Fortunately, there are good prospects for combining utilization with conservation.

Production forest

The possibility exists to manage rain forests for multiple purposes, to meet the needs of conservation as well as to produce useful products. But to retain the long-term benefits implied by conservation it is necessary to forgo some immediate cash profit. Multiple use involves compromises.

It was demonstrated in Chapter 7 that sustainable production of timber from a forest is certainly possible while maintaining most of the natural richness and diversity, so long as one operates within the limits of natural dynamics.

Low-volume selective logging on a polycyclic system alters the forest least. The full plant species complement is likely to be retained, but the proportion of light-demanding gap species, including pioneers, will increase. It is essential to enforce laws to minimize the damage caused by logging and to prevent hunters, collectors of minor products, or farmers from entering along roads and causing damage, depletion or destruction. Logging at this intensity is unlikely to eliminate many animal species (section 10.7). Knowledge of their ecology can be utilized to favour them. Inviolate patches should be deliberately retained. Hollow trees used as nesting sites, and plant species which are important for food, should also be retained, especially keystone species which provide food in times of famine (pp. 71–2), for example, strangling figs and woody climbers. More research is needed to identify these groups. It has been found in both East Kutai, Kalimantan, and Kuala Lompat, Malaya, that strangling figs grow most commonly on timber tree species,[427] so to maintain enough of them means forgoing some of the timber harvest.

Rural people commonly get much of their meat by hunting wild animals, so-called 'bushmeat' in West Africa. It has been shown in Sarawak that the rain forest provides *c.* 18 000 tonne of wild meat every year, which is equivalent to about 12 kg per person. The cost of replacing this from domestic livestock and fishponds would be prohibitive, about £25 million per year.[428] Production forest is a good potential source of wild meat.

Forest plantations

Timber is produced cheaply in natural rain forests. The trees can be left to continue growth until market conditions are good. The numerous species have different properties and as the market changes different products can be sold. This cheapness and flexibility mean that there will always be a strong case for a tropical nation to retain part of its rain forest for timber production.[429] The growth rate of commercial timber in natural forests is commonly 1–2 m^3 $year^{-1}$, but can be boosted for a few years to 3 m^3 $year^{-1}$ or more by silviculture.

There is also a case for plantations.[430] These are much more productive (Fig. 7.37) (indeed the humid tropics have an ideal climate for tree growth,[431]), but much less flexible than natural forest. They require a substantial initial cash investment and a prediction that there will be a suitable specific market at a particular future date. Moreover, timber plantations are in essence a form of long-term agriculture and, like agricultural crops, sooner or later fertilization is required (see pp. 165–7). Timber sold from plantations takes the pressure off natural forests as a source of foreign exchange. They should only be established on already degraded sites, never at the expense of good natural forest.

Most plantations in the humid tropics have been of pioneer species which produce low density, pale, soft timber. Only a handful of species are planted extensively and sooner or later most that have been tried have succumbed to a pest or disease and been dropped from the foresters' repertoire. There are very few quality hardwoods which have proved successful in plantations. The experience of Malaysia illustrates these points. There, plantations are being established to supplement timber production from the second cut of dipterocarp rain forests,

Table 11.4 Areas of tropical forest plantations in 1990 (km^2)

	Africa	Asia and Pacific	Tropical America	Total
Industrial wood*	14 000	91 000	57 000	156 000
Fuelwood etc.†	16 000	231 000	35 000	282 000
Total	30 000	322 000	86 000	438 000
Annual area planted	1 300	21 100	3 700	26 100

* Pulpwood and lumber

† Fuelwood, soil protection, domestic wood, resin

From FAO (1993) table 20

The effectively planted area is estimated to be 70 per cent of these figures. The largest areas planted were India 189 000 km^2, Indonesia 88 000 km^2 and Brazil 70 000 km^2. The total area planted by 1990 is 2.5 times the 1980 area of only 180 000 km^2 (Prebble 1997).

which will be less than the first cut even assuming silvicultural rules have been closely followed.[432] Quality hardwoods (mahogany, *Swietenia macrophylla*), and lower value *Acacia mangium*, *Gmelina arborea* and *Paraserianthes (Albizia) falcataria* have performed satisfactorily grown on 20–30- and 7–10-year rotations, respectively.[433] Many other species have been commercial failures, for example, *Anthocephalus chinensis*, *Araucaria cunninghamii*, *A. hunsteinii* (kadam, hoop pine, klinki pine), and because of the perhumid rain-forest climate, pines (mainly *P. caribaea*, *P. oocarpa*) also. Far more thought and more trials are needed to find other quality hardwood species suitable for plantations. There will be a huge future demand for this sort of timber as the virgin forests become exhausted.[434] Botanists and foresters need to get their heads together. One species ripe for investigation is *Pterocarpus indicus* (angsana, narra, sena) of Malesia, which has a very valuable, finely figured cabinet timber and fast growth, as witnessed in the garden city of Singapore where it is the most extensively planted shade tree.[435]

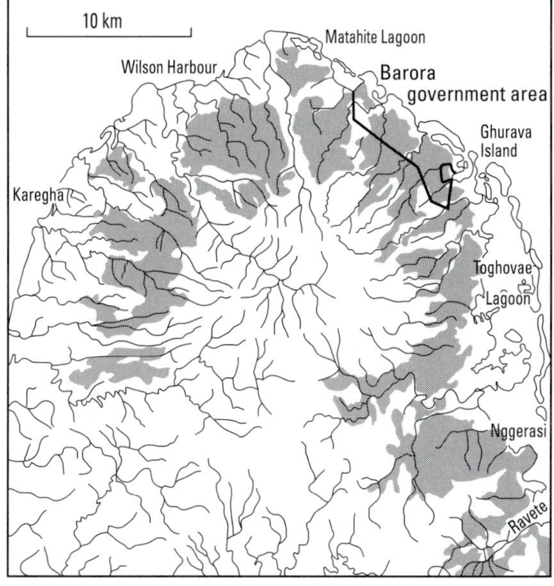

Fig. 11.4. Plantation opportunity areas. New Georgia, Solomon Islands. (Shield 1992, exhibit 8.) The north cape of New Georgia island (8°S, 157°30′E) is an extinct volcano. There are scattered villages with their associated gardens and coconut plantations around the coast. Inland the land rises increasingly steeply to the crater rim. The broad ridge crests on the lower slopes between the radial rivers are ideal sites for tree plantations. Steep valley sides and the uppermost slopes would be left under natural rain forest.

The rate of establishment of tree plantations in the tropics for both fuelwood and industrial wood is increasing (Table 11.4). For example, the plans of Indonesia to become a major producer of paper have already been mentioned. Such plantations should be emplaced in the

most suitable parts of the landscape (Fig. 11.4). Belts of natural forest should be conserved along streams to act as barriers to fire, pests and diseases and to harbour insectivorous birds (Fig. 11.7). These belts should also connect the blocks of retained natural conservation and production forest, thus providing contact between different populations of plants and animals.

Looking ahead, fuelwood grown in industrial plantations is likely to become increasingly important for power generation as fossil fuels become depleted, and to reduce overall increase in atmospheric CO_2.[436] There has been much discussion about creating plantations for fuelwood and constructional timber for local people, as buffer zones around National Parks, to reduce the pressure of traditional usages on the rain forest. Buffer zone planting has began at Korup, Cameroon.

Timber plantations play a role in biodiversity conservation by providing part of the living space and food source for some animals (see below p. 233). Agroforests (p. 180) which are intermediate in composition between plantations and natural forest also play a part (Fig. 11.5). Plantations deliberately not cleaned develop a rich understorey. For example, the ground layer of tree plantations at Omu, Nigeria, were found to have one-third of the species of the rain forest, including many used by the rural population, such as chewing sticks (*Garcinia* spp.), cordage and medicines[437] (see Fig 11.6).

Restoration of forest via plantations could be an important tool for the land manager of tomorrow. It has been observed in Puerto Rico that once plantations have restored a forest canopy native species invade and eventually natural vegetation is restored. Species that take up larger amounts of mineral nutrients best facilitate this succession.

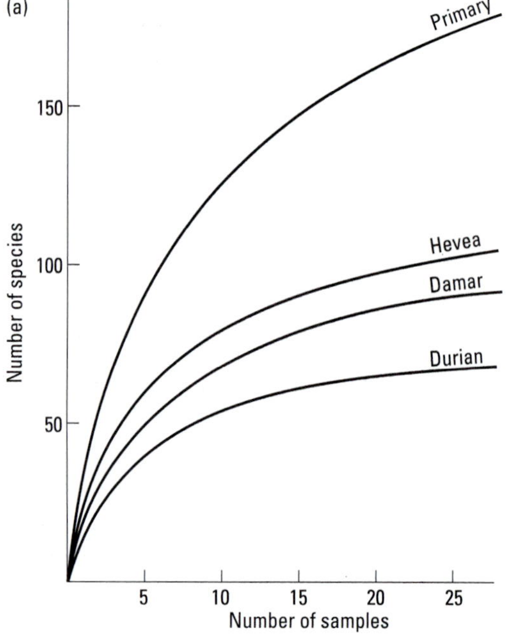

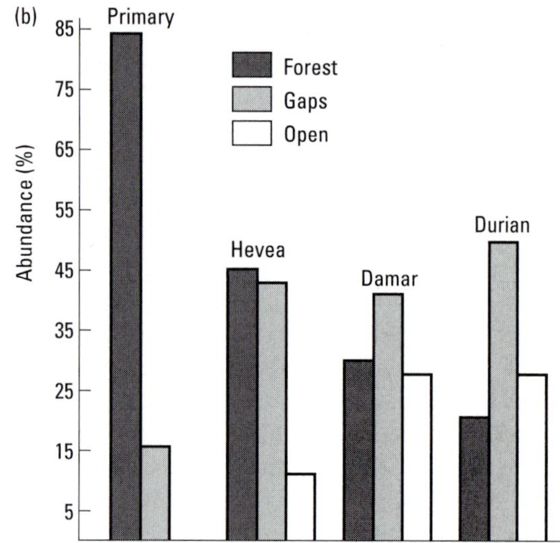

Fig. 11.5. Bird species numbers in Sumatran agroforests. (Thiollay 1995, figs. 3, 5.)
(a) Species numbers are reduced, and (b) the proportion of species of canopy gaps and open woodlands are increased.

Species survival in forest fragments[438]

Increasingly in the future, tropical rain forest will only remain as fragments. Some such already exist, for example, Barro Colorado Island in Gatun Lake, Panama, Pasoh and Sungai Menyala in Malaya, Bukit Timah in Singapore, Semengo in Sarawak and the tiny

Fig. 11.6. Plantation of Meliaceae (big tree, right) in Nigeria, which has developed a rich understorey of native species.

Natural forest can be restored via plantations and in Nigeria the understorey provides villagers with many useful plants. The plantation shown here, one of the Kennedy plots, was illegally felled in the late 1980s.

surviving parts of the Brazil Atlantic Coast rain forest (Fig. 10.11). There is the problem that such fragments may break the 50/500 rule (p. 227) and contain too few individuals of a species for its long-term genetic integrity. Species that occur at low density are especially vulnerable to genetic erosion, to chance extinction when numbers fall (p. 72), or to inbreeding depression. In particular, many trees live several centuries and may be persisting today but unable to breed, so the species is 'living but dead', doomed to extinction. Moreover, small forest remnants may be too small to support certain species and this may have repercussions on other components of the ecosystem. For example, tiny islets (< 2 ha) near to Barro Colorado are too small to support large vertebrate seed predators (agouti, paca, peccary, and squirrels) and have a much higher abundance than the nearby mainland of big-seeded palms (*Astrocaryum*, *Oenocarpus*, *Scheelea*) and trees (*Dipteryx*, *Protium*).[439] Pasoh forest has lost the animals that once dispersed the seeds of *Chrysophyllum roxburghii*, and, as was described in section 5.5, today piles of rotting fruit accumulate below the parent trees.

Besides reduction in area, forest fragmentation also increases the proportion of edge relative to interior (Table 10.13) and if the fragments are surrounded by open land this will result in a change of microclimate.

Continual local extinction and immigration is probably a feature of species of plants and animals which occur at low density in tropical

rain forest. This mechanism is likely to break down in isolated fragments which become cut off from immigration.

Despite the poor long-term outlook for species survival in forest fragments there are several places where a surprisingly high proportion have persisted for decades or longer. Singapore island (544 km^2) still has 71 per cent of its inland vascular plants despite the loss of 99.8 per cent of the inland primary rain forest.[440] At least half the birds have been lost, and the surviving species have become more abundant. Many of the plants now exist only on Bukit Timah, an 81-ha relict patch of mainly primary forest, and in specks of primary forest that lie within the regrowth forest of the central catchment area.[441] Bukit Timah still has dipterocarp forest and has persisted for over 100 years, long enough for there to have been a complete turnover amongst most of its big trees (section 7.7). It has over 800 plant species, but has lost about half of its mammals and birds.[442] It seems probable that ecosystem functioning will have changed, for example, birds now disperse fruits previously dispersed by mammals. Until 1985 when a motorway was built, Bukit Timah was contiguous with the mainly old secondary forest central catchment area, which perhaps contributed to its resilience.

There are several examples of a rich vertebrate fauna enduring in small forest fragments. At Sungai Bernam, Malaya, a 75-ha patch left intact within an oil palm plantation and protected from hunting was found after 'more than 60 years' still to be extremely species-rich in primary forest birds and mammals. In Sabah, small relict strips and patches of primary forest that had been logged and subsequently accidentally burned also retain a similar rich fauna, including in this case leopard. Deer roam out to feed in the surrounding timber-tree plantations. Birds feed on insect pests. Bee populations have exploded, feeding on the copious nectar provided by the plantation trees.[443] Results being published from studies of small blocks of forest of sizes 1 to 10 000 ha, left when pastures were created north of Manaus, Brazil, are also showing high survival, although some groups have died out from the smallest patches. For example, three bird species that are obligatory feeders on insects fleeing from swarming army ants (*Eciton burchelli*) and a further 13 insectivorous birds that feed in mixed-species flocks have disappeared from the 1- and 10-ha fragments.[444]

The most dramatic instance of long-term survival is in the Brazil Atlantic Coast rain forests, of which only about 5 per cent still exists (Fig. 10.11). All the vascular plants, mammals, birds, and butterflies ever recorded from these forests are still there. This is a region of rugged mountains and has a gradient of increasing dryness and seasonality inwards from the coast. It is likely always to have been a mosaic of small patches of different habitat and forest, so its component species always have occurred in small populations. The difference today is that only a few of them remain, so despite amazingly high persistence, the biota are highly vulnerable to extinction.[445]

The surprisingly long survival of species in all these various fragments is good news because it provides the breathing space within which conservation measures can be devised.

11.4 PRIORITIES FOR ACTION

From this picture of the situation of many tropical rain forests today, which will apply to more in the future, it is possible to identify priorities.

Biodiversity conservation in cultural landscapes

Firstly, in any scheme for managing a tract of forest for sustainable production, it is absolutely essential that some patches are left inviolate in their virgin state. Secondly, production forest also has an important contribution to make. The extent to which animal and plant populations thrive in retained fragments of these two kinds of forest, as well as in tree plantations, needs to be monitored. All these three kinds of

Fig. 11.7. Industrial tree plantations of *Acacia mangium* planted to retain belts and blocks of the former rain forest. Riau, Sumatra.

The very uneven canopy of the lowland evergreen dipterocarp rain forest shows it has been logged. The riparian strip in the middle distance is probably too narrow to be ideal for forest conservation.

forest need to be contiguous or joined by corridors (Figs. 11.4, 11.7), and much remains to be discovered about the real value of rain-forest corridors, their minimum effective width and their long-term health.[446] The way animals use different parts of a landscape needs to be understood and allowed for in land-use plans.[447] Much remains to be learned about the ecology of rain-forest fragments. To what extent are plant–animal interactions closely knit and to what extent opportunistic but with the same or similar end result? For example, if a fruit-dispersing mammal goes extinct can it be replaced by a bird?

Management of previously logged forest

Foresters need to look ahead. For much of rain-forest Asia the days of giant trees are over and to refine low-impact logging techniques for virgin forest is largely to fight yesterday's battles. Low-impact logging of primary forest remains top priority for tropical America and central Africa but in Asia the focus needs now to be changed. Most of the Asian dipterocarp rain forests have now been logged once and some several times, and there are vast areas of regenerating forest. Future harvests will take out much smaller volumes of timber and smaller trees. The priority today should be to devise logging, management and silvicultural techniques for these regrowth forests.[448] The temperate broadleaf forests of Europe or North America provide likely models. The rules for sustainability currently being developed in the drive towards forest certification need to include rules for these relogged forests.

Rural people and the rain forest

If the forest frontier is to be stabilized, if fragments of the rain forest and its biota are to survive, rural people must live in balance with the landscape they inhabit. Their exclusion from the forest is unrealistic and may involve annulling traditional rights. The difficulty is the inexorable, and in many places rapid, increase in human population (Fig. 10.10). Ultimately this has to be controlled. The forest and the people who depend on it need to be considered as a single ecosystem, managed to maintain a continual but changing stream of goods and services.[449] This is a tall order, and in the face of really high human population pressure, forests are likely to be sacrificed. There are moves throughout the tropics to weave the people's interests into forest management and a wide range of views on how to do this.[450] In some

cases this involves the state giving up claims to ownership, often made in colonial times. To give just two examples: forest control is being returned to rural communities under the new forest policy of India;[451] and Colombia has returned ownership of its Amazonian forests to the Amerindian inhabitants. National sustainable timber exploitation may be best retained under Forest Department control, but power to the people by participatory management, especially of minor forest products and in the establishment of plantations, is a welcome recoil from control by central government. It remains for viable mechanisms to be explored and developed.

Priorities for campaigners

NGOs will have most leverage if they take note of the real situation of tropical rain forests and not some idealized fantasy. They can have as vital a function in the future as they have had in the past, but should not fight yesterday's battles, regrettable though some losses have been. A look at likely impacts on the world's rain forests tomorrow suggests several key issues that need to be pushed hard.

National parks require active management, not the proliferation of plans on paper. Parks must be seen to contribute to human welfare.

The rights and roles of rural people, in both conservation and production forests, need to be reconciled with biodiversity conservation and the necessity for production to be sustainable.

Low-impact logging of primary rain forest, the realization of ITTO's goal of sustainable management by AD 2000, is a high priority for tropical America and the parts of Asia and Africa not yet logged. For much of the Asian rain forests, especially the dipterocarp forests, pressure is now needed to stimulate the development of different techniques that will use much smaller machines and take out lower timber volumes.

Biologists need to be pressed to research and monitor the effectiveness of different landuse patterns and to develop active management tools for biodiversity conservation in cultural landscapes of which rain forest is only a part. As rural landscapes evolve it is known in general terms to be important to maintain contact between different sorts of forest, either in contiguity or by corridors, but the details of effective patterns remain unknown. The long-term viability of small populations in forest fragments and the health of the ecosystem as measured by species interactions needs to be monitored and manipulative tools may need to be developed.

11.5 TROPICAL RAIN FORESTS AT THE CUSP OF THE NEW MILLENNIUM—CHAPTER SUMMARY

1. It is increasingly accepted that tropical rain forests have many values, and are the source of many goods and services, not simply timber quarries. There is a general awareness that careless logging or the reduction of forests to fragments damages their biology and the livelihood of rural people.

2. ITTO has set the target that by AD 2000 all tropical forests will be managed sustainably. Sustainability will need to be certified, and bodies are evolving to make the necessary inspections. The basic requirements for sustainability are simple, but some are proving difficult to codify.

3. Some rain forests today, and increasingly more in the future, exist as fragments in cultural landscapes, mixed with tree plantations and various sorts of agriculture, and this is the setting for the long-term survival of their biodiversity.

4. The genetic resources of rain forests can be conserved *ex situ* in seed banks, or living collections, but for many species, especially those with recalcitrant seeds, conservation in the forest *in situ* is the only realistic possibility. An adequate population must be conserved, and a *very* rough rule is 50 breeding individuals short-term, 500 long-term.

5. For most species, except some birds and mammals, detailed information on occurrence and abundance does not exist. Given the nature

of variation from place to place within rain forests the only rational means to conserve them is by preserving samples of forest across their geographical range.

6. Conservation means, in practice, caring for rare species, such as many carnivores. Adequate habitat for these will encompass larger populations of other more abundant species.

7. The different parts of the landscape mosaic collectively contribute to species conservation. Timber production forests can be managed for multiple uses, to grow timber and minor forest products, to continue as animal habitat, and with the full complement of plant species. Inviolate bench-mark reserves are essential. Forest plantations take the pressure off natural forest as a timber source and also are useful to some animals. These forest patches need to be linked by forest corridors.

8. There is evidence that some species survive for decades or longer in fragments of forest. Where populations are too low and unable to breed successfully they are doomed eventually to extinction. This long survival gives breathing space within which to devise conservation measures.

9. Priorities for research are the biology of species survival in forest fragments and in production forests, and then if necessary to devise active management procedures. Secondly, the design and the role of connecting corridors needs investigation. Low-impact logging needs refinement and further demonstration, especially in central Africa and tropical America. A higher priority in Asia is to devise methods to log low-volume regrowth forests.

10. People need to live in balance with the landscape they inhabit. Means need to be devised to increase their participation in forest management.

Epilogue

Ultimately we need a new paradigm.[452] The human species must develop a new philosophy for interaction with the biosphere of which it is part. Population has to stabilize. Environmentally sustainable development means reaching sustainable levels of production and consumption.

Natural forest management has to become the practical application of a land ethic. The sum of the values of a forest may outweigh alternative land uses but they are diffuse. The logger does not benefit from the soil. Farmers do not own biodiversity. The intrinsic worth of non-human species is virtually ignored by economics. With how many other species are we willing to share the earth?

Worldwide there is a growing acceptance of the limits to growth, and recognition that ultimately the limit will be set by the supply and distribution of resources. Concern over population increase and resource depletion is becoming an accepted part of the dialogue between government and industry, scientists and the general public.

For some tropical rain forests this awakening is coming too late, and for others the response is still too little. Logging proceeds as fast as ever and moves on to fresh countries. The rain-forest resource is being liquidated, while simultaneously, under scrutiny from pressure groups, the new dialogue on sustainable utilisation is being engaged: foresters and sociologists (but as yet few scientists) pursue the labyrinths of defining and measuring sustainability in all its senses.

By AD 2000 ITTO has called for all tropical rain forest to be under sustainable management. What does this mean in reality?[453] The primary rain forests of Asia will have been almost completely carelessly logged by then, with 10 years or less left in Kalimantan and New Guinea. The timber industry there will have shifted to logged forest or plantations. Sustainability will be based on logging a seriously damaged, very reduced resource. Will virgin forest logging in tropical America and Africa be causing less damage by then?

The picture is no different from human impact on other 'forest frontiers' in the past, except that change is much more rapid. Those other forests have grown back, so one can have some sympathy with the protests of tropical politicians. However, tropical rain forests are qualitatively different from all others. If systematically and almost totally eliminated from a landscape, or very heavily damaged by excessive and careless logging, species become extinct and the same forests and their animals will never come back. In this way current rapine, aided by late 20th century technology, really does cause an impact never previously seen elsewhere.

What then is the future of tropical rain forests? Large parts (but not all) of the Eastern rain forests are lost, fragmented or damaged beyond full recovery. The future as I see it lies mainly in the great blocks of the Amazonian and central African rain forests. There are glimmerings of hope these will not suffer the same fate. We

largely have the technical and biological knowledge for sustainable utilization, and are a fair way to knowing how this can be made compatible with the maintenance of biodiversity. The political climate is just starting to change, and the many values of tropical forests to be appreciated.

Let us hope the forest frontier stabilizes in the East in time to save just a fraction of the grandest of all rain forests, the dipterocarp forests of the Sunda Shelf, and that exploitation in central Africa and tropical America takes heed of the lessons learned so hard in Asia.

Envoi, the end of the saga. Lowland evergreen dipterocarp rain forest transmuted to poorly established rubber, 1997.

All saleable big trees were removed by logging. This reduced the biomass by up to 30 per cent. The rest was then burned during a dry period, causing part of the massive smoke haze that has periodically plagued western Malesia since the 1980s (p. 148).

Text notes

1. There is also a torrent of new scientific publications. For fuller reading than in this book the whole subject of tropical rain forests is covered in two comprehensive accounts. Richards P.W. (1996) is a new edition of the classic text, originally published (in 1952) as a synthesis of the work of European scientists in tropical rain forests up to about 1940. The new edition uses the same framework and adds much new information. Whitmore (1984*a*) is the only modern comprehensive book on rain-forest ecology. It is built around contemporary interests in forest dynamics. In it most of the subjects covered in the present book are discussed in great detail and with examples drawn from the Eastern rain forests. Some topics have developed fast in the 1980s and 1990s and an attempt is made to give reference to these in the present *Introduction.* On a more popular level Ayensu (1980) is one of the best of the plethora of quasi-scientific books aimed at the concerned layman. Mitchell (1986) and Richards (1970) are beautifully illustrated general introductions. Other sources of more detailed information are referred to later in these text notes.
2. See Putz and Holbrook in Denslow and Padoch (1988) for an analysis of the influence tropical rain forests have had on European culture.
3. Stearn (1977) as paraphrased by Mabberley (1988).
4. *Antiaris toxicaria*, in modern Malay called ipoh.
5. Burkill (1935).
6. Quoted in Merrill (1945).
7. *Caesalpinia pulcherrima.*
8. A.R. Wallace quoted by Richards (1952).
9. *Verbena officinalis.*
10. *Aesculus hippocastanum.*
11. *Polygala vulgaris.*
12. *Vinca.*
13. *Viola.*
14. R. Spruce writing on the Amazon and quoted by Richards (1952). He uses as examples plants of northwest Europe likely to be known to his readers.
15. In this book the term canopy is used for the whole forest, not just its upper surface.
16. Corner (1964).
17. Whitmore (1989*c*).
18. Richards (1952) with a second edition in 1996.
19. A remarkable 20th century recapturing of 19th century science, recorded amongst other plants by: *Whitmorea grandiflora*, *Allowoodsonia*, *Anodendron*, *Aporusa*, *Bridielia*, *Calamus*, *Cyathea*, *Endiandra*, *Freycinetia*, *Heliciopsis*, *Horsfieldia*, *Licuala*, *Macaranga*, *Pandanus*, *Physokentia*, *Piper*, *Rhododendron*, *Swintonia*, *Terminalia* and *Vaccinium whitmorei.*
20. Sommer (1976).
21. Walter and Lieth (1967).
22. For modern detailed maps of tropical moist forests see Sayer *et al.* (1992) for Africa; Harcourt and Sayer (1996), and Dinerstein *et al.* (1995) for the Americas; Whitmore (1984*b*), Collins *et al.* (1991), and Blasco *et al.* (1996) for Asia.

23. These figure are for potential extent, viz. the area covered before humans came on to the scene.
24. Brown and Brown in Whitmore and Sayer (1992).
25. MacKinnon *et al.* (1996), Monk *et al.* (1997), and Whitten *et al.* (1984, 1987, 1996) are important regional accounts published in both English and Indonesian on the ecology of Borneo, Maluku, Sumatra, Sulawesi, and Java and Bali, respectively. They are at the same time student textbooks and source books for researchers.
26. Webb (1968).
27. As a belt in suitable sites between *c.* 26° and 28°N; see Academica Sinica (1979) for China; Proctor *et al.* (1998) for Nampalapha National Park, Assam (now Arunachal Pradesh); and Kingdon-Ward (1945) for Burma.
28. White (1983).
29. White (1983).
30. Baur (1968), White (1983).
31. Nepstad *et al.* (1994). It is possible that some of the rain forests of Indo-China are on similarly deep soils, but in India north of 14° 30′, where the dry season is 6–7 months (Pascal 1986) this explanation is unlikely to hold.
32. Proctor *et al.* (1998), Kingdon-Ward (1945), Academia Sinica (1979), Whitmore (1982*b*).
33. See Whitmore (1984*a*), plus, on upper montane forest, in two symposium volumes Churchill *et al.* (1995), and Hamilton *et al.* (1995).
34. Van Steenis (1972) is an elegant illustrated account.
35. See White (1983).
36. Brunig in Ashton (1971).
37. Davis and Richards (1933–4). This study has been repeated in full modern detail by ter Steege *et al.* (1993).
38. Ramin (*Gonystylus bancanus*), with timber pre-eminent for mouldings, grows here.
39. A scheme started in 1996 to convert 1.46 × 10^6 ha of Central Kalimantan is doomed to create a gigantic fern and shrub-covered wasteland. The environmental impact analysis, conducted after work started, found only 40 per cent of the area to be suitable for agriculture. (*Down to Earth* 29/30, 1996, 1–5; 33, 1997, 8–9.)
40. Driessen (1978).
41. Watt (1947), Whitmore (1978, 1982*a*, 1988).
42. Kramer (1933).
43. Swaine and Whitmore (1988).
44. Some authors have talked about 'mature-phase species' but that is to confuse the forest growth cycle, which concerns forest structure only, with the species of which a growth cycle is composed.
45. Bormann and Likens (1979).
46. Johns (1986).
47. Garwood *et al.* (1979).
48. Whitmore (1988, 1989*a*).
49. *Diospyros virginiana* (America), *D.kaki* (Orient).
50. Gentry (1988*a*) gives a comprehensive analysis.
51. Ashton (1982).
52. See Richards in Meggers *et al.* (1973).
53. Turner (1993).
54. Hall and Swaine (1976).
55. A 52-ha plot at Lambir, Sarawak, with more diverse topography and soil, has *c.* 1300 species (T. Yamakura, personal communication).
56. Grubb (1987), excluding 'critical' species of *Sorbus* in Europe and including 5 and 23 species of conifers, respectively.
57. Ng and Low (1982) document this for Malaya.
58. Gentry and Dodson (1987).
59. P.J. Grubb (personal communication).
60. Whittaker (1972).
61. Whitmore (1984*a*) discusses this problem more fully, using Far Eastern examples.
62. Richards (1952).
63. Whitmore (1974).
64. We do not know why rheophytes are commoner in some places, or why only a few families have evolved them (van Steenis 1981). One family, Podostemaceae, consists entirely of rheophytes; it has 50 genera and 275 species and is largely tropical. Members of Podostemaceae are remarkable moss-like or thalloid plants with minimal or no vascular tissues, which cling by adhesive roots to fast-flowing rocky cataracts. They die back at low water. Most are tiny, but a few develop seaweed-like flowering scapes to 0.6 m tall. It is always interesting to search for podostems while floundering up a rocky tropical stream. By keeping his eyes open,

Dr. J. Dransfield in 1968 made the first discovery of the family in Malaya, at a locality that had been much visited by botanists for 80 years.

65. Whitmore (1973*b*).
66. Newbery *et al.* (1986, 1988).
67. Evans (1972).
68. Sometimes known as screwpines.
69. Jacobs (1988).
70. Gentry (1988*a*).
71. Holttum (1955).
72. Hallé and Ng (1981).
73. Hallé *et al.* (1978).
74. Kursar and Coley (1992).
75. Henwood (1973) in Whitmore (1984*a*).
76. Corner (1988).
77. Formerly *Tristania*.
78. The Far East leads, e.g. Symington (1943); Whitmore (1972, 1973*a*); Ng (1978, 1989); recently introducing computer-based multi-access keys (Newman *et al.* 1996*a*,*b*,*c*). An important American example is Gentry (1993).
79. Klinge (1973).
80. For an excellent review see Ng (1984).
81. Ng (1984) and Longman and Jenik (1987) give clear accounts.
82. Holttum (1940) in Ng (1984).
83. Medway (1972*a*).
84. Borchert (1994*b*).
85. Borchert (1994*a*).
86. Wright and Cornejo (1990).
87. Borchert (1992), Reich (1995).
88. Borchert (1996), p. 79.
89. Janzen (1976).
90. R.T. Corlett, personal communication.
91. Holttum (1954).
92. Ashton *et al.* (1988) found that 5–8 nights with uniquely low minimum temperature of 18.9 °C, not the usual 21–23 °C, had occurred 8–9 weeks before a series of mass flowerings in Malaya and deduced that this chilling caused young, undifferentiated buds to become inflorescences not leaf flushes. However, this trigger did not precede the 1987 mass flowerings in Singapore and Danum, Sabah (R.T. Corlett and N.D. Brown, personal communication).
93. Bourlière (1983, 1985), Leigh *et al.* (1983), Montgomery (1978), and Terborgh (1983) give general accounts.
94. Erwin (1982). See also May (1988).
95. Others have subsequently repeated or questioned Erwin's study (e.g. Basset *et al.* 1996, Stork 1988) and argued the figure lies between 7 and 80 million depending on assumptions and the validity of extrapolations. J.B.S. Haldane, on being asked what he inferred of the Creator from a study of the natural world, replied that God must have an inordinate fondness for beetles. Erwin's study confirms this.
96. Robinson (1992) describes several bizarre adaptations and interdependencies of insects in rain forests, including moths that feed on mammals' tears or live in the fur of sloths.
97. Terborgh and van Schaik (1987) describe these differences more fully and relate them to differences in the phenology of leaf and fruit production in the three regions.
98. Robinson (1992).
99. Powell and Stradling in Huxley and Cutler (1991).
100. Terborgh (1983, 1986). Charles-Dominique *et al.* (1981) made an equally elegant study for nine nocturnal frugivorous mammals in a secondary forest in French Guiana.
101. Terborgh (1983).
102. Leighton and Leighton in Sutton *et al.* (1983).
103. Gautier-Hion and Michaloud (1989).
104. See Faegri and van der Pijl (1979) and Howe and Westley (1988).
105. Analysis has shown that species differ substantially in the composition of their nectar. Sometimes it contains amino acids as well as sugars. Intricate relationships have been discovered between nectar composition, amount and time of production, and the particular animal pollinators to which it is offered as an attractant and reward.
106. For recent discussions of beetle pollination see Schatz, and Irvine and Armstrong in Bawa and Hadley (1990).
107. Appanah and Chan (1981).
108. Gould (1978).
109. Start and Marshall (1976).
110. See van der Pijl (1982) and Howe and Westley (1988).
111. Snow (1981).
112. Gautier-Hion in Bawa and Hadley (1990).
113. Terborgh (1983).
114. Goulding (1980).

115. Snow (1966).
116. Wheelwright (1986).
117. To a plant-eating animal 'the plant world is not coloured green; it is coloured morphine, caffeine, tannin, phenol, terpene, canavanine, latex, phytohaemagglutinin, oxalic acid, saponin, and L-dopa ... a many textured landscape of scents and tastes, the distinction of which may be a life and death matter' (Janzen in Montgomery 1978). Howe and Westley (1988) give a broad review of herbivory, the various defences plants have set up, and how animals overcome them.
118. For example, the pretty Apocynaceous garden herb *Catharanthus roseus*, the rosy periwinkle, contains the alkaloids vincristine and vinblastine, valuable in the treatment of two previously intractable cancers. This species is a native of semi-arid Madagascar but is now pantropical. It commonly escapes and becomes naturalized on sandy shores.
119. Gilbert (1980). He used the term 'food web' not 'plant web' which is confusing to zoologists who use the term in a different sense in discussions of trophic levels.
120. The indications from those comprehensive studies that have been made are that fewer than 10 per cent of all plant-eating insects feed on more than three different plant families (Bernays and Graham 1988 in Strong 1988).
121. Gilbert (1980).
122. Appanah (1985). This pattern must have developed since human activity reduced Pasoh to an isolated patch. *Melastoma* dominated scrub is neither extensive nor common in the primary forest landscape of Malaya.
123. Prance (1986), but the fact that fruits form on trees planted in Malaya shows that as, so often, new interconnections can develop.
124. See Ehrlich and Raven (1964), Futuyma and Slatkin (1983), Thompson (1986), Strong (1988), and Howe and Westley (1988) for overviews.
125. The hypothesis takes its name from the Red Queen who made this statement in *Alice through the looking glass*, a famous children's story by Lewis Carroll.
126. Coley (1983).
127. For fuller details than given here see Huxley and Cutler (1991).
128. Huxley (1978).
129. Kleinfeldt (1986); Davidson (1988).
130. Fiala *et al.* (1989).
131. Maschwitz *et al.* (1984).
132. Benson *et al.* (1976).
133. Ng in Sutton *et al.* (1983).
134. Bawa and Hadley (1990) review the state of knowledge for rain forests.
135. Fedorov (1966); Ashton (1969). See also Whitmore (1984*a*) for a detailed account of the subsequent observations sketched here which have been made to flesh out the speculations.
136. Kaur *et al.* (1978).
137. Ashton (1969); Schatz in Bawa and Hadley (1990); Kress and Beach in McDade *et al.* (1994).
138. Gan *et al.* (1977, 1981).
139. See Hallam (1973) for a lively survey.
140. Smith and Briden (1977).
141. *Pseudomonotes tropenbosii.*
142. Thorne in Meggers *et al.* (1973); Thorne (1977).
143. Hallam and Audley Charles (1988); Whitmore (1981, 1987).
144. Of these only dipterocarps occur in India today, the others have been found as fossil pollen (Morley 1998).
145. See Hall (1996).
146. Morley (1998).
147. Charles (1997).
148. Flenley (1979) gives a comprehensive synthesis.
149. Haffer in Whitmore and Prance (1987) reviews geoscientific evidence for past different climates.
150. Fig. 6.10 shows only the upper part of this record.
151. Thorne (1977), van der Hammen and van der Cleef (1983).
152. Richards in Meggers *et al.* (1973).
153. Haffer (1967, 1974) for birds, Vanzolini (1970) for lizards: cited by Brown in Whitmore and Prance (1987).
154. Nelson *et al.* (1990).
155. See Brown in Whitmore and Prance (1987) for a critical review of neotropical Pleistocene refugia.
156. Bush (1994); Colinvaux (1987); Nelson *et al.* (1994); Colinvaux *et al.* (1996).
157. Hamilton (1976, 1982); Mayr and O'Hara (1986); Maley (1987); Maley in Alexander *et al.* (1996).
158. Bahuchet in Hladik *et al.* (1993).
159. Hu (1980).

160. Medway (1972*b*).
161. van Steenis (1972).
162. Wong and Phillips (1997).
163. For overviews on tropical rain forest dynamics see Whitmore (1978, 1982*a*, 1988, 1989*a*, 1996); Platt and Strong (1989).
164. But in two out of the five studies made gap soils were moister (Whitmore 1996), presumably due to transpiration of soil moisture from below closed canopy.
165. Brown (1993) analyses the Danum forest; Whitmore (1996) reviews the topic.
166. Lieberman *et al.* (1989).
167. Swaine and Whitmore (1988).
168. Some pioneers have now been found below closed canopy as miniscule seedlings which soon die; they have germinated, presumably triggered by sunflecks, but cannot establish and grow, most probably because they receive insufficient light to maintain a positive carbon balance (Whitmore 1996).
169. See Finegan (1992), who calls them 'long-lived intolerant' species.
170. Budowski (1965).
171. Egler (1954).
172. See Swaine (1996) for recent reviews.
173. Symington (1933).
174. Keay (1960).
175. Whitmore (1983); Vazquez-Yanes and Orozco-Segovia (1984).
176. Roberts (1973).
177. Which may be caused by high irradiance, viz. exposure to high light; see Swaine *et al.* (1997) for this effect in Ghanaian pioneers.
178. Roberts (1973).
179. With the exception of the pioneer mangrove *Avicennia*.
180. Burgess in Whitmore (1984*a*).
181. Itoh (1995).
182. Howe and Smallwood (1982); Clark and Clark (1984). At Barro Colorado Island during 1982–5 the commonest tree *Trichilia tuberculata* had least recruits near parents, more species showed the reverse, and most showed no significant spatial pattern (Condit *et al.* in Whitmore 1996).
183. Sometimes also known as 'Janzen-Connell spacing' after its discoverers.
184. Chan (1980).
185. Burgess (1969).
186. Riera (1985); Raich (1987).
187. Brandani *et al.* (1988) and Nũnez-Farfán and Diezo (1988) in Whitmore (1996).
188. Denslow *et al.* (1990) in Whitmore (1996).
189. Connell (1978).
190. See Mulkey *et al.* (1996) and Press *et al.* in Swaine (1996) for recent reviews.
191. For example, mesh shade-houses create an unnatural microclimate (Whitmore 1996).
192. Chazdon and Pearcy (1986*a*,*b*).
193. Burslem *et al.* (1995); review in Whitmore (1996); Swaine *et al.* 1997, section 5).
194. 11.3 cm diameter in the biggest gap (range 7.6–16.7 cm, $n = 10$).
195. Thompson *et al.* (1988) in Whitmore (1996).
196. Hawthorne (1993); this guild was predicted by Oldeman and Dijk (1991); both in Whitmore (1996).
197. Clark and Clark (1992).
198. Lee *et al.* (1997); Scholes *et al.* (1997); Press *et al.* in Swaine (1996); Whitmore (1996); Zipperlen and Press (1996, 1997).
199. Ashton and Berlyn (1991). In Ghana species that occupy different places along a rainfall gradient have seedlings with different sensitivity to drought (Swaine in Swaine 1996).
200. Today gutta-percha is used only for temporary dental stoppings for which it is well suited by its unusual property of swelling as its sets from the liquid to solid state.
201. The term ecology is usually dated to the German E.H. Haeckel, 1866.
202. See Dawkins (1988) and Schmidt in Gomez Pompa *et al.* (1991) for reviews.
203. Technically termed the 'dysgenic effect'.
204. Bruijnzeel and Critchely (1994), fig. 7.
205. Reviewed by Bruijnzeel and Critchely (1994).
206. For example, Baharuddin (1988) demonstrated this in central Malaya. In Queensland such cross-drains were evocatively named 'whoaboys'.
207. Respectively, Pinard *et al.* (1995), Bertault and Sist (1995), Precious Woods Ltd and Tropical Forest Foundation.
208. Schmidt (1987) reviews silviculture throughout the humid tropics.
209. For fuller description see Newman *et al.* (1996*a*,*b*).
210. See Whitmore (1984*a*) and Salleh in Cranbrook (1988) for fuller descriptions of silviculture in Malaya. In its full form the

Malayan Uniform System included repeated interventions during the first decade after felling to assist growth of the next timber crop.

211. The so-named Selective Management System (SMS). A modified MUS with a rotation of 55 years was retained for peat swamp forests, poorly regenerated logged-over forest, lower montane and degraded forest (Anon, 1996.)
212. In Sarawak a highly modified form called Liberation Thinning has been introduced in which after logging very selective thinning is applied just around potential crop trees.
213. Philippine Selective Logging System (introduced 1955) and Indonesian Selective Cutting and Planting System (Tebang Pilih Tanaman Indonesia), respectively.
214. Fearnside (1989) reviews the handful of examples.
215. De Graaf (1986); Gradwohl and Greenberg (1988).
216. Whitmore and Silva (1990).
217. Silva *et al.* (1995).
218. See MacLellan (1996) for a review of the patchy knowledge.
219. Parren and de Graaf (1995); Hardcastle (1997); and, for the underlying seedling ecology, Swaine *et al.* (1997).
220. Nicholson (1965) and McWhirter (1980) in Whitmore (1984*a*).
221. Failure to allow for varying growth rate with forest phase is a major weakness of rain forest productivity studies, see Whitmore (1984*a*). Thus, it is also a weakness of scenarios that extrapolate from these data to compute how much CO_2 rain forests sequester, and hence how they might contribute to amelioration of global warming.
222. See review by Whitmore and Burslem in (1998).
223. Johns (1986).
224. Salo *et al.* (1986).
225. Balée (1989).
226. Anderson *et al.* in Lugo *et al.* (1987); Balée (1988).
227. Nelson (1994).
228. Nelson *et al.* (1994).
229. The ocean-surface warming of the eastern Pacific altered the currents across the Pacific and that is what altered the rainfall; see Gill and Rasmusson (1983) for a full description. One result was the changed position or oscillation of the current which strikes the coast of Peru about Christmas time. This change reduced the harvest of anchovies, mainstay of the local fisheries. The term El Niño is Spanish for 'The Child' and is an allusion to the Christ child, who was born at Christmas time.
230. Beaman *et al.* (1985).
231. Leighton and Wirawan (1986). In logged forest, by contrast, more of the small trees were killed.
232. Walsh (1996).
233. Goldammer and Seibert (1990) in Whitmore and Burslem (1998).
234. Condit *et al.* (1995).
235. Sanford *et al.* (1985).
236. Piperno and Becker (1996) in Whitmore and Burslem (1998).
237. At 54–57°W, 2–3°S (Nelson 1994).
238. And once opened up by burning becomes more susceptible to further fires (Swaine *et al.* 1998).
239. Unpublished data of Burslem and Whitmore.
240. Newbery in Alexander *et al.* (1996).
241. Guevara *et al.* (1986).
242. Kochummen and Ng (1977).
243. Hartshorn (1980) in Whitmore (1996). In Scotland and Chile, respectively, birch (*Betula*) and *Nothofagus* are self-perpetuating pioneers in locations where shade-tolerants are absent.
244. Hall and Swaine (1976); Swaine *et al.* (1997); Veenendaal and Swaine in Newbery *et al.* (1998).
245. Swaine and Hall (1988).
246. Whitmore (1974).
247. Jones (1955, 1956).
248. Manokaran and Swaine (1994).
249. Whitmore (1989*b*). The observations now continue for 30 years, analyses are still in progress.
250. Swaine *et al.* (1987). A perplexing discovery (Phillips 1996), which has caused considerable interest, is an increase in turnover since the 1950s shown by 67 sample plots spread across the humid tropics and believed all to sample mature-phase forests. There are difficulties in sampling multi-aged forests and this result could perhaps be an artefact (Sheil and May 1996); and 17 plots measured over 25 years in Venezuela do not show increasing mortality (Carey *et al.* 1994).

251. Primack and Hall (1992).
252. Zinke *et al.* (1978).
253. And was in fact also practised in prehistoric Europe.
254. D.U.U. Okali (personal communication).
255. Chin (1985).
256. Boerboom and Wiersum (1983); Denslow and Padoch (1988).
257. Kunstadter in Denslow and Padoch (1988); Schmidt-Vogt (1995).
258. Uhl and Jordan (1984).
259. Ramakrishnan in Proctor (1989).
260. Jordan (1985) is the only comprehensive text and Proctor (1989) reports an important symposium.
261. The common assumption that feeding roots in tropical rain forests are shallow is challenged by the discovery that in the lower Amazon they go down to 8 m (Nepstad *et al.* 1994, and see p. 17).
262. Bruijnzeel (1989).
263. Baillie and Ashton (1983); Baillie in Proctor (1989).
264. Vitousek and Sanford (1986); Bruijnzeel (1990) pointed out the need for work on all facets of the nitrogen cycle to secure sounder generalizations than those based on individual processes.
265. Proctor (1984).
266. viz. k the decay factor ranges from 3 to 1, a turnover thrice to once a year (Table 8.3).
267. Collins in Cranbrook (1988).
268. Bruijnzeel (1995).
269. See Proctor in Primack and Lovejoy (1996) for a critical discussion.
270. Gillman *et al.* (1985).
271. Gradwohl and Greenberg (1988) give an account.
272. See Lamb (1990) for a critical review of the Gogol venture.
273. Ecosystem nutrient capital in a plantation changes with age but is also likely to vary across the plantation due to intrinsic site differences. This makes it difficult to interpret data on nutrients against age put together from different sites, so-called 'false time series' studies. Spanenberg *et al.* (1996) documented serious nutrient depletion in the soils under the Jari plantations in Amazonian Brazil after one to three rotations with wood plus bark being harvested; deficiency, especially of calcium, will soon develop. Bruijnzeel (1990) reviewed other studies.
274. Nykvist *et al.* (1994).
275. Fearnside (1988) gives a good review of the Jari plantation forest project which covers *c.* 100 000 ha.
276. P. Lavelle (personal communication).
277. Kellman (1970).
278. The Yurimaguas experiment has been critically appraised by Fearnside and others (Fearnside 1987). The Nigerian work is described by Juo and Kang in Proctor (1989).
279. Jordan (1987); Gradwohl and Greenberg (1988).
280. Juo and Kang in Proctor (1989). Alley-cropping has been a 'nine days' wonder'. It gave the new agro-forestry agencies an apparent success, so was widely trumpeted, but has not lived up to its early promise. It requires repeated, intensive, and unusual labour-input. Yields drop as soil-organic matter of the original forest decomposes, and it has proved much less easy on oxisols and ultisols than on the alfisols on which it was pioneered.
281. Fuller discussion and references to the research literature mentioned here can be found in Whitmore (1984*a*, 1991).
282. Turner (1994) argues that low resource availability, whatever may cause it, selects for long leaf life, hence for sclerophylly which protects leaves from vicissitudes of the environment and from herbivory. See also Reich *et al.* (1991).
283. Medina and Cuevas in Proctor (1989).
284. i.e. the cation exchange capacity drops to nearly zero.
285. Experiments by J. Proctor and colleagues have shown this.
286. Franco and Dezzeo (1994) give a detailed analysis.
287. There is also the complication that a swampy facies of heath forest exists, especially common in Borneo where it is called kerapah, see Whitmore (1984*a*, section 12.3).
288. See Whitmore (1984*a*, 1989*d*) for fuller details.
289. The unusual absence of a sharp boundary between lower and upper montane forests on Volcán Barva, Costa Rica, 2906 m (Lieberman *et al.* 1996), may be a consequence of young, fertile volcanic soils and the absence of a regular cloud cap. By contrast, the ecotone

between lowland and lower montane forest is always gradual. In Brunei it was deduced that, in the absence of evidence of high winds, drought, waterlogging, high soil acidity, or nutrient limitation, the change in forest with altitude, 200 to 850 m, was due to a reduction in temperature (Pendry and Proctor 1996).

290. However nutrient budget computations have shown an excess of supply in precipitation over uptake by plants in three of four forests analysed (Bruijnzeel and Proctor in Hamilton *et al.* 1995).
291. At Gunung Silam, Sabah, the stunted summit forest at 870 m lies within the cloud cap, nutrients seem adequate and drought not to occur. It was suggested that the high concentrations of polyphenols in the soil may be interfering with growth (Bruijnzeel *et al.* 1993), as has been suggested also for heath forests.
292. See also Whitmore (1984*a*) and Diamond (1988).
293. Whittaker (1977).
294. Who have not discovered how to make fire and have to keep naturally occurring fires alight.
295. Foresta and Michon in Hladik *et al.* (1993).
296. Brookfield (1993).
297. Balée (1987).
298. See also Peluso (1983) on the breakdown of social control on rattan exploitation in East Kalimantan after commercial logging began.
299. Dunn (1975) is the best account.
300. Gaharu is immensely valuable and was being traded in the late 1980s at \$200–300 kg^{-1} in Sarawak, where it is also produced by some other species of the same family (Caldecott 1996). Throughout its wide range in the Eastern forests the trees are always rare and highly sought after. Many trees are felled that prove not to have the dark, dense, fragrant, diseased heartwood which is gaharu.
301. Richards, P. (1996).
302. Keay (1996).
303. All these examples lie in seasonal climates. One may speculate that permanent agriculture was easier and diseases less serious than in the perhumid tropics which have no dry season.
304. Gomez Pompa *et al.* (1987).
305. See Rambo in Cranbrook (1988) on Malaya.
306. Lucidly analysed by Schmidt-Vogt (1995).
307. For a few years in the early 17th century King James I was 'by the Grace of God King of England, Scotland, Ireland, France, Puloway and Puloroon' till the Dutch dislodged the East India Company traders from Pulau Run and P.Ai, their toehold specks of land in the Banda group of Maluku (Keay 1991).
308. Monk *et al.* (1997).
309. Merrill (1945).
310. Whitmore (1977).
311. And in the Dutch East Indies a network of Nature Reserves was created half a century before these became widely used as a conservation measure.
312. Layrisse (1992) estimates that the Amazon had 10×10^6 inhabitants at the time of European contact, when Europe itself had 12×10^6 and that it has declined to 0.25×10^6 today.
313. These are less biodegradable, hence the permanent mounds of rubbish which today disfigure so many tropical settlements.
314. The estimated plant-based prescription drug market of OECD countries (which make up most of the developed Western world) was \$$43 \times 10^9$ in 1985, 41 per cent of the total market (Collins in McDermott 1988).
315. Even more remarkable, several species of *Brosimum* in Amazonia have a white, freely flowing latex which can be drunk and is used as a substitute for milk.
316. Cocoa entered cultivation in Central America where it was believed to be of divine origin (the scientific name *Theobroma* is Greek for food of the Gods). Amongst the Aztecs only the wealthy drank cocoa.
317. Recently wild-type coconuts have been found in both Australia and the Philippines, finally settling the old conundrum of where this species, 'nature's greatest gift to man', originated (Uhl and Dransfield 1987).
318. In fact only 25 of the *c.* 119 plant-based drugs in worldwide use come from tropical rain forests. However, the value of tropical forest-derived pharmaceutical products in the United States in 1990 is *c.* \$$5 \times 10^9$ (Soejarto 1997). For a wide ranging perspective on rain-forest drugs see Balick *et al.* (1996).
319. Swanson (1995); Principe in Akerele *et al.* (1991); for AIDS see Soejarto (1997).

320. Corner (1966).
321. Peluso (1983).
322. Already meranti rods are being substituted for the large diameter rotan manau (*C. manan*) as the straight parts of chairs and tables.
323. Rotan irit, *Calamus trachycoleus*, is especially suitable, and has been grown by villagers in the remote interior of central Kalimantan for over a century; over 7000 ha are now planted in Sabah (Dransfield 1988).
324. Peters *et al.* (1989) started this hare running in a notorious analysis of a forest exceptionally rich in perishable fruits (see Phillips 1993), and only 30 km from Iquitos, a big urban market. They compared retail price of the fruit with the price of lumber at the sawmill, far from the point of retail sale. Not surprisingly minor forest products fared better.
325. Godoy *et al.* (1993).
326. See, for example, Bawa (1992).
327. The forests of northeast Siberia and the great valleys which penetrate the Tibetan plateau westwards from China are the others.
328. Plato described Attica in the 5th century BC as 'like the skeleton of a sick man, all the fat and soft earth having been washed away, only the bare framework of the land being left' (quoted in Aiken and Leigh 1992).
329. Fischer and Heilig in Greenland *et al.* (1997).
330. Browder (1988); Uhl *et al.* (1988).
331. Hecht *et al.* (1988).
332. Cattle raised on tropical American pastures have lean, tough meat. Some of it is sold to North America. It is very suitable for hamburgers. For this reason rain forest destruction has been evocatively ascribed to 'the hamburger connection'. It is, however, totally misleading to regard the 'fast food' market as a major engine of tropical deforestation. In 1982 hamburger-bound beef imports to the USA from nine Latin American countries were only 7900 t (*c.* 26 000–32 000 carcasses representing at the most 100 000 ha of pasture). This was 1.2% of total beef imported to USA where total beef consumption was 24 million t (Browder 1988).
333. These can also produce sawlogs but *A. mangium* is rendered useless by dark discoloration of the heartwood where water gets in through branch snags. This defect can be avoided by growing trees whose lateral branches are horizontal or drooping rather than erect. This trait is apparently inherited.
334. Fearnside (1988).
335. Federal Land Development Authority.
336. See Davis in Denslow and Padoch (1988) for an account of transmigration and its effect on the forests of Indonesia. In the 1990s governmental emphasis has switched to rehabilitation of poorly functioning transmigration schemes and the private sector has begun to fund new schemes, as a source of labour for plantation development, etc.
337. See Vadya in Denslow and Padoch (1988).
338. Lanly (1982).
339. McCully (1996).
340. Aiken and Leigh (1992), pp. 127–8.
341. Stone in Denslow and Padoch (1988) gives a full account.
342. Consumption is 1×10^6 tonne charcoal year^{-1} to fuel 7 pig iron, 2 iron alloy and 2 cement factories (Gradwohl and Greenwood 1988).
343. Roughly speaking, 50 per cent of the human race gets 90 per cent of its energy from fuelwood.
344. Full discussion is beyond the scope of this book but it can be noted that in much of the seasonal tropics fuelwood is becoming hard to get. In Karnataka, India, its cost increased eleven-fold from 1970 to 1988, compared to an overall rise in the cost of living of under four-fold.
345. With the major exception of teak.
346. Gillis in Denslow and Padoch (1988) describes the exploitation of the west Malesian dipterocarp rain forests.
347. A country endowed with a natural resource does not necessarily have a comparative advantage in processing it. In both Malaysia and Indonesia jobs in timber processing were created at a very high price, once the costs of machines and the reduced value of logs sold to a captive market were allowed for. A protected internal market also suppresses signals of impending shortage so log prices do not rise, nor does extraction become more efficient as resources dwindle (Vincent 1992).
348. International Tropical Timber Organization, see Table 10.14.
349. Mittermeier (1995) and Sizer (1996) describe the interest being shown in Suriname by Asian

loggers, but in late 1996 the government postponed any decision for three years (*Plant Talk* 8, 1997, 12).

350. Pleydell (1995). Medium density fibreboard (MDF) and oriented strandboard (OSB) make use of what was previously waste material. Small pieces of wood, finger-jointed and stuck together with modern glues, also utilize fragments that were previously unusable.
351. *Diospyros*; *Chloroxylon*, *Distemonanthus*, *Fagara*; *Swietenia*.
352. Uhl and Viera (1989).
353. The mahogany which was so important and fashionable as a cabinet timber in 19th-century Europe came from the Caribbean (*Swietenia mahagoni*) and Central America (*S. macrophylla*).
354. Hutan Tanaman Industri, HTI.
355. 1 ha forest has 200–300 tonne of wood, 4 tonne make 1 tonne of charcoal and this fuels a smelter for 5 minutes.
356. Most recently synthesized in Johns (1997).
357. Crome in Gomez Pompa *et al.* (1991); Thiollay (1992).
358. Struhsaker (1997).
359. Holloway *et al.* in Marshall and Swaine (1992); Hill *et al.* (1995).
360. So-called lesser known species (LKS) or less used species (LUS). Their utilization is often promoted by aid agencies without considering the implication that, were extra exploitation successful, it could seriously increase damage to the forest.
361. FAO (1993, 1995); for commentary see Whitmore (1997).
362. Lanly (1982), see ed. 1 for details.
363. FAO (1995), p. 8.
364. Brown *et al.* (1989).
365. Myers (1980), (1990).
366. Strong (1972) in Whitmore (1984*a*), referring to photographs taken by the first men on the moon.
367. Goodland (1995).
368. World Commission on Environment and Development (1987). This report, as Goodland (1995) has pointed out, achieved the brilliant feat of garnering almost worldwide political consensus for the urgent need for sustainability.
369. This is often called the concept of 'intergenerational equity'.
370. IUCN was formerly called International Union for Conservation of Nature, WWF was formerly World Wildlife Fund.
371. For the uneasy relationship between the Malaysian government and local NGOs see Aiken and Leigh (1992), pp. 121–2.
372. Well-expressed in *Proceedings of the conference on forest resources crisis in the Third World 6–8 September, 1986*, Sahabat Alam Malaysia, Penang (1987), and evocatively in Waiko (1980).
373. Aiken and Leigh (1992), pp. 126–7.
374. The prolonged struggle since 1972 to create a national park is analysed by Aiken (1993).
375. IUCN (1980), IUCN/UNEP/WWF (1991).
376. Attended by 117 Heads of State, more than ever before assembled.
377. Chapter 11 Agenda 21 was all that could be agreed. This is a 'non legally binding authoritative statement of principles for a global consensus on the management, conservation and sustainable development of all types of forest and combating deforestation.'
378. Hadley and Kartawinata (1993).
379. See Groombridge (1992) for a full analysis.
380. May (1988).
381. See Heywood and Stuart in Whitmore and Sayer (1992). One example is cocoa, *Theobroma cacao*, an understorey tree of lowland rain forest, which is endemic to northeast Peru and adjacent Ecuador in the upper Amazon, is currently losing diversity as oil prospecting, logging and settlement are eating away its habitat.
382. Simberloff in Whitmore and Sayer (1992).
383. Reid in Whitmore and Sayer (1992).
384. Anon. (1988). It is not possible to update these figures because the next assessment (Collar *et al.* 1994) used a new set of criteria of endangerment, making the point that Indonesia and Brazil together with Philippines were the top three countries in the world for number of threatened bird species.
385. Wilson (1992). The most recent mass extinction event was the so-called K/T boundary event at the end of the Cretaceous 65×10^6 years ago when *inter alia* dinosaurs and ammonites became extinct.
386. Balick *et al.* (1996).
387. Iltis *et al.* (1979).
388. Kahn (1988).

389. There are now numerous studies. See, for example, Denslow and Padoch (1988); Padoch and Vayda in Golley (1983); Aiken and Leigh (1992), pp. 97–107; Sponsel *et al.* (1996).
390. Hart and Petrides in Lugo *et al.* (1987).
391. Bevis (1995), pp. 184–5, 189–90. See Caldecott (1996) for the context in which logging of the Baram basin took place.
392. Bruijnzeel (1990).
393. Conversion of rain forest to grassland increases water yield by 300–400 mm $year^{-1}$ and to arable crops by 600 mm $year^{-1}$ (L.J. Bruijnzeel personal communication).
394. The Rio Nima in Colombia is an example of a formerly forested water catchment where re-establishment of vegetation and the control of erosion has improved water quality, in this case for the downstream city of Palmira (Gradwohl and Greenberg 1988).
395. Bruijnzeel (1990).
396. Salati and Vose (1984).
397. Gash *et al.* (1996).
398. Watson *et al.* (1996).
399. For a general account of the 'greenhouse effect' see Bolin *et al.* (1986).
400. Lugo and Brown (1993).
401. Nepstad *et al.* (1994).
402. Pearce (1989); 33 l = 200 g at room temperature.
403. The claim for tropical rain forests is made from heroic extrapolation of measurements of CO_2 fluxes in and above the canopy for a few weeks at two places in the Amazon, by Grace *et al.* (1996). For a critical discussion see Keller *et al.* (1996) who point out the very large heterogeneity in both space and time in any large area of rain forest (see Chapters 1 and 7), which renders the extrapolation highly speculative. Moreover, the net uptake of carbon is a small difference between photosynthesis and respiration, of which soil respiration is very temperature sensitive, such that a 1 °C change alters sink to source (Nobre *et al.* in Gash *et al.* 1996). Given these caveats it was perhaps unfortunate that discovery of this putative sink was highlighted by NERC in 1996 as one of the major findings of its interdisciplinary TIGER research programme.
404. Hulme and Viner in Markham (1998).
405. Numerous aspects of rain forest biology were considered, see Markham (1998).
406. Whitmore in Markham (1998).
407. Watson *et al.* (1996).
408. Set up in 1991 and administered by UNDP, UNEP and the World Bank. For a summary of its activities up to 1997 see *Nature* 385 (1997), pp. 106–7.
409. Vincent (1992).
410. ITTO (1990, 1992).
411. Poore *et al.* (1989). Unfortunately campaigners seized on the one million ha, and ignored the important corollary.
412. For example, the Virgin Jungle Reserves of Malaya.
413. So-called usufruct.
414. Japan, USA, France, Germany, Holland, Ireland, Italy, Portugal, Spain, UK (Thompson in McDermott 1988). See also Secrett (1987).
415. So-called accreditation. By 1997 five bodies had been accredited and FSC had launched an 'eco-label' to be attached to timber and timber products derived from certified forests.
416. Lembaga Eco-Label Indonesia, Green Label initiative of the African Timber Organisation, and the Amazon Co-operation Treaty countries, respectively. Outside the tropics there are other national and regional schemes.
417. Specifically to ISO 14 000.
418. Analysis for the World Bank by Varangis *et al.* (1995).
419. All once clothed with dipterocarp rain forests, which now remains extensive only on Borneo (Brunei, Kalimantan, Sabah and Sarawak).
420. Prescott Allen and Prescott Allen (1983).
421. See Bramwall *et al.* (1987).
422. Bibby *et al.* (1992).
423. Just such a rational approach to pinpointing priorities for conservation has been published for tropical America (Dinerstein *et al.* 1996).
424. Whitmore (1973*a*).
425. Soulé and Wilcox (1980), Frankel and Soulé (1981), Soulé (1986, 1987).
426. In Malaya 72 virgin jungle reserves, totalling 21 095 ha remain (Thang 1995) of a former 130 totalling 106 500 ha (Putz 1978); Sabah has 47 of total area 88 299 ha (Thang 1995).
427. Leighton and Leighton in Sutton *et al.* (1983); F. Lambert (unpublished).
428. Caldecott (1988).

429. Kio (1978), Leslie (1987).
430. Evans (1984).
431. *The Economist*, 31 August 1996.
432. Areas planted by the mid-1990s were 1918 ha teak, 5681 ha pines and *Araucaria*, 153 124 ha *Acacia mangium*, *Paraserianthes*, *Gmelina* and others (Thang 1995).
433. *Swietenia macrophylla* is probably the most extensively planted quality hardwood. Plantations have been established in some 40 countries, the largest being 116 000, 42 000 and 25 000 ha in Indonesia, Fiji and the Philippines respectively (Mayhew 1997.) As a strongly light-demanding climax species (section 7.6) it is easy to grow fast in plantations. Some attempts have been made to plant dipterocarps but their low-value timber and tendency to develop broad, spreading crowns (except *Shorea platyclados*) makes them unsuitable.
434. This is a currently empty niche for a far-sighted entrepreneur looking for a rich return in 20–25 years' time.
435. Sena sometimes suffers from a fungal disease (Corner 1988). Ironically when one is felled in Singapore its timber goes to waste. Other fast-growing quality hardwood species which deserve fuller trial are *Acrocarpus fraxinifolius* of India and Burma, and *Cybistax donnell-smithii*, *Hibiscus elatus* and *Tabebuia* spp. of tropical America (L.J. Whitmore personal communication). A race of teak which grows well in rain-forest climates has recently been developed by Solomon Islands Forestry Division; previously this species only thrived in seasonal climates.
436. Already biomass is the fourth most important source of energy globally, and at 13 per cent is equivalent to 20×10^6 barrels of oil day^{-1}, ten times the mid-1990s production from the North Sea (D.O. Hall personal communication).
437. Okali (1992).
438. Turner (1996); Turner and Corlett (1996); Laurance and Bierregaard (1997).
439. Terborgh (1992).
440. Corlett and Turner in Laurance and Bierregaard (1997); Turner *et al.* (1994).
441. Only a few tree species have moved out of these specks to colonize the central catchment since it was created 70 years ago; most remain confined to sites from which primary forest was never cleared.
442. Corlett, (1988); 16 out of 30 mammals, excluding bats, and 80 out of 150 bird species, have probably disappeared. Chin *et al.* (1995) collates all data on Bukit Timah.
443. Bennett and Caldecott (1981) and Duff *et al.* (1984) in Whitmore (1997).
444. Bierregaard *et al.* (1992).
445. Brown and Brown in Whitmore and Sayer (1992).
446. The state of knowledge is reviewed in Laurance and Bierregaard (1997).
447. For example, Stiles (in Almeda and Pringle 1988) has demonstrated that on the Caribbean slope of the mountains of Costa Rica 20–25 per cent of the rain forest birds migrate daily or seasonally from lowlands to highlands and can therefore only persist if forest is conserved at all elevations.
448. Deramakot forest Sabah, 50 000 ha, yielded *c.* 4×10^6 m^3 timber (80 m^3 ha^{-1}) at first cut and 20×10^3 m^3 is anticipated from the second cut (0.4 m^3 ha^{-1}). (M.Kleine personal communication.)
449. Sayer (1995).
450. Jeanrenaud (1997).
451. See several papers in Hladik *et al.* (1993).
452. Goodland (1995); Kimmins (1987); Vincent (1992); Whitten *et al.* (1996).
453. Kemp and Phantumvanit (1995) thought that, amongst major producers of internationally traded tropical timber, only Ghana, Indonesia, and Malaysia had any realistic prospect of bringing their entire permanent production forest under sustainable management by 2000. Nevertheless, the ITTO 2000 goal has created an awareness and a commitment to sustainability even if it is not met in full.

References

Academia Sinica (1979). [*Vegetation map of China*] (in Chinese). Map Publisher of the People's Republic of China, Beijing.

Aiken, S.R. (1993). Struggling to save Malaysia's Endau-Rompin rain forest. *Environmental Conservation* **20**, 157–62.

Aiken, S.R. (1994). Peninsular Malaysia's protected areas' coverage, 1903–92: creation, rescission, excision, and intrusion. *Environmental Conservation* **21**, 49–56.

Aiken, S.R. and Leigh, C.H. (1992). *Vanishing rain forests, the ecological transition in Malaysia.* Clarendon Press, Oxford.

Akerele, O., Heywood, V. and Synge, H. (1991). (eds.) *The conservation of medicinal plants.* Cambridge University Press, Cambridge.

Alexander, I.J., Swaine, M.D. and Watling, R. (1996) (eds.). Essays on the ecology of the Guinea-Congo rain forest. *Royal Society of Edinburgh Proceedings* **B 104**, 1–365.

Almeda, F. and Pringle, C.M. (1988) (eds.). *Tropical rain forests: diversity and conservation.* California Academy of Sciences and Pacific Division, American Association for the Advancement of Science, San Francisco.

Anon. (1985). *Tropical forests: a call for action. Part 1. The plan.* World Resources Institute, Washington D.C.

Anon. (1988). Birds to watch: the ICBP world checklist of threatened birds. *ICBP Technical Publication*, 6.

Anon. (1996). *Forestry in Peninsular Malaysia.* Forest Department, Kuala Lumpur.

Appanah, S. (1985). General flowering in the climax rain forest of south east Asia. *Journal of Tropical Ecology* **1**, 225–40.

Appanah, S. and Chan, H.T. (1981). Thrips: the pollinators of some dipterocarps. *Malaysian Forester* **44**, 234–52.

Ashton, P.M.S. and Berlyn, G.P. (1991). Leaf adaptations of some *Shorea* species in sun and shade. *New Phytologist* **121**, 587–96.

Ashton, P.S. (1969). Speciation amongst tropical forest trees: some deductions in the light of recent evidence. *Biological Journal of the Linnean Society, London* **1**, 155–96.

Ashton, P.S. (1971). The plants and vegetation of Bako National Park. *Malayan Nature Journal*, 24, 151–62.

Ashton, P.S. (1982). Dipterocarpaceae. *Flora Malesiana, Ser.I*, **9**, 237–552.

Ashton, P.S., Givnish, T.J., and Appanah, S. (1988). Staggered flowering in the Dipterocarpaceae: new insights into floral induction and the evolution of mast fruiting in the aseasonal tropics. *American Naturalist* **132**, 44–66.

Ayensu, E.S. (1980) (ed.). *Jungles.* Jonathan Cape, London.

Baharuddin, Kasran (1988). Effects of logging on sediment yield in a hill dipterocarp forest in Peninsular Malaysia. *Journal of Tropical Forest Science* **1**, 56–66.

Baillie, I.C. and Ashton, P.S. (1983). Some soil aspects of the nutrient cycle in the mixed dipterocarp forests of Sarawak, East Malaysia. In *Tropical forest ecology and management* (ed. S.L. Sutton, T.C. Whitmore, and A.C. Chadwick). Blackwell, Oxford.

Balée, W. (1987). Cultural forests of the Amazon. *Garden* **11** (6), 12–14, 32.

Balée, W. (1988). Indigenous adaptation to Amazonian palm forests. *Principes* **32**, 47–54.

Balée, W. (1989). The culture of Amazonian forests. In *Resource management in Amazonia: indigenous and folk strategies* (eds. D.A. Posey and W. Balée). Advances in Economic Botany **7**.

Balick, M.J., Elisabetsky, E. and Laird, S.A. (1996) (eds.). *Medicinal resources of the tropical forest: biodiversity and its importance*. Columbia University Press, New York.

Basset, Y., Samuelson, G.A., Allison, A. and Miller, S.E. (1996). How many species of host-specific insects feed on a species of tropical tree? *Biological Journal of the Linnean Society* **59**, 201–16.

Baur, G.N. (1968). *The ecological basis of rain forest management*. Forestry Commission of New South Wales, Sydney.

Bawa, K.S. (1992). The riches of tropical forests: non-timber products. *Trends in Research in Ecology and Evolution* **7**, 361–3.

Bawa, K.S. and Hadley, M. (1990) (eds.). *Reproductive ecology of tropical forest plants*. UNESCO and Parthenon, Paris and Carnforth.

Beaman, R.S., Beaman, J.H., Marsh, C.W., and Woods, P.V. (1985). Drought and forest fires in Sabah in 1983. *Sabah Society Journal* **8**, 10–30.

Benson, W.W., Brown, K.S.Jr., and Gilbert, L.E. (1976). Co-evolution of plants and herbivores: passion flower butterflies. *Evolution* **29**, 659–80.

Bertault, J-G. and Sist, P. (1995). Impact de l'exploitation en forêt naturelle. *Bois et Forêts Tropiques* **245**, 5–20.

Bevis, W.W. (1995). *Borneo log*. University of Montana Press, Seattle.

Bibby, C.J. and 8 others (1992). *Putting biodiversity on the map: priority areas for global conservation*. International Council for Bird Preservation, Cambridge.

Bierregaard, R.O., Lovejoy, T.E., Kapos, V., Augusto dos Santos, A. and Hutchings, R.W. (1992). The biological dynamics of tropical rain forest fragments. *BioScience* **42**, 859–66.

Blasco, F., Bellan, M.F. and Aizpura, M. (1996). A vegetation map of tropical continental Asia at scale 1:5 million. *Journal of Vegetation Science* **7**, 623–34.

Blume, C.L. (1835–48). *Rumphia*, 4 Vols. Lugduni Batavorum.

Boerboom, J.H.A. and Wiersum, K.F. (1983). Human impact on tropical moist forest. In *Man's impact on vegetation* (eds. W. Holzner, M.J.A. Werger, and I. Ikusima), pp. 83–106. Junk, The Hague.

Bolin, B., Döös, B.R., Jäger, J. and Warrick, R.A. (1986) (eds.). *The greenhouse effect, climatic change and ecosystems*. Wiley, Chichester.

Borchert, R. (1992). Computer simulation of tree growth periodicity and climatic hydroperiodicity in tropical forests. *Biotropica* **24**, 385–95.

Borchert, R. (1994*a*). Induction of rehydration and bud break by irrigation or rain in decid[u]ous trees of a tropical dry forest in Costa Rica. *Trees* **8**, 198–204.

Borchert, R. (1994*b*). Soil and stem water storage determine phenology and distribution of tropical dry forest trees. *Ecology* **75**, 1437–49.

Borchert, R. (1996). Phenology and flowering periodicity of neotropical dry forest species: evidence from herbarium collections. *Journal of Tropical Ecology* **12**, 65–80.

Bormann, F.H. and Likens, G.E. (1979). *Pattern and process in a forested ecosystem*. Springer, New York.

Bourlière, F. (1983). Animal species diversity in tropical forests. In *Tropical rain forest ecosystems, structure and function* (ed. F.B. Golley), pp. 77–92. Elsevier, Amsterdam.

Bourlière, F. (1985). Primate communities: their structure and role in tropical ecosystems. *Internatonal Journal of Primatology* **6**, 1–26.

Bramwall, D., Hamman, O., Heywood, V. and Synge, H. (1987) (eds.). *Botanic gardens and the World Conservation Strategy*. Academic Press, London.

Brandani, A., Hartshorn, G.S. and Orians, G.H. (1988). Internal heterogeneity of gaps and tropical tree species richness. *Journal of Tropical Ecology* **4**, 99–119.

Brookfield, H.C. (1993). Farming the forests in south east Asia. In *Environmental change in rain forests and dry lands. UNU Global Environmental Forum II*, pp. 51–70. United Nations University, Tokyo.

Browder, J.O. (1988). The social cost of rain forest destruction: a critique and economic analysis of the 'hamburger debate'. *Interciencia* **13**, 115–20.

Brown, N.D. (1993). The implications of climate and gap microclimate for seedling growth conditions in a Bornean lowland rain forest. *Journal of Tropical Ecology* **9**, 153–68.

Brown, N.D. (1996). A gradient of seedling growth from the centre of a tropical rain forest canopy gap. *Forest Ecology and Management* **82**, 239–44.

Brown, S., Gillespie, A.J.R. and Lugo, A.E. (1989). Biomass of tropical forests of south and southeast Asia. *Canadian Journal of Forest Research* **21**, 11–7.

Bruijnzeel, L.A. (1984). Immobilization of nutrients in plantation forests of *Pinus merkusii* and *Agathis dammara* growing on volcanic soils in central Java, Indonesia. In *International conference on soils and nutrition of perennial crops* (eds. E. Pusparaja and Tachib) pp. 19–29. Malayan Soil Science Society, Kuala Lumpur.

Bruijnzeel, L.A. (1989). Nutrient content of bulk precipitation in south-central Java, Indonesia. *Journal of Tropical Ecology* **5**, 187–202.

Bruijnzeel, L.A. (1990). *Hydrology of moist tropical forests and effects of conversion: a state of knowledge review*. UNESCO, Paris.

Bruijnzeel, L.A. (1991). Nutrient input-output budgets of tropical forests: a review. *Journal of Tropical Ecology* **7**, 1–24.

Bruijnzeel, L.A. (1992). Managing tropical watersheds for production: where contradictory theory and practice co-exist. In *Wise management of tropical forests* (eds. F.R. Miller and K.L. Adam), pp. 37–75. Oxford Forestry Institute, Oxford.

Bruijnzeel, L.A. (1995). Soil chemical and hydrochemical responses to tropical forest disturbance and conversion: a hydrologist's perspective. In *Soils of tropical forest ecosystems 3 Soil and water relationships* (eds. A.S. Schute and D. Ruhiyat, pp. 5–47). Mulawarmam University Press, Samarinda.

Bruijnzeel, L.A. and Critchley, W.R.S. (1994). *Environmental impacts of logging tropical moist forests* IHP humid tropics programme series **7**, UNESCO-MAB.

Bruijnzeel, L.A. and Wiersum, K.F. (1985). A nutrient balance sheet for *Agathis dammara* Warb. plantation forest under various management conditions in central Java, Indonesia. *Forest Ecology and Management* **10**, 195–208.

Bruijnzeel, L.A., Waterloo, M.J., Proctor, J., Kuiters, A.T. and Kotterink, B. (1993). Hydrological observations in montane rain forests on Gunung Silam, Sabah, Malaysia, with special reference to the 'Massenerhebung' effect. *Journal of Ecology* **81**, 145–67.

Brunig, E.F. (1996). *Conservation and management of tropical rain forests*. CAB International, Wallingford.

Budowski, G. (1965). Distribution of tropical American rain forest species in the light of successional processes. *Turrialba* **15**, 40–2.

Burgess, P.F. (1969). Preliminary observations on the autecology of *Shorea curtisii* Dyer ex King in the Malay Peninsula. *Malayan Forester* **32**, 438.

Burkill, I.H. (1935). *A dictionary of the economic products of the Malay Peninsula*, Vols 1 and 2. Crown Agents for the Colonies, Oxford.

Burslem, D.F.R.P., Grubb, P.J. and Turner, I.M. (1995). Responses to nutrient addition among shade-tolerant tree seedlings of lowland tropical rain forest in Singapore. *Journal of Ecology* **83**, 113–22.

Bush, M.B. (1994). Amazonian speciation: a necessarily complex model. *Journal of Biogeography* **21**, 5–17.

Caldecott, J. (1988). *Hunting and wildlife management in Sarawak*. IUCN, Gland and Cambridge.

Caldecott, J. (1996). *Designing conservation projects.* Cambridge University Press, Cambridge.

Carey, E.V., Brown, S., Gillespie, A.J.R. and Lugo, A.E. (1994). Tree mortality in mature lowland tropical moist and tropical lower montane moist forests of Venezuela. *Biotropica* **26**, 255–65.

Chan, H.T. (1980). Reproductive biology of some Malaysian dipterocarps. In *Tropical ecology and development* (ed. J.I. Furtado), pp. 169–75. International Society of Tropical Ecology, Kuala Lumpur.

Charles, C. (1997). Cool tropical punch of the ice ages. *Nature* **385**, 681–3.

Charles-Dominique, P., Atramentowicz, M., Charles-Dominique, M., Gérard, H., Hladik, A., Hladik, C.M., and Prevost, M.F. (1981). Les mammifères frugivores arboricoles nocturnes d'une forêt guyanaise: inter-relations plantes-animaux. *Revue Ecologie (Terre et Vie)*, 35, 341–35.

Chazdon, R.L. and Fetcher, N. (1984). Light environments of tropical forests. In *Physiological ecology of plants of the wet tropics* (eds. E. Medina *et al.*), pp. 27–36. Junk, The Hague.

Chazdon, R.L. and Pearcy, R.W. (1986*a*). Photosynthetic responses to light variation in rain forest species. I. Induction under constant and fluctuating light climates. *Oecologia (Berlin)* **69**, 517–23.

Chazdon, R.L. and Pearcy, R.W. (1986*b*). Photosynthetic responses to light variation in rain forest species. II. Carbon gain and photosynthetic efficiency during sunflecks. *Oecologia (Berlin)* **69**, 524–31.

Chin, S.C. (1985). Agriculture and resource utilization in a lowland rain forest Kenyah community. *Sarawak Museum Journal, Special Monograph* **4**.

Chin, S.C., Corlett, R.T., Wee, Y.C. and Geh, S.Y. (1995) (eds.). Rain forest in the city: Bukit Timah

nature reserve Singapore. *Garden's Bulletin, Supplement* **3**, 1–168.

Churchill, S.P., Balslev, H. Forero, E. and Luteyn, J.L. (1995) (eds.). *Biodiversity and conservation of neotropical montane forests.* New York Botanic Garden, Bronx.

Clark, D.A. and Clark, D.B. (1984). Spacing dynamics of a tropical rain forest tree: evaluation of the Janzen-Connell model. *American Naturalist* **124**, 769–88.

Clark, D.A. and Clark, D.B. (1992). Life history diversity of canopy and emergent trees in a neotropical rain forest. *Ecological Monographs* **62**, 315–44.

Clough, B. (1982) (ed.). *Mangrove ecosystems in Australia: structure, function and management.* Australian National University Press, Canberra.

Coley, P.D. (1983). Herbivory and defensive characteristics of tree species in a lowland tropical forest. *Ecological Monographs* **53**, 209–33.

Colinvaux, P.A. (1987). Amazon diversity in the light of the paleoecological record. *Quaternary Science Review* **6**, 93–114.

Colinvaux, P.A., de Oliviera, P.E., de Moreno, J.E., Miller, M.C. and Bush, M.B. (1996). A long pollen record from lowland Amazonia: forest and cooling in glacial times. *Science* **274**, 85–8.

Collar, N.J., Crosby, M.J. and Stattersfield, A.J. (1994). *Birds to watch 2. The world list of threatened birds.* Birdlife International, Cambridge.

Collins, N.M., Sayer, J.A. and Whitmore, T.C. (1991) (eds.). *The conservation atlas of tropical forests. Asia and the Pacific.* IUCN, Gland & Macmillan, London.

Condit, R., Hubbell, S.P. and Foster, R.B. (1995). Mortality rates of 205 neotropical tree and shrub species and the impact of a severe drought. *Ecological Monographs* **65**, 419–39.

Connell, J.H. (1978). Diversity in tropical rain forests and coral reefs. *Science* **199**, 1302–10.

Corlett, R.T. (1988). Bukit Timah: the history and significance of a small rain-forest reserve. *Environmental Conservation* **15**, 37–44.

Corner, E.J.H. (1964). *The life of plants.* Wiedenfeld and Nicolson, London.

Corner, E.J.H. (1966). *The natural history of palms.* Wiedenfeld and Nicholson, London.

Corner, E.J.H. (1988). *Wayside trees of Malaya* (3rd edn.). Malayan Nature Society, Kuala Lumpur.

Cranbrook, Lord (1988) (ed.). *Key environments, Malaysia.* Pergamon, Oxford.

Davidson, D.W. (1988). Ecological studies of neotropical ant gardens. *Ecology* **69**, 1138–52.

Davis, T.A.W. and Richards, P.W. (1933–4). The vegetation of Moraballi Creek, British Guiana; an ecological study of a limited area of tropical rain forest. I, II. *Journal of Ecology* **21**, 350–84; **22**, 106–55.

Dawkins, H.C. (1988). The first century of tropical silviculture: successes forgotten and failures misunderstood. In *The future of the tropical rain forest* (ed. M.J. McDermott), pp. 4–8. Oxford Forestry Institute, Oxford.

De Graaf, N.R. (1986). *A silvicultural system for natural regeneration of tropical rain forest in Suriname.* Agricultural University, Wageningen.

Denslow, J.S. and Padoch, C. (1988) (eds.). *People of the tropical rain forest.* University of California Press, Berkeley.

Diamond, J. (1988). Factors controlling species diversty: overview and synthesis. *Annals of the Missouri Botanic Garden* **75**, 117–29.

Dinerstein, E. and 6 others. (1996). *A conservation assessment of the terrestrial ecoregions of Latin America and the Caribbean.* World Bank, Washington.

Dransfield, J. (1988). Prospects for rattan cultivation. *Advances in Economic Botany* **6**, 190–200.

Driessen, P.M. (1978). Peat soils. In IRRI (ed.) *Soils and rice* IRRI, Los Banos.

Dunn, F.L. (1975). Rain forest collectors and traders. *Malayan Branch Royal Asiatic Society Monograph* **5**.

Egler, F.E. (1954). Vegetation science concepts. 1. Initial floristic composition, a factor in old field vegetation development. *Vegetatio* 4, 412–7.

Ehrlich, P.R. and Raven, P.H. (1964). Butterflies and plants: a study in co-evolution. *Evolution* **18**, 586–608.

Erfurth, T. and Rusche, H. (1978). The marketing of tropical wood in South America. *FAO Forestry Paper*, 5.

Erwin, T.L. (1982). Tropical forests: their richness in Coleoptera and other arthopod species. *The Coleopterists' Bulletin* **36**, 74–5.

Evans, G.C. (1972). *The quantitative analysis of plant growth.* Blackwell, Oxford.

Evans, J. (1984). *Plantation forestry.* Clarendon Press, Oxford. (2nd edn.).

FAO (1985). *Dipterocarps of South Asia.* FAO Regional Office for Asia and the Pacific, Bangkok.

FAO (1989). *Yearbook of forest products*, 1987. FAO, Rome.

FAO (1993). Forest resources assessment 1990. Tropical countries. *FAO Forestry Paper* **112**.

FAO (1995). Forest resources assessment 1990. Global synthesis. *FAO Forestry Paper* **124**.

FAO (1997). *State of the world's forests*. FAO, Rome.

Faegri, K. and Pijl, L. van der (1979). *Principles of pollination ecology*. (3rd edn.). Pergamon, Oxford.

Fearnside, P.M. (1986). Spatial concentration of deforestation in the Brazilian Amazon. *Ambio* **15**, 74–81.

Fearnside, P.M. (1987). Rethinking continuous cultivation in Amazonia. *BioScience* **37**, 209–14, 638–40; **38**, 525–7.

Fearnside, P.M. (1988). Jari at age 19: lessons for Brazil's silvicultural plans at Carajas. *Interciencia* **13**, 12–24.

Fearnside, P.M. (1989). Forest management in Amazonia: the needs for new criteria in evaluating development options. *Forest Ecology and Management* **27**, 61–79.

Fedorov, A.A. (1966). The structure of the tropical rain forest and speciation in the humid tropics. *Journal of Ecology* **54**, 1–11.

Fiala, B., Maschwitz, U., Tho, Y.P. and Helbig, A.J. (1989). Studies of a south-east Asian ant-plant association: protection of *Macaranga* trees by *Crematogaster borneensis*. *Oecologia* **79**, 463–70.

Finegan, B. (1992). The management potential of neotropical secondary lowland rain forest. *Forest Ecology and Management* **47**, 295–321.

Flenley, J.R. (1979). *The equatorial rain forest: a geological history*. Butterworth, London.

Fölster, H., de Las Salas, G, and Khama, P. (1976). A tropical evergreen forest site with perched water table, Magdalena valley, Colombia. *Oecologica Plantarum* **11**, 297–320.

Franco, W. and Dezzeo, N. (1994). Soils and soil water regime in the terra firme—caatinga forest complex near San Carlos de Rio Negro, state of Amazonas, Venezuela. *Interciencia* **19**, 305–16.

Frankel, O.H. and Soulé, M.E. (1981). *Conservation and evolution*. Cambridge University Press, Cambridge.

Futuyma, D.J. and Slatkin, M. (1983). *Co-evolution*. Sinauer, Sunderland.

Gan, Y.Y., Robertson, F.W., Ashton, P.S., Soepadmo, E., and Lee, D.W. (1977). Genetic variation in wild populations of rain forest trees. *Nature* **269**, 323–5.

Gan, Y.Y., Robertson, F.W. and Soepadmo, E. (1981). Isozyme variation in some rain forest trees. *Biotropica* **13**, 20–8.

Garwood, N.C., Janos, D.P. and Brokaw, N. (1979). Earthquake-caused landslides: a major disturbance to tropical forests. *Science* **205**, 997–9.

Gash, J.H.C., Nobre, C.A., Roberts, J.M. and Victoria, R.L. (1996) (eds.). *Amazonian deforestation and climate*. Wiley, Chichester, etc.

Gautier Hion, A. and Michaloud, G. (1989). Are figs always keystone resources for tropical frugivorous vertebrates? A test in Gabon. *Ecology* **70**, 1826–33.

Gentry, A.H. (1988*a*). Changes in plant community diversity and floristic composition on environmental and geographical gradients. *Annals Missouri Botanical Garden* **75**, 1–34.

Gentry, A.H. (1988*b*). Tree species richness of upper Amazonian forests. *Proceedings National Academy of Sciences USA* **85**, 156–9.

Gentry, A.H. (1990) (ed.). *Four neotropical rain forests*. Yale University Press, New Haven and London.

Gentry, A.H. (1993). *A field guide to the families and genera of woody plants of northwest South America (Colombia, Ecuador, Peru) with supplementary notes on herbaceous taxa*. Conservation International, Washington, D.C.

Gentry, A.H. and Dodson, C. (1987). Contributions of nontrees to species richness of a tropical forest. *Biotropica* **19**, 149–56.

Gentry, A.H. and Vasquez, R. (1988). Where have all the *Ceibas* gone? A case history of mismanagement of a tropical forest resource. *Forest Ecology and Management* **23**, 73–6.

Gilbert, L.E. (1980). Food web organisation and conservation of neotropical diversity. In *Conservation biology* (eds. M.E. Soulé and B.A. Wilcox), pp. 11–33. Sinauer, Sunderland.

Gill, A.E. and Rasmusson, E.M. (1983). The 1982–83 climate anomaly in the equatorial Pacific. *Nature* **306**, 229–34.

Gillman, G.P., Sinclair, D.F., Knowlton, R. and Keys, M.G. (1985). The effect on some soil chemical properties of the selective logging of a north Queensland rain forest. *Forest Ecology and Management* **12**, 195–214.

Godoy, R., Lubowski, R. and Markandya, A. (1993). A method for the economic valuation of non-timber tropical forest products. *Economic Botany* **47**, 220–33.

Golley, F.B. (1983) (ed.). *Tropical rain forest ecosystems—structure and function*. Ecosystems of the world 14 A. Elsevier, Amsterdam, etc.

Gomez-Pompa, A., Salvador, J. and Sosa, V. (1987). The 'pet-kot': a man-made tropical forest of the Maya. *Interciencia* **12**, 10–15.

Gomez-Pompa, A., Whitmore, T.C. and Hadley, M. (1991) (eds.). *Rain forest regeneration and management*. UNESCO and Parthenon, Paris and Carnforth.

Goodland, R. (1995). The concept of environmental sustainability. *Annual Reviews of Ecology and Systematics* **26**, 1–24.

Gould, E. (1978). Foraging behaviour of Malaysian nectar-feeding bats. *Biotropica* **10**, 184–93.

Goulding, M. (1980). *The fishes and the forest*. University of California Press, Berkeley.

Grace, J. and 6 others (1996). The use of eddy covariance to infer the net carbon dioxide uptake of Brazilian rain forest. *Global Change Biology* **2**, 209–17.

Gradwohl, J. and Greenberg, R. (1988). *Saving the tropical forest*. Earthscan, London.

Greenland, D.J., Gregory, P.J. and Nye, P.H. (eds.). (1997). Land resources: on the edge of a Malthusian precipice? *Philosophical Transactions of the Royal Society of London* B **352**, 859–1033.

Gressitt, J.L. (1982) (ed.). *Biogeography and ecology of New Guinea*. Junk, The Hague.

Groombridge, B. (1992) (ed.). *Global biodiversity*. Chapman and Hall, London, etc.

Grubb, P.J. (1987). Global trends in species-richness in terrestrial vegetation: a view from the northern hemisphere. In *Organisation of communities, past and present* (eds. J.H.R. Gee and P.S. Gilles). Blackwell, Oxford.

Grubb, P.J., Lloyd, J.R., Pennington, T.D. and Whitmore, T.C. (1963). A comparison of montane and lowland rain forest in Ecuador. I. The forest structure, physiognomy and floristics. *Journal of Ecology* **51**, 567–601.

Guevara, S., Purata, S.E. and van der Maarel, E. (1986). The role of remnant forest trees in tropical secondary succession. *Vegetatio* **66**, 77–84.

Hadley, M. and Kartawinata, K. (1993). Issues and opportunities in managing renewable natural resources in the humid tropics: insights from Asia. In *Economically sound socio-economic development in the humid tropics: perspectives from Asia and Africa* (eds. J.I. Uitto and M. Clüsener-Godt). United Nations University, Tokyo.

Hall, R. (1996). Reconstructing Cenozoic SE Asia. *Geological Society Special Publication* **106**, 153–84.

Hall, J.B. and Swaine, M.D. (1976). Classification and ecology of closed canopy forest in Ghana. *Journal of Ecology* **64**, 913–51.

Hallam, A. (1973). *A revolution in the earth sciences*. Clarendon Press, Oxford.

Hallam, A. and Audley Charles, M.G. (1988) (eds.). *Gondwana and Tethys*. Geological Society Special Publication, **37**.

Hallé, F. and Ng, F.S.P. (1981). Crown construction in mature dipterocarp trees. *Malaysian Forester* **44**, 222–33.

Hallé, F, Oldeman, R.A.A. and Tomlinson, P.B. (1978). *Tropical trees and forests*. Springer, Berlin.

Hamilton, A. (1976). The significance of patterns of distribution shown by forest plants and animals in tropical Africa for the reconstructions of upper Pleistocene palaeoenvironments. In *Palaeoecology of Africa* (ed. Zinderen Bakker Sr., E.M. van). Balkema, Cape Town.

Hamilton, A.C. (1982). *Environmental history of East Africa. A study of the Quaternary*. Academic Press, London.

Hamilton, L.S., Juvik, J.O. and Scatena, F.N. (1995) (eds.). *Tropical montane cloud forests*. Springer, New York.

Hammen, T. van der, and Cleef, A.M. van der (1983). *Trigonobalanus* and the tropical amiphipacific element in the North Andean forest. *Journal of Biogeography* **10**, 437–40.

Harcourt, C.S. and Sayer, J.A. (1996) (eds.). *The conservation atlas of tropical forests. The Americas*. Simon and Schuster, New York, etc.

Hardcastle, P.D. (1997). Silvicultural options for the rain forest zone of Cameroon. DFID *Forestry Series* (**in press**).

Hawthorne, W.D. (1995). Ecological profiles of Ghanaian forest trees. *Tropical Forestry Papers* **29**.

Heaney, A. and Proctor, J. (1989). Chemical elements in litter in forests on volcán Barva, Costa Rica. In *Mineral nutrients in tropical forest and savanna ecosystems* (ed. J. Proctor), pp. 255–72. Blackwell, Oxford.

Hecht, S.B., Norgaard, R.B., and Possio, G. (1988). The economics of cattle ranching in eastern Amazonia. *Interciencia* **13**, 233–40.

Hill, J.K., Hamer, K.C., Lace, L.A. and Banhay, W.M.T. (1995). Effects of selective logging on

tropical forest butterflies on Buru, Indonesia. *Journal of Applied Ecology* **32**, 754–60.

Hladik, C.M., Hladik, A., Linares, O.F., Pagezy, H., Semple, A. and Hadley M. (1993) (eds.). *Tropical forests, people and food.* UNESCO and Parthenon, Paris and Carnforth.

Holttum, R.E. (1954). *Plant life in Malaya.* Longman, London.

Holttum, R.E. (1955). Growth habits of Monocotyledons—variation on a theme. *Phytomorphology* **5**, 399–413.

Houghton, R.A. (1991) Biomass burning from the perspective of the global carbon cycles. In *Global biomass burning* (ed. J.S.Levine), pp. 321–5. MIT Press, Cambridge, Mass.

Houghton, R.A. (1995*a*). Tropical forests and climate. In *Ecology, conservation, and management of Southeast Asian rainforests* (eds. R.B. Primack and T.E. Lovejoy), pp. 263–89. Yale University Press, Newhaven and London.

Houghton, R.A. (1995*b*). Land-use and the carbon cycle. *Global Change Biology* **1**, 275–87.

Howe, H.F. and Smallwood, J. (1982). Ecology of seed dispersal. *Annual Review of Ecology and Systematics* **13**, 201–28.

Howe, H.F. and Westley, L.C. (1988). *Ecological relationships of plants and animals.* Oxford University Press, Oxford.

Hu, S.Y. (1980). The *Metasequoia* flora and its phytogeographic significance. *Journal of the Arnold Arboretum* **60**, 41–94.

Huxley, C.R. (1978). The ant-plants *Myrmecodia* and *Hydnophytum* (Rubiaceae) and the relationships between their morphology, ant occupants, physiology and ecology. *New Phytologist* **80**, 231–68.

Huxley, C.R. and Cutler, D.F. (1991) (eds.). *Ant-plant interactions.* Oxford University Press, Oxford.

Iltis, H.H., Doebley, J.F. Guzman, R.M. and Pazy, B. (1979). *Zea diploperennis* (Graminae), a new teosinte from Mexico. *Science* **203**, 186 -8.

Itoh, A. (1995). Effects of forest floor environment on germination and seedling establishment of two Bornean rain forest emergent species. *Journal of Tropical Ecology* **11**, 517–27.

ITTO (1990). *ITTO guidelines for the sustainable management of natural tropical forests.* ITTO, Technical Series **5**, Yokohama.

ITTO (1992). *ITTO guidelines on the conservation of biodiversity in tropical production forests.* ITTO, Yokohama.

ITTO (1993*a*). Global context of tropical timber. *Tropical Forest Management Update* **3** (3), 10.

ITTO (1993*b*). Top timber exporters. *Tropical Forest Management Update* **3** (2), 8.

IUCN (1980). *World conservation strategy.* IUCN, UNEP, WWF, Morges.

IUCN/UNEP/WWF (1991). Caring for the earth, a strategy for sustainably living. IUCN, UNEP, WWF, Gland.

Jacobs, M. (1988). *The tropical rain forest, a first encounter.* Springer Verlag, Berlin.

Jaffré, T. (1985). Composition minérale et stocks de bioéléments dans la biomasse épigée de recrûs forestiers en Côte d'Ivoire. *Acta Oecologia, Oecologica Plantarum* **20**(6), 233–46.

Janzen, D.H. (1976). Why do bamboos wait so long to flower? In *Tropical trees: variation breeding and conservation* (eds. J. Burley and B.T. Styles), pp. 135–9. Academic Press, London.

Jeanrenaud, S. (1997). Perspectives in people-oriented conservation. *Arborvitae Supplement* February 1997, pp. 6.

Johns, A.D. (1986). Effects of selective logging on the bahavioural ecology of West Malaysian primates. *Ecology* **67**, 684–94.

Johns, R.J. (1986). The instability of the tropical ecosystem in New Guinea. *Blumea* **31**, 341–71.

Johns, A.G. (1997). *Timber production and biodiversity conservation in tropical rain forests.* Cambridge University Press, Cambridge.

Johns, A.G. and Johns, B.G. (1995). Tropical forests and primates: long term co-existence? *Oryx* **29**, 205–11.

Johnson, S. (1997). Production and trade of tropical logs. *ITTO Tropical Forest Update* **7**(**2**), 18–20.

Jones, E.W. (1955, 1956). Ecological studies on the rain forest of southern Nigeria. IV. The plateau forest of the Okomu Forest Reserve (contd.). *Journal of Ecology* **43**, 564–94; **44**, 83–117.

Jordan, C.F. (1985). *Nutrient cycling in tropical forest ecosystems.* Wiley, Chichester.

Jordan, C.F. (1987) (ed.). *Amazonian rain forests ecosystem disturbance and recovery—case studies of ecosystem dynamics under a spectrum of land use intensities.* Springer, New York.

Kahn, F. (1988). Ecology of economically important palms in Peruvian Amazonia. *Advances in Economic Botany* **6**, 42–9.

Kaur, A., Ha, C.O., Jong, K., Sands, V.E., Chan, H.T., Soepadmo, E. and Ashton, P.S. (1978).

Apomixis may be widespread among trees of the climax rain forest. *Nature* **271**, 440–2.

Keay, J. (1991). *The honourable company, a history of the English East India Company*. Harper Collins.

Keay, R.W.J. (1960). Seeds in forest soils. *Nigerian Forestry Department Information Bulletin (New Series)* **4**, 1–12.

Keay, R.W.J. (1996). Foreword. *Proceedings of the Royal Society of Edinburgh* **B 104**, 1–4.

Keller, M., Clark, D.A., Clark, D.B., Weitz, A.M. and Veldkamp, E. (1996). If a tree falls in the forest... *Science* **273**, 201.

Kellman, M.C. (1970). *Secondary plant succession in tropical montane Mindanao*. Research School for Pacific Studies, Australian National University, Publication BG\2.

Kemp, R.H. and Phantumvanit, D. (1995). 1995 mid-term review of progress towards the achievement of the year 2000 objective. *Consultants' report for International Tropical Timber Council 19th session*. ITTO, Yokohama.

Kerner, A.K. and Oliver, F.W. (1895). *The natural history of plants*. Blackie, London.

Kimmins, J.P. (1987). *Forest ecology*. Macmillan, New York, Collier MacMillan, London.

Kingdon-Ward, F. (1945). A sketch of the botany and geography of north Burma. *Journal of the Bombay Natural History Society* **45**, 16–30.

Kio, P.R.O. (1978). What future for natural regeneration of tropical high forest? An appraisal with examples from Nigeria and Uganda. *Commonwealth Forestry Review* **55**, 309–18.

Kleinfeldt, S. (1986). Ant gardens: mutual exploitation. In *Insects and the plant surface* (eds. B. Juniper and T.R.E. Southwood), pp. 283–94. Arnold, London.

Klinge, H. (1973). Root mass estimation in lowland rain forests of central Amazonia, Brazil. 1. Fine root masses of pale yellow latosols and giant humus podzol. *Tropical Ecology* **14**, 29–38.

Kochummen, K.M. and Ng, F.S.P. (1977). Natural plant succession after farming at Kepong. *Malaysian Forester* **40**, 61–78.

Kohyama, T. (1987). Significance of architecture and allometry in saplings. *Functional Ecology* **1**, 399–404.

Kramer, F. (1933). De natuurlijke verjonging in het Goenoeng Gedeh complex. *Tectona* **26**, 156–85.

Kursar, T.A. and Coley, P.D. (1992). Delayed greening in tropical leaves: an antiherbivore defence? *Biotropica* **24**, 256–62.

Lamb, D. (1990). *Exploiting the tropical rain forest; an account of pulpwood logging in Papua New Guinea*. UNESCO and Parthenon, Paris and Carnforth.

Lanly, J.P. (1982). Tropical forest resources. *FAO Forestry Paper*, **30**.

Laumonier, Y., Gadrinab, A. and Purnajaya. (1983). *Southern Sumatra. International map of the vegetation and of environmental conditions*. Institut de la Carte de Tapis Végétal and SEAMEO/BIOTROP, Toulouse.

Laumonier, Y., Purnadjaja and Setiabudhi. (1986*a*). *Central Sumatra. International map of the vegetation and of environmental conditions*. Institut de la Carte de Tapis Végétal and SEAMEO/BIOTROP, Toulouse.

Laumonier, Y., Purnadjaja and Setiabudi. (1986*b*). *Northern Sumatra. International map of the vegetation and of environmental conditions*. Institut de la Carte du Tapis Végétal and SEAMEO/ BIOTROP, Toulouse.

Laurance, W.F. and Bierregaard, R.O.Jr. (1997) (eds.). *Tropical forest remnants: ecology, management, and conservation of fragmented communities*. University of Chicago Press, Chicago.

Layrisse, M. (1992). The 'holocaust' of the Amerindians. *Interciencia* **17**, 274.

Ledec, G. and Goodland, R. (1988). *Wildlands, their protection and management in economic development*. The World Bank, Washington D.C.

Lee, D.W., Oberbauer, S.F., Krishnapilay, B., Mansor, M., Mohamad, H. and Yap, S.K. (1997). Effects of irradiance and spectral quality on seedling development of two southeast Asian *Hopea* species. *Oecologia* **110**, 1–9.

Leigh, E.G., Rand, A.S. and Windsor, D.M. (1983) (eds.). *The ecology of a tropical forest*. Oxford University Press, Oxford.

Leighton, M. and Wirawan, N. (1986). Catastrophic drought and fire in Borneo tropical rain forest associated with the 1982–83 El Niño southern oscillation event. In *Tropical rain forests and the world atmosphere* (ed. G.T. Prance), pp. 75–102. Westview, Boulder.

Lennertz, R. and Panzer, K.F. (1983). Preliminary assessment of the drought and forest fire damage in Kalimantan Timur. DFS German Forest Inventory Service.

Leslie, A.J. (1987). A second look at the economics of natural management system in tropical mixed forests. *Unasylva* **39**(155), 46–58.

Lieberman, D., Lieberman, M., Hartshorn, G.S., and Peralta, R. (1985). Growth rates and age-size relationships of tropical wet forest trees in Costa Rica. *Journal of Tropical Ecology* **1**, 97–109.

Lieberman, M., Lieberman, D., and Peralta, R. (1989). Forests are not just swiss cheese: canopy stereogeometry of non-gaps in tropical forests. *Ecology* **70**, 550–2.

Lieberman, M., Lieberman, D., Peralta, R. and Hartshorn, G.S. (1996). Tropical forest structure and composition on a large scale altitudinal gradient in Costa Rica. *Journal of Ecology* **84**, 137–52.

Longman, K.A. and Jenik, J. (1987). *Tropical forest and its environment* (2nd edn.). Longman, London.

Lugo, A.E. and Brown, S. (1993). Management of tropical soils as sinks or sources of atmospheric carbon. *Plant and Soil* **149**, 27–41.

Lugo, A.E. and Snedaker, S.C. (1974). The ecology of mangroves. *Annual Review of Ecology and Systematics* **5**, 39–64.

Lugo, A.E. *et al.* (1987) (eds.). *People and the tropical forest*. U.S. Man and Biosphere Program, Washington D.C.

Lutz, W., Sanderson, W. and Scherbov, S. (1997). Doubling of world population unlikely. *Nature* **387**, 803–5.

Mabberley, D.J. (1988). The living past: time state of the tropical rain forest. In *Forests, climate and hydrology: regional impacts* (eds. E.R.C. Reynolds and F. Thompson), pp. 6–15. United Nations University.

McCully, P. (1996). *Silenced rivers. The ecology and politics of large dams*. Zed Books, London and New Jersey.

McDade, L.A., Bawa, K.S., Hespenheide, H.A. and Hartshorn, G.S. (1994) (eds.). *La Selva: ecology and natural history of a neotropical rain forest*. University of Chicago Press, Chicago.

McDermott, M.J. (1988) (ed.). *The future of the tropical rain forest*. Oxford Forestry Institute, Oxford.

MacKinnon, K.S., Hatta, G., Halim, H. and Mangalik, A. (1996). *The ecology of Kalimantan, Indonesian Borneo*. Periplus Editions, Hong Kong.

MacLellan, A. (1996) (ed.). Is there a future for mahogany? *Botanical Journal of the Linnean Society* **122**, 1–87.

MacLellan, A.J. and Frankland, F. (1985). A simple field method for measuring light quality: seasonal changes in a temperate deciduous wood. *Photochemistry and Photobiology* **42**, 689–95.

Maley, J. (1987). Fragmentation de la forêt dense humide Africaine. In *Palaeoecology of Africa and the surrounding islands* 18. Balkema, Rotterdam.

Manokaran, N. and Swaine, M.D. (1994). *Population dynamics of trees in dipterocarp forests of Peninsular Malaysia*. Malayan Forest Records **40**.

Markham, A. (1998) (ed.). Climatic change and tropical forests. *Climatic Change* (**in press**).

Marshall, A.G. and Swaine, M.D. (1992) (eds.). Tropical rain forest: disturbance and recovery. *Philosophical Transactions of the Royal Society of London* **B 335**, 323–457.

Martius, C.F.P. von (1840–65). *Flora Brasiliensis* 1 (1) *Tabulae Physiognomicae Brasiliae*. Monachii, Lipsioe.

Maschwitz, U., Schroth. M, Hänel, H. and Tho, Y.P. (1984). Lycaenids parasitizing symbiotic plant-ant relationships. *Oecologia (Berlin)* **64**, 78–80.

May, R.M. (1988). How many species are there on earth? *Science* **241**, 1441–9.

Mayhew, J. (1997). The future of mahogany plantations. *ITTO Tropical Forest Update* **7(2)**, 14.

Mayr, E. and O'Hara, R.J. (1986). The biogeographic evidence supporting the Pleistocene refuge hypothesis. *Evolution* **40**, 55–67.

Medway, Lord. (1972*a*). Phenology of a tropical rain forest in Malaya. *Biological Journal of the Linnean Society, London* 4, 117–46.

Medway, Lord. (1972*b*). The Quaternary mammals of Malesia: a review. In *The Quaternary era in Malesia* (eds. P. Ashton and M. Ashton), Miscellaneous Series, **13**, 63–83. Department of Geography, University of Hull,

Meggers, B.J., Ayensu, E.S. and Duckworth, W.D. (1973) (eds.). *Tropical forest ecosystems in Africa and South America: a comparative review*. Smithsonian Institution, Washington.

Merrill, E.D. (1945). *Plant life of the Pacific world*. Macmillan, New York.

Metcalfe, I. (1996). Pre-Cretaceous evolution of SE Asian terranes. *Geological Society Special Publication* **106**, 97–122.

Mitchell, A.W. (1986). *The enchanted canopy, secrets from the rain forest roof*. Collins, London.

Mittermeier, R.A. (1995). Suriname crisis illustrates global threat to biodiversity. *Tropinet* **6**(4), 1.

Monk, K.A., de Fretes, Y. and Reksodiharjo-Lilley, G. (1997). *The ecology of Nusa Tenggara and Maluku*. Periplus Editions, Hong Kong.

Montgomery, G.G. (1978) (ed.). *The ecology of arboreal folivores*. Smithsonian Institution, Washington.

Morley, R.J. (1998). Palynological evidence for Tertiary plant dispersals in the southeast Asia region in relation to plate tectonics and climate. In Hall, R. and Holloway, J.D. (eds.) *Biogeography and geological evolution in southeast Asia.* Backhuys, Amsterdam.

Mulkey, S.S., Chazdon, R.L. and Smith, A.P. (1996) (eds.). *Tropical forest plant ecophysiology.* Chapman and Hall, London.

Myers, N. (1980). *Conversion of tropical moist forests.* National Academy of Sciences, Washington.

Myers, N. (1990). Tropical forests. In *Global warming—the Greenpeace report* (ed. J. Leggatt). Oxford University Press, Oxford.

Nectoux, F. and Kuroda, Y. (1989) *Timber from the South Seas.* WWF International, Gland.

Nelson, B.W. (1994). Natural forest disturbance and change in the Brazilian Amazon. *Remote Sensing Reviews* **10**, 105–25.

Nelson, B.W., Ferreira, C.A.C., da Silva, M.F. and Kawasaki, M.L. (1990). Endemism centres, refugia and botanical collection density in Brazilian Amazonia. *Nature* **345**, 714–6.

Nelson, B.W., Kapos, V., Adams, J.B., Oliviera, W.J., Braun, O.P.G. and do Amaral, I.L. (1994). Forest disturbance by large blow downs in the Brazilian Amazon. *Ecology* **75**, 853–8.

Nepstad, D.C. and 9 others (1994). The role of deep roots in the hydrological and carbon cycles of Amazonian forests and pastures. *Nature* **372**, 666–9.

Newbery, D.M., Prins, H.H.T. and Brown, N.D. (1998) (eds.). *Dynamics of tropical communities.* Blackwell Science, Oxford.

Newbery, D.M., Gartlan, J.S., Thomas, D. and Waterman, P.G. (1986). The influence of topography and soil phosphorus on the vegetation of Korup Forest Reserve, Cameroun. *Vegetatio* **65**, 131–48.

Newbery, D.M., Alexander, I.J., Thomas, D.W. and Gartlan, J.S. (1988). Ectomycorrhizal rain-forest legumes and soil phosphorus in Korup National Park, Cameroon. *New Phytologist* **109**, 433–50.

Newman, M.F., Burgess, P.F. and Whitmore, T.C. (1996*a*). *Manuals of dipterocarps for foresters, Philippines.* Royal Botanic Garden and CIFOR, Edinburgh and Jakarta.

Newman, M.F., Burgess, P.F. and Whitmore, T.C. (1996*b*). *Manuals of dipterocarps for foresters, Borneo Island Light Hardwoods.* Royal Botanic Garden and CIFOR, Edinburgh and Jakarta.

Newman, M.F., Burgess, P.F. and Whitmore, T.C. (1996*c*). *Manuals of dipterocarps for foresters, Sumatra Light Hardwoods.* Royal Botanic Garden and CIFOR, Edinburgh and Jakarta.

Ng, F.S.P. (1978) (ed.). *Tree Flora of Malaya*, Vol. 3. Longman, Kuala Lumpur and London.

Ng, F.S.P. (1984). Plant phenology in the humid tropics. *Malaysian Forest Department Research Pamphlet* **96**, 129–62.

Ng, F.S.P. (1989) (ed.). *Tree Flora of Malaya*, Vol. 4. Longman, Kuala Lumpur and London.

Ng, F.S.P. and Low, C.M. (1982). Check List of endemic trees of the Malay peninsula. *Malaysian Forestry Department Research Pamphlet*, **88**.

Nykvist, N., Grip, H., Sim, B.L., Malmer, A. and Wong, F.K. (1994). Nutrient losses in forest plantations in Sabah, Malaysia. *Ambio* **23**, 210–5.

Oberbauer, S.F. and Donnelly, M.A. (1986). Growth analysis and successional status of Costa Rican rain forest trees. *New Phytologist* **104**, 517–21.

Oberbauer, S.F. and Strain, B.R. (1984). Photosynthesis and successional status of Costa Rican rain forest trees. *Photosynthesis Research* **5**, 227–32.

O'Brien, K.L. (1996). Tropical deforestation and climate change. *Progress in Physical Geography* **20**, 311–35.

Okali, D.U.U. (1992). Sustainable use of West African moist forest lands. *Biotropica* **24**, 335–44.

Orozco-Segovia, A., Vazquez-Yanes, C., Coates-Estrada, R., and Pérez-Nasser, N. (1987). Ecophysiological characteristics of the seed of the tropical forest pioneer *Urera caracasana* (Urticaceae). *Tree Physiology* **3**, 375–86.

Parren, M.P.E. and de Graaf, N.R. (1995). *The quest for natural forest management in Ghana, Côte d'Ivoire and Liberia.* Tropenbos Series **13**. Tropenbos Foundation, Wageningen.

Pascal, J.P. (1986). *Explanatory booklet on the forest map of south India.* Institut Français de Pondichery. Travaux de la Section Scientifique et Technique, Hors Série **18**.

Pearce, F. (1989). Methane: the hidden greenhouse gas. *New Scientist* **122**(1663), 37–41.

Peluso, N.L. (1983). Networking in the commons: a tragedy for rattan. *Indonesia* **35**, 95–108.

Pendry, C.A. and Proctor, J. (1996). The causes of altitudinal zonation of rain forests on Bukit Belalong, Brunei. *Journal of Ecology* **84**, 407–18.

Peters, C.M., Gentry, A.H. and Mendelsohn, R.O. (1989). Valuation of an Amazonian rain forest. *Nature* **339**, 655–6.

Phillips, O. (1993). The potential of harvesting fruits in tropical rain forests: new data from Amazonian Peru. *Biodiversity and Conservation* **2**, 18–38.

Phillips, O.L. (1996). Long-term environmental change in tropical forests: increasing tree turnover. *Environmental Conservation* **23**, 235–48.

Pijl, L. van der (1982). *The principles of dispersal in higher plants*. (3rd edn.). Springer, Berlin.

Pinard, M.A., Putz, F.E., Tay, J. and Sullivan, T.E. (1995). Creating timber harvest guidelines for a reduced-impact logging project in Malaysia. *Journal of Forestry* **93**, 41–5.

Platt, W.J. and Strong, D.R. (1989) (ed.). Treefall gaps and forest dynamics. *Ecology* **70**, 537–76.

Pleydell, G. (1995). The substitution of tropical timbers. *ITTO Tropical Forest Update* 5(2), 18.

Pleydell, G. and Johnson, S. (1997). Tropical timber trade trends 1987–1996. *Tropical Forest Update* 7(**1**), 13–15.

Poore, M.E. D., Burgess, P.F., Palmer, J.R., Rietbergen, S. and Synnott, T.J. (1989). *No timber without trees*. Earthscan, London.

Prance, G.T. (1986). Introduction to tropical rain forests. In *Tropical rain forests and world atmosphere* (ed. G.T. Prance) pp. 1–8. Westview Press, Boulder.

Prebble, C. (1997). A plantation perspective. *ITTO Tropical Forest Update* 7(**2**), 1.

Prescott Allen, R. and Prescott Allen, C. (1983). *Genes from the wild: using wild genetic resources for food and raw materials*. International Institute for Environment and Development, London.

Primack, R.B. and Hall, P. (1992). Biodiversity and forest change in Malaysian Borneo. *BioScience* **42**, 829–37.

Primack, R.B. and Lovejoy, T.E. (1996) (eds.). *Ecology, conservation, and management of Southeast Asian rain forests*. Yale University Press, Newhaven and London.

Proctor, J. (1984). Tropical forest litterfall. II. The data set. In *Tropical rain forest. The Leeds symposium* (eds. A.C. Chadwick and S.L. Sutton), pp. 83–113. Leeds Philosophical and Literary Society, Leeds.

Proctor, J. (1989) (ed.). *Mineral nutrients in tropical forest and savanna ecosystems*. Blackwell, Oxford.

Proctor, J., Haridasan, K. and Smith, G.W. (1998). How far north does lowland evergreen tropical rain forest go? *Global Ecology and Biogeography Letters* (in press).

Purseglove, J.W. (1968). *Tropical crops, dicotyledons*. Longman, London.

Putz, F.E. (1978). A survey of virgin jungle reserves in Peninsular Malaysia. *Malaysian Forestry Department Research Pamphlet*, **73**.

Raich, J.W. (1987). *Canopy openings, seed germination, and tree regeneration in Malaysian coastal hill dipterocarp forest*. Ph.D. Thesis, Duke University.

Raunkiaer, C. (1934). *The life forms of plants and statistical plant geography*. Oxford University Press, Oxford.

Reich, P.B. 1995. Phenology of tropical forests; patterns, causes and consequences. *Canadian Journal of Botany* **73**, 164–74.

Reich, P.B., Uhl, C., Walters, M.B. and Ellsworth, D.S. (1991). Leaf lifespan as a determinant of leaf structure and function among 23 Amazonian species. *Oecologia* **86**, 16–24.

Richards, P. (1996). Forest indigenous peoples: concept, critique and cases. *Proceedings of the Royal Society of Edinburgh* B **104**, 349–65.

Richards, P.W. (1952). *The tropical rain forest* (2nd edn. 1996). Cambridge University Press, Cambridge.

Richards, P.W. (1970). *The life of the jungle*. McGraw Hill, New York and London.

Richter, D.D. and Babbar, L.I. (1991). Soil diversity in the tropics. *Advances in Ecological Research* **21**, 316–90.

Riera, B. (1985). Importance des buttes de racinement dans la regeneration forestière en Guyane Française. *Revue Ecologie (Terre Vie)* **40**, 321–9.

Rijksen, H.D. (1978). *A field study on Sumatran orang utans* (Pongo pygmaeus abelii *Lesson 1827*). Mededelingen Landbouwhogeschool, Wageningen, 78–2.

Roberts, E.H. (1973). Predicting the storage life of seeds. *Seed Science and Technology* **1**, 499–574.

Robinson, M.H. (1992). The uniqueness of tropical natural history. *Malayan Nature Journal* **45**, 273–309.

Sader, S.A. and Joyce, A.T. (1988). Deforestation rates and trends in Costa Rica, 1940–1983. *Biotropica* **20**, 11–9.

Salati, E. and Vose, P.D. (1984). Amazon basin: a system in equilibrium. *Science* **225**, 129–38.

Salo, J., Kalliola, R., Häkkinen, I., Mäkinen, Y., Niemalä, P., Puhakka, M. and Coley, P.D. (1986). River dynamics and the diversity of Amazon lowland forest. *Nature* **322**, 254–8.

Sanchez, P.A. (1976). *Properties and management of soils in the tropics*. Wiley, New York.

Sanford, R.L., Saldarriaga, J., Clark, K.E., Uhl, C. and Herrera, R. (1985). Amazon rain forest fires. *Science* **227**, 53–5.

Sayer, J. (1995). Forestry research, a way forward to sustainable development. *CIFOR News* **6**, 1–2.

Sayer, J.A., Harcourt, C.S. and Collins, N.M. (1992) (eds.). *The conservation atlas of tropical forests: Africa*. IUCN and Macmillan, Gland and Basingstoke.

Schmidt, R. (1987). Tropical rain forest management. *Unasylva* **39**(156), 2–17.

Schmidt-Vogt, D. (1995). Swidden farming in secondary vegetation, two case studies from northern Thailand. In *Counting the costs: economic growth and environmental change in Thailand* (ed. J. Rigg), pp. 47–64. Institute of Southeast Asian Studies, Singapore.

Scholes, J.D., Press, M.C. and Zipperlen, S.W. (1997). Differences in light energy utilisation and dissipation between dipterocarp rain forest tree seedlings. *Oecologia* **109**, 41–8.

Secrett, C. (1987). Friends of the Earth UK and the hardwood compaign. In *Proceedings of the conference on forest resources crisis in the Third World, 6–8 September 1986*, pp. 348–56. Sahabat Alam Malaysia, Penang.

Sheil, D. and May, R.M. (1996). Mortality and recruitment rate evaluation in heterogeneous tropical forests. *Journal of Ecology* **84**, 91–100.

Shield, E.D. (1992). Plantation opportunity areas in the Solomon Islands. *AIDAB-MNR Solomon Islands National Forest Resources Inventory Working Paper* **13**.

Silva, J.N.M and 7 others (1995). Growth and yield of a tropical rain forest in the Brazilian Amazon 13 years after logging. *Forest Ecology and Management* **71**, 267–74.

Sizer, N. (1996). Out on a limb in Suriname. *Interciencia* **21**, 147–53.

Skole, D. and Tucker, C. (1993). Tropical deforestation and habitat fragmentation in the Amazon: satellite data from 1978 to 1988. *Science* **260**, 1905–10.

Smith, A.G. and Briden, J.C. (1977). *Mesozoic and Cenozoic palaeocontinental maps*. Cambridge University Press, Cambridge.

Snow, D.W. (1966). A possible selective factor in the evolution of flowering seasons in tropical forest. *Oikos* **15**, 274–81.

Snow, D.W. (1981). Tropical frugivorous birds and their food plants: a world survey. *Biotropica* **13**, 1–14.

Soejarto, D.D. (1997). Forest biodiversity and the continuing search for new pharmaceuticals. *Flora Malesiana Bulletin* 12(1), 4–6.

Sommer, A. (1976). Attempt at an assessment of the world's tropical forests. *Unasylva* **28**(112/113), 5–25.

Soulé, M.E. (1986) (ed.). *Conservation biology*. Sinauer, Sunderland.

Soulé, M.E. (1987) (ed.). *Viable populations for conservation*. Cambridge University Press, Cambridge.

Soulé, M.E. and Wilcox, B.A. (1980) (eds.). *Conservation biology*. Sinauer, Sunderland.

Spangenberg, A., Grimm, U., Sepeda da Silva, J.R. and Fölster, H. (1996). Nutrient stores and export rates of *Eucalyptus urograndis* plantations in eastern Amazonia (Jari). *Forest Ecology and Management* **80**, 225–34.

Sponsel, L.E., Headland, T.N. and Bailey, R.C. (1996) (eds.). *Tropical deforestation, the human dimension*. Columbia University Press, New York.

Start, A.N. and Marshall, A.G. (1976). Nectarivorous bats as pollinators of trees in West Malaysia. In *Tropical trees: variation, breeding and conservation* (eds. J. Burley and B.T. Styles), pp. 141–50. Academic Press, London.

Steege, H. ter, Jetten, V.G., Polak, A.M. and Werger, M.J.A. (1993). Tropical rain forest types and soil factors in a watershed area in Guyana. *Journal of Vegetation Science* **4**, 705–16.

Steenis, C.G.G.J. van (1962). The land bridge theory in botany. *Blumea* **11**, 235–372.

Steenis, C.G.G.J. van (1972). *The mountain flora of Java*. Brill, Leiden.

Steenis, C.G.G.J. van (1981). *Rheophytes of the world*. Sijthoff and Noordhoff, Alpen aan den Rijn.

Stiles, F.G. (1977). Co-adapted pollinators: the flowering seasons of hummingbird-pollinated plants in a tropical forest. *Science* **198**, 1177–8.

Stork, N.E. (1988). Insect diversity: facts, fiction and speculation. *Biological Journal of the Linnean Society London* **35**, 321–7.

Strong, D.R. (1988) (ed.). Special feature: insect host range. *Ecology* **69**, 885–915.

Struhsaker, T.T. (1997). *Ecology of an African rain forest*. University Press of Florida, Gainesville.

Sutton S.L., Whitmore T.C., Chadwick A.C. (1983) (eds.). *Tropical forest ecology and management*. Blackwell, Oxford.

Swaine, M.D. (1996) (ed.). *The ecology of tropical forest tree seedlings.* UNESCO and Parthenon, Paris and Carnforth.

Swaine, M.D. and Hall, J.B. (1988). The mosaic theory of forest regeneration and the determination of forest composition in Ghana. *Journal of Tropical Ecology* **4**, 253–69.

Swaine, M.D. and Whitmore, T.C. (1988). On the definition of ecological species groups in tropical rain forests. *Vegetatio* **75**, 81–6.

Swaine, M.D., Lieberman, D. and Putz, F.E. (1987). The dynamics of tree populations in tropical forests: a review. *Journal of Tropical Ecology* **3**, 359–66.

Swaine, M.D., Agyeman, V.K., Kyereh, B., Orgle, T.K., Thompson, J. and Veenendaal, E.M. (1997). *Ecology of forest trees in Ghana.* ODA Forestry Series **7**.

Swanson, T. (1995) (ed.). *Intellectual property right and biodiversity conservation: an interdisciplinary analysis of the values of medicinal plants.* Cambridge University Press, Cambridge.

Symington, C.F. (1933). The study of secondary growth on rain forest sites in Malaya. *Malayan Forester* **2**, 107–17.

Symington, C.F. (1943). Foresters' manual of dipterocarps. *Malayan Forest Records*, **16**.

Terborgh, J. (1983). *Five New World primates.* Princeton University Press, Princeton.

Terborgh, J. (1986). Keystone plant resources in the tropical forest. In *Conservation biology* (ed. M. Soulé), pp. 330–44. Sinauer, Sunderland.

Terborgh, J.H. (1992). Maintenance of diversity in tropical forests. *Biotropica* **24**, 283–92.

Terborgh, J. and Schaik, C.P. van (1987). Convergence vs nonconvergence in primate communities. In *Organisation of communities past and present* (eds. J.H.R. Gee and P.S. Giller), pp. 205–26. Blackwell, Oxford.

Thang, H.C. (1995). Forest resources monitoring systems in Malaysia. Expert consultation on forest resources monitoring systems, Bangkok 27 Feb–3 March 1995. FAO, Bangkok.

Thiollay, J.M. (1992). Influence of selective logging on bird species diversity in a Guianan rain forest. *Conservation Biology* **6**, 47–63.

Thiollay, J.M. (1995). The role of traditional agroforests in the conservation of rain forest bird diversity in Sumatra. *Conservation Biology* **9**, 335–53.

Thompson, J.N. (1986). Constraints on arms race in co-evolution. *Trends in Ecology and Evolution* **1**, 105–7.

Thorne, R.F. (1977). Plate tectonics and angiosperm distributions. *Notes Royal Botanic Garden, Edinburgh* **36**, 297–316.

Toledo, V.M., Batis, A.I., Becerra, R., Martinez, E. and Ramos, C.H. (1995). La Selva util: etnobotánica cuantitativa de los grupos indígenas del trópico húmedo de México. *Interciencia* **20**, 177–87.

Tomlinson, P.B. (1986). *The botany of mangroves.* Cambridge University Press, Cambridge.

Turner, I.M. (1993). The names used for Singapore plants since 1900. *Garden's Bulletin Singapore* **45**, 1–287.

Turner, I.M. (1994). Sclerophylly: primarily protective? *Functional Ecology* **8**, 669–75.

Turner, I.M. (1996). Species loss in fragments of tropical rain forest: a review of the evidence. *Journal of Applied Ecology* **33**, 200–9.

Turner, I.M. and Corlett, R.T. (1996). The conservation value of small, isolated fragments of lowland tropical rain forest. *Trends in Ecology and Evolution* **11**, 330–3.

Turner, I.M., Tan, H.T.W., Wee, Y .C., Ibrahim, A., Chew, P.T. and Corlett, R.T. (1994). A study of plant species extinction in Singapore: Lessons for the conservation of tropical biodiversity. *Conservation Biology* **8**, 705–12.

Uhl, N.W. and Dransfield, J. (1987). *Genera Palmarum.* International Palm Society, Lawrence.

Uhl, C. and Jordan, C.F. (1984). Succession and nutrient dynamics following forest cutting and burning in Amazonia. *Ecology* **65**, 1476–90.

Uhl, C., and Vieira, I.C.G. (1989). Ecological impacts of selective logging in the Brazilian Amazon: a case study from the Paragominas region of the state of Pará. *Biotropica* **21**, 98–106.

Uhl, C., Clark, K., Clark, H. and Murphy, P. (1981). Early plant succession after cutting and burning in the upper Rio Negro region of the Amazon basin. *Journal of Ecology* **69**, 631–49.

Uhl, C., Buschbacher, R. and Serrão, E.A.S. (1988). Abandoned pastures in eastern Amazonia. I Patterns of plant succession. *Journal of Ecology* **76**, 663–81.

Valencia, R., Balslev, H. and Paz Y Miño, C.G. (1994). High tree alpha-diversity in Amazonian Ecuador. *Biodiversity and Conservation* **3**, 21–28.

Varangis, P.N., Crossley, R. and Trimo Braga, C.A. (1995). Is there a commercial case for tropical

timber certification? *Policy Research Working Paper* **1479**. World Bank, Washington.

Vazquez-Yanes, C. and Orozco-Segovia, A. (1984). Ecophysiology of seed germination. In *Physiological ecology of plants of the wet tropics* (eds. E.Medina, H.A. Mooney and C. Vazquez-Yanes), pp. 37–50. Junk, The Hague.

Vincent, J.R. (1992). The tropical timber trade and sustainable development. *Science* **256**, 1651–5.

Vitousek, P.M. and Sanford, R.L. (1986). Nutrient cycling in moist tropical forest. *Annual Review of Ecology and Systematics* **17**, 137–68.

Waiko, J.D. (1980). Yu no buggarapim em tasol. (Leave the trees alone). *Uniterra* 5(2), 4–5.

Wallace, A.R. (1876). *The geographical distribution of animals*. Macmillan, London.

Walsh, R.P.D. (1996). Drought frequency changes in Sabah and adjacent parts of northern Borneo since the late nineteenth century and possible implications for tropical rain forest dynamics. *Journal of Tropical Ecology* **12**, 385–407.

Walter, H. and Lieth, H. (1967). *Klimadiagramm Weltatlas*. Gustav Fischer, Jena.

Watson, R.T., Zinyowere, M.C., Moss, R.H. (1996) (eds.). *Climatic change 1995; impacts, adaptations and mitigation of climatic change: scientific-technical analysis*. Cambridge University Press, Cambridge.

Watt, A.S. (1924). On the ecology of British beechwoods with species reference to their regeneration. II. *Journal of Ecology* **12**, 145–204.

Watt, A.S. (1947). Pattern and process in the plant community. *Journal of Ecology* **35**, 1–22.

Webb, L.J. (1959). A physiognomic classification of Australian rain forest. *Journal of Ecology* **47**, 551–70.

Webb, L.J. (1968). Environmental relationships of the structural types of Australian rain forest vegetation. *Ecology* **49**, 296–311.

Wheelwright, N.T. (1986). Competition for dispersers and the timing of flowering and fruiting in a guild of tropical trees. *Oikos* **44**, 465–77.

White, F. (1983). The vegetation of Africa: a descriptive memoir to accompany the UNESCO/ AETFAT/ UNSO vegetation map of Africa. UNESCO, Paris.

Whitmore, T.C. (1972) (ed.). *Tree flora of Malaya*, Vol. 1. Longman, Kuala Lumpur and London.

Whitmore, T.C. (1973*a*) (ed.). *Tree flora of Malaya*, Vol. 2. Longman, Kuala Lumpur and London.

Whitmore, T.C. (1973*b*). Frequency and habitat of tree species in the rain forest of Ulu Kelantan. *Garden's Bulletin Singapore* **26**, 195–210.

Whitmore, T.C. (1974). Change with time and the role of cyclones in tropical rain forest on Kolombangara, Solomon Islands. *Commonwealth Forestry Institute Paper*, **46**.

Whitmore, T.C. (1975). *Tropical rain forests of the Far East* (1st edn.). Clarendon Press, Oxford.

Whitmore, T.C. (1977). A first look at *Agathis*. *Tropical Forestry Papers*, **11**.

Whitmore, T.C. (1978). Gaps in the forest canopy. In *Tropical trees as living systems* (eds. P.B. Tomlinson and M.H. Zimmermann), pp. 639–55. Cambridge University Press, Cambridge.

Whitmore, T.C. (1980). Utilization, potential and conservation of *Agathis*, a genus of tropical Asian conifers. *Economic Botany* **34**, 1–12.

Whitmore, T.C. (1981) (ed.). *Wallace's line and plate tectonics*. Clarendon Press, Oxford.

Whitmore, T.C. (1982*a*). On pattern and process in forests. In *The plant community as a working mechanism* (ed. E.I. Newman), pp. 45–60. Blackwell, Oxford.

Whitmore, T.C. (1982*b*). Fleeting impressions of some Chinese rain forests. *Commonwealth Forestry Review* **61**, 51–8.

Whitmore, T.C. (1983). Secondary succession from seeds in tropical rain forests. *Forestry Abstracts* **44**, 767–79.

Whitmore, T.C. (1984*a*). *Tropical rain forests of the Far East* (2nd edn.). Clarendon Press, Oxford.

Whitmore, T.C. (1984*b*). A new vegetation map of Malesia at scale l:5 million. *Journal of Biogeography* **11**, 461–71.

Whitmore, T.C. (1987). *Biogeographical evolution of the Malay archipelago*. Clarendon Press, Oxford.

Whitmore, T.C. (1988). The influence of tree population dynamics on forest species composition. In *Population biology of plants* (eds. A.J. Davy, M.J. Hutchings and A.R. Watkinson), pp. 271–91. Blackwell, Oxford.

Whitmore, T.C. (1989*a*). Canopy gaps and the two major groups of forest trees. *Ecology* **70**, 536–8.

Whitmore, T.C. (1989*b*). Changes over 21 years in the Kolombangara rain forests. *Journal of Ecology* **77**, 469–83.

Whitmore, T.C. (1989*c*). Forty years of rain forest ecology. *Geojournal* **19**, 347–60.

Whitmore, T.C. (1989*d*). Tropical forest nutrients, where do we stand? A *tour de horizon*. In *Mineral nutrients in tropical forest and savanna ecosystems* (ed. J. Proctor), pp. 1–13. Blackwell, Oxford.

Whitmore, T.C. (1991). Tropical rain forest dyanics and is implications for management. In *Rain forest regeneration and management* (eds. A. Gomez Pompa, T.C. Whitmore, and M.J. Hadley), pp. 67–89. UNESCO and Parthenon, Paris and Carnforth.

Whitmore, T.C. (1996). A review of some aspects of tropical rain forest seedling ecology with suggestions for further enquiry. In *Ecology of tropical tree seedlings* (ed. M.D. Swaine), pp. 3–39. UNESCO and Parthenon, Paris and Carnforth.

Whitmore, T.C. (1997). Tropical forest disturbance, disappearance and species loss. In *Tropical forest remnants: ecology, management and conservation of fragmented communities* (eds. W.F. Laurance and R.O. Bierregaard Jr.), pp. 3–12. University of Chicago Press, Chicago.

Whitmore, T.C. and Brown, N.D. (1996). Dipterocarp seedling growth in rain forest canopy gaps during six and a half years. *Philosophical Transactions of the Royal Society of London* B **351**, 1195–1203.

Whitmore, T.C. and Burslem, D.F.R.P. (1998). Major disturbances in tropical rain forests. In *Dynamics of tropical communities*. (eds. D.M. Newbery, N.D. Brown and H.H.T. Prins). Blackwell Science, Oxford.

Whitmore, T.C. and Prance, G.T. (1987) (eds.). *Biogeography and Quaternary history in tropical America*. Clarendon Press, Oxford.

Whitmore, T.C. and Sayer J.A. (1992). Deforestation and species extinction in tropical moist forests. In *Tropical deforestation and species extinction* (eds. T.C. Whitmore and J.A. Sayer), pp. 1–14. Chapman and Hall, London.

Whitmore, T.C. and Sidiyasa, K. (1986). Composition and structure of a lowland rain forest at Toraut, northern Sulawesi. *Kew Bulletin* **41**, 747–56.

Whitmore, T.C. and Silva, J.N.M. (1990). Brazilian rain forest timbers are mostly very dense. *Commonwealth Forestry Review* **69**, 87–90.

Whitmore, T.C., Peralta, R. and Brown, K. (1986). Total species count in a Costa Rican tropical rain forest. *Journal of Tropical Ecology* **1**, 375–8.

Whitmore, T.C., Sidiyasa, K. and Whitmore, T.J. (1987). Tree species enumeration of 0.5 ha on Halmahera. *Garden's Bulletin, Singapore* 40, 31–4.

Whittaker, R.H. (1972). Evolution and measurement of species diversity. *Taxon* **21**, 213–51.

Whittaker, R.H. (1977). Evolution of species diversity in land communities. *Evolutionary Biology* **10**, 1–67.

Whitten, A.J., Damanik, S.J., Anwar, J. and Hisyam, N. (1984). *The ecology of Sumatra*. Gadjah Mada University Press, Yogyakarta.

Whitten, A.J., Mustafa, M. and Henderson, G. (1987). *The ecology of Sulawesi*. Gadja Mada University Press, Yogyakarta.

Whitten, A.J., Soeriaatmadja, R.E. and Afiff, S. (1996). *The ecology of Java and Bali*. Periplus Editions, Hong Kong.

Wilson, E.O. (1992). *The diversity of life*. Belknap, Cambridge.

Wong, K.M. and Phillips, A. (1997) (eds.). *Kinabalu: summit of Borneo* (2nd edn.). Sabah Society, Kota Kinabalu.

World Commission on Environment and Development (1987). *Our common future*. Oxford University Press, Oxford.

Wright, J.S. and Cornejo, F.H. (1990). Seasonal drought and leaffall in a tropical forest. *Ecology* **71**, 1165–75.

Zinke, P.J., Sabhasri, S. and Kunstadter, P. (1978). Soil fertility aspects of the Lua forest fallow system of shifting cultivation. In *Farmers in the forest* (eds. P. Kunstadter, E.C. Chapman, and S. Sabhasri), pp. 134–59. University Press of Hawaii, Honolulu.

Zipperlen, S.W. and Press, M.C. (1996). Photosynthesis in relation to growth and seedling ecology of two dipterocarp rain forest tree species. *Journal of Ecology* **84**, 863–76.

Zipperlen, S.W. and Press, M.C. (1997). Photosynthetic induction and stomatal oscillations in relation to light environment of two dipterocarp rain forest tree species. *Journal of Ecology* **85**, 491–503.

Glossary

When the reader encounters an unfamiliar word or concept he should first consult the index, because many of them are defined or described somewhere in the text. This glossary gives definitions of a number of specialized terms which it was not appropriate to cover in the main text. For standard scientific terms which are not familiar and which are not explicitly defined somewhere in the book the reader should consult a science dictionary. Two very useful ones are:

Allaby, M. (1985). *The Oxford dictionary of natural history*. Oxford University Press, Oxford.

Mabberley, D.J. (1997). *The plant book, a portable dictionary of the higher plants* (2nd edn.). Cambridge University Press, Cambridge.

albedo The proportion of the incoming radiation that is reflected back from the Earth's surface; albedo varies for different wavelengths and types of land surface cover.

alfisol A relatively fertile tropical clay-rich soil, high in calcium and magnesium, of only limited occurrence (Table 8.1), well suited for agriculture.

allopatry, allopatric Refers to related species with different non-overlapping geographical areas of occurrence. See **sympatry.**

Amerindian The original inhabitants of the New World.

andesite A fine-grained volcanic rock which weathers to give a fertile soil.

andosol A fairly fertile soil produced by the initial weathering of volcanic ash; of very limited extent (Table 8.1).

apomixis Reproduction that does not involve the sexual process, with the consequence that the progeny have the same genetic constitution as their parents.

aroid A member of the family Araceae.

association A plant community with several or many co-dominant species. See **consociation.**

autotroph An organism that uses CO_2 as its main or sole source of carbon.

basal area (BA) The cross-sectional area of the trees of a block of forest (above a specified minimum diameter). BA is proportional to timber volume (BA × bole height) and so is a useful measure in forestry.

biomass The total weight of the living components of an ecosystem, usually expressed as dry weight per unit area.

biome Biological subdivision of the Earth that reflects the ecological and physiognomic character of the vegetation, e.g. the tropical forest biome.

bromeliad A member of the New World plant family Bromeliaceae which are mainly xeromorphic epiphytes.

CAM See crassulacean acid metabolism.

canopy Used here to mean the whole of a forest from the ground upwards. Some scientists use canopy to mean just the top of the forest, here called canopy top.

cline Gradual change across its distribution, without sharp disjunctions, of the gene frequencies or characters of a species.

consociation A plant community with a single dominant species. See **association.**

coppice shoot Shoot that arises from a bud at the base of a tree, usually one that has been cut down.

coupe The area a concessionaire is permitted to fell in a given period, usually annual.

crassulacean acid metabolism (CAM) Special kind of metabolism which allow CO_2 to be absorbed at night when water loss is minimal and to be stored and used next day for photosynthesis. Common in Crassulaceae and other xerophytic succulent plants.

cultivar Cultivated variety, e.g. Malayan dwarf is a high-yielding, disease-resistant cultivar of coconut.

dipterocarp Member of the Dipterocarpaceae, pre-eminent tree family of lowland rain forests of western Malesia.

diurnal (daily) Applied to the rhythms found during a 24 hour night-day time period.

diurnal range The range expressed during the 24-hour night-day cycle.

ecosystem A community of plants and animals plus their physical environment.

ecotone A narrow and fairly sharply defined transition between two plant communities.

ectotrophic mycorrhiza (ectomycorrhiza) A type of mycorrhiza in which the fungal hyphae do not penetrate the root cells but grow between them. Confined to rather few families of trees.

emergent A tree whose crown stands above the general level of the canopy top.

endemism Situation in which a species or other taxonomic group is confined to a particular geographic region which may be 'narrow', e.g. the beautiful ornamental tree *Maingaya malayana* is narrowly endemic to north-west Malaya, or 'wide', e.g. *Swietenia*, true mahogany, is a genus endemic to the neotropics.

endotrophic mycorrhiza (endomycorrhiza) Usually called vesicular arbuscular mycorrhiza; the fungal hyphae penetrate the root cells as little tree-like growths. Found in numerous plant families.

eutrophic Rich in nutrients. See **oligotrophic.**

frugivore A fruit eater.

FAO Food and Agricultural Organization (of the United Nations).

folivore A leaf eater.

Forest Estate The area of a nation set permanently aside for forestry; may include untouched forest, managed natural forest, and plantations, and be in public or private ownership or both.

genotype Genetic constitution of an organism. See **phenotype.**

Glacial maximum The height of an Ice Age when glaciation is at its maximum extent.

greenhouse effect Solar radiation which enters a greenhouse is reflected as long wavelengths which cannot pass through the glass, so the greenhouse heats up. Carbon dioxide, methane, oxides of nitrogen, and certain other gases in the lower atmosphere behave in the same way as glass. As these gases increase so does the amount of trapped solar energy, and this leads to climatic warming.

hardwood Wood of a flowering plant, technically recognized by its possession (with rare exceptions) of vessels. Hardwoods in fact range from hard and dense (e.g. lignum vitae) to soft (e.g. balsa). See **softwood.**

hemiparasitic Plants that are partial parasites, able to fix CO_2 by photosynthesis but dependent on other plants for mineral nutrients. Sometimes misused with reference to stranglers (see text).

heterotroph An organism that uses organic carbon as its main or sole source of carbon.

insectivore An insect eater.

irradiance Amount of radiation received on a surface per unit area per unit time.

ITTO International Tropical Timber Organisation.

kebun (Malay garden) A term used in Indonesia and Malaysia for orchards and fields of mixed crops, home gardens.

kunkar nodules Nodules of concretionary calcium carbonate, formed within soil as a result of evaporation of mineral-rich water under a strongly seasonal or arid climate.

leaf size spectrum The division of leaves into a range of size classes (see Fig. 3.26), from leptophyll (very tiny) through nanophyll, microphyll, notophyll, mesophyll, and macrophyll to megaphyll (huge).

leptophyll See **leaf size spectrum.**

lithology The gross features of rocks.

Malaya The geographical region that encompasses the political states of Peninsular Malaysia and Singapore.

Malesia The phytogeographical region that stretches from the isthmus of Kra in south peninsular Thailand, throughout the Malay archipelago to the Bismarck archipelago northeast of New Guinea.

Melanesia The biogeographical region of the islands of the western Pacific Ocean, extending from the Solomon islands southwards to Vanuatu, New Caledonia, Fiji, Samoa, and Tonga.

mesophyll See **leaf size spectrum.**

microphyll See **leaf size spectrum.**

monopodial Tree crown with a single leading stem (leader), the branches lateral and often eventually falling off; commonly deep and narrow. See **sympodial.**

monotypic A genus that has only one species.

mutualism Interaction of two or more species that benefits both partners. See **symbiosis.**

mycorrhiza Close physical association between a fungus and the roots of a plant. See **endotrophic** and **ectotrophic mycorrhiza.**

mymecophyte Ant-plant.

nanophyll See **leaf size spectrum.**

nectarivore Nectar eating.

neotropics New World (American) tropics.

niche (ecological niche) The functional position of an organism in a community.

notophyll See **leaf size spectrum.**

ochrea A swollen enlarged ligule.

OECD Organization for Economic Co-operation and Development.

oligotrophic Poor in nutrients. See **eutrophic.**

oxisol An infertile loamy and clayey tropical soil, of very wide occurrence (Table 8.1).

palaeotropics Old World tropics.

pantropical Throughout the tropics.

PAR Photosynthetically active radiation of wave-lengths 400–700 nm, i.e. the visible spectrum. See Fig. 7.2.

perhumid climate A climate with no dry season.

phenotype Physical appearance of an organism, produced by the influence of the environment on the genotype (q.v.).

photophyte A sun-loving plant commonly a xerophyte. See **skiophyte.**

phytochrome A photoreversible pigment found in plants which exists in two interchangeable forms, a red form in the presence of visible red light (660 nm) and a far-red form in the presence of far-red light (730 nm, beyond the visible spectrum). The amount of each form depends on the ratio of those two wave-lengths in the light received by the plant.

palynology The study of pollen, and especially its use in the reconstruction of vegetation history.

primary succession The sequential change in vegetation commencing from the colonization of a new site.

profile diagram A conventional way of depicting the canopy of a forest, by a drawing of the trees larger than a specified small size on a long narrow strip, conventionally 60×6 m in size.

radiocarbon dating Atmospheric CO_2 contains a mixture of ^{12}C and the unstable radioactive isotope ^{14}C which has a half life of 5600 years. In dead plants the amount of ^{14}C progressively decreases, so the ratio of ^{14}C to ^{12}C can be used to measure the age of a piece of wood or other dead plant tissue.

rattan (Malay rotan) A climbing palm of sub-family Calamoideae. Rattans occur from Africa to Fiji and are abundant and economically very important in western Malesia.

sclerophyll A tough leathery leaf, in the humid tropics characteristic of heath forest.

secondary succession The sequential change in vegetation recolonizing a previously vegetated site.

sere The characteristic sequence of developmental stages occuring in primary or secondary succession (q.v.).

skid trail A foresters' term for the track in a forest along which logs are dragged or skidded. Sometimes called a snig track.

skiophyte A shade-loving plant. See **photophyte.**

snig track A foresters' term. See **skid trail.**

softwood Wood of a conifer, technically recognized by the absence of vessels. Softwoods have abundant fibres and make good paper. See **hardwood.**

stand table A foresters' term for the population of a tree species, usually divided into diameter or girth classes. See Fig. 7.12.

sunflecks Patches of direct sunlight which move across the floor of a forest.

symbiosis Said of species that live in close physical contact to their mutual benefit; a special kind of mutualism (q.v.).

sympatry The occurrence of species together in the same area. See **allopatry.**

sympodial Tree crown with several main leading stems, commonly broader than deep. See **monopodial.**

synusia A life form community, e.g. big woody climbers.

testa Seed coat.

ultisol An infertile loamy and clayey tropical soil, of very wide occurrence (Table 8.1).

UNESCO United Nations Educational Scientific and Cultural Organization.

unit leaf rate (E) The rate at which a plant gains dry weight per unit leaf area.

xeromorphic A plant with morphological and anatomical characters which appear adapted to withstand drought. See **xerophyte.**

xerophyte A plant that withstands drought.

zygomorphic Bilaterally symmetrical, as some flowers.

Index to plants and minor forest products

General Index